Flow diagram

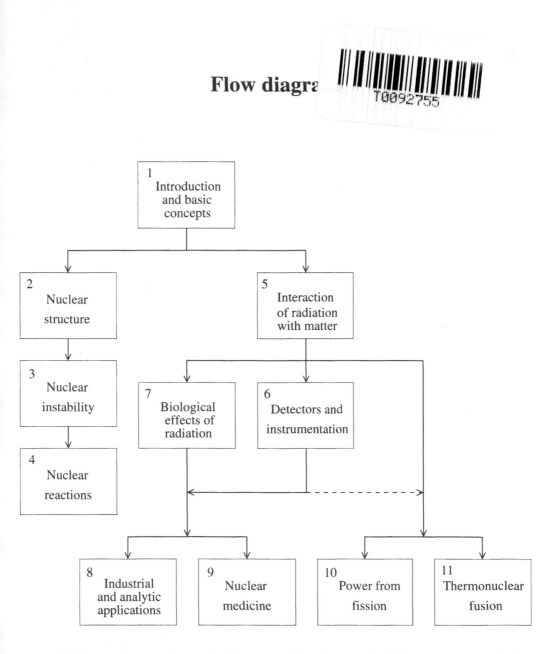

The solid lines leading to a chapter indicate the earlier chapters of which substantial knowledge is presupposed.

The dashed line indicates that Chapter 6 contains some material referred to in Chapters 10 and 11, but is not a prerequisite for those later chapters.

Nuclear Physics

The Manchester Physics Series

General Editors
D. J. SANDIFORD: F. MANDL: A. C. PHILLIPS

Department of Physics and Astronomy,
University of Manchester

Properties of Matter: B. H. Flowers and E. Mendoza

Statistical Physics: F. Mandl
Second Edition

Electromagnetism: I. S. Grant and W. R. Phillips
Second Edition

Statistics: R. J. Barlow

Solid State Physics: J. R. Hook and H. E. Hall
Second Edition

Quantum Mechanics: F. Mandl

Particle Physics: B. R. Martin and G. Shaw
Second Edition

The Physics of Stars: A. C. Phillips
Second Edition

Computing for Scientists: R. J. Barlow and A. R. Barnett

Nuclear Physics: J. S. Lilley

Nuclear Physics

Principles and Applications

J. S. Lilley

Department of Physics and Astronomy
The University of Manchester

John Wiley & Sons, Ltd

CHICHESTER · NEW YORK · WEINHEIM · BRISBANE · SINGAPORE · TORONTO

Copyright © 2001 by John Wiley & Sons Ltd,
Baffins Lane, Chichester,
West Sussex PO19 1UD, England

National 01243 779777
International (+44) 1243 779777
e-mail (for orders and customer service enquiries): cs-books@wiley.co.uk
Visit our Home Page on http://www.wiley.co.uk
or http://www.wiley.com

Other Wiley Editorial Offices

John Wiley & Sons, Inc., 605 Third Avenue,
New York, NY 10158-0012, USA

Wiley-VCH Verlag GmbH, Pappelallee 3,
D-69469 Weinheim, Germany

John Wiley, Australia, 33 Park Road, Milton,
Queensland 4064, Australia

John Wiley & Sons (Asia) Pte Ltd, 2 Clementi Loop #02-01,
Jin Xing Distripark, Singapore 0512

John Wiley & Sons (Canada) Ltd, 22 Worcester Road,
Rexdale, Ontario M9W 1L1, Canada

British Library Cataloguing in Publication Data

A catalogue record for this book is available from the British Library

ISBN 978-0-471-97936-4

Typeset by Kolam Information Systems, Pondicherry, India

Printed and bound by CPI Group (UK) Ltd, Croydon, CR0 4YY

C9780471979364_131223

Contents

Editors' preface to the Manchester Physics Series

The Manchester Physics Series is a series of textbooks at first degree level. It grew out of our experience at the Department of Physics and Astronomy at Manchester University, widely shared elsewhere, that many textbooks contain much more material than can be accommodated in a typical undergraduate course; and that this material is only rarely so arranged as to allow the definition of a short self-contained course. In planning these books we have had two objectives. One was to produce short books: so that lecturers should find them attractive for undergraduate courses; so that students should not be frightened off by their encyclopaedic size or price. To achieve this, we have been very selective in the choice of topics, with the emphasis on the basic physics together with some instructive, stimulating and useful applications. Our second objective was to produce books which allow courses of different lengths and difficulty to be selected with emphasis on different applications. To achieve such flexibility we have encouraged authors to use flow diagrams showing the logical connections between different chapters and to put some topics in starred sections. These cover more advanced and alternative material which is not required for the understanding of latter parts of each volume.

Although these books were conceived as a series, each of them is self-contained and can be used independently of the others. Several of them are suitable for wider use in other sciences. Each Author's Preface gives details about the level, prerequisites, etc., of that volume.

The Manchester Physics Series has been very successful with total sales of more than a quarter of a million copies. We are extremely grateful to the many students and colleagues, at Manchester and elsewhere, for helpful criticisms and stimulating comments. Our particular thanks go to the authors for all the work they have done, for the many new ideas they have contributed, and for discussing patiently, and often accepting, the suggestions of the editors.

Finally we would like to thank our publishers, John Wiley & Sons, Ltd, for their enthusiastic and continued commitment to the Manchester Physics Series.

D. J. Sandiford
F. Mandl
A. C. Phillips
February 1997

Author's preface

This book deals with the basic principles of nuclear physics and their many applications in the modern world. Much of the book is based on a course "Applications of Nuclear Physics", which has been offered as a one-semester, third-year option in physics at the University of Manchester for a number of years.

The book is in two parts. Part I is a brief general introduction to the principles of nuclear physics. However, the emphasis of the book is on applications. These form Part II, which presupposes only a knowledge of the basic concepts developed in Chapter 1 of Part I. The aim of Part II is to introduce the reader to a wide diversity of applications and the underlying physics rather than to attempt a complete coverage of the subject.

The book is addressed mainly to science and engineering students, who require knowledge of the fundamental principles of nuclear physics and its applications. Some of these students may wish later to take advanced courses in nuclear physics or specialized courses in different areas of nuclear science and technology such as nuclear chemistry, nuclear engineering, instrumentation, radiation biology and nuclear medicine.

The level of the book is suitable for undergraduates in their second and third years of physics education and a corresponding grounding in introductory physics and mathematics is assumed.

The approach to the different topics is mainly from an experimental point of view, with illustrative examples. Complex and extensive mathematical treatments generally are avoided. However, where possible, an attempt is made to give a proper understanding based on fundamental physics principles. Derivations of formulae are given or outlined with a minimum of mathematical complexity. A bibliography contains references where much more extensive coverage can be found on all topics. Problem solving is an integral part of learning and understanding physics and, to that end, a set of problems is attached to each chapter. Hints and outlined solutions to all of them are collected together in an appendix.

The material in the book is designed to be used flexibly for a range of different courses. Part I could comprise a one-semester core course on the principles of nuclear physics, which introduces the main elements of nuclear structure, radioactivity and nuclear reactions. Part II can form the basis of different courses on applications allowing choice of topics and courses of different lengths. The flow diagram inside the front cover shows the logical connections between chapters and how material may be used selectively. In addition, much of the material in Part II can be tailored to meet specific needs. For example, if only the basic principles of a nuclear reactor are wanted, the sections in Chapter 10 dealing with the finite reactor, reactor operation and future uses may be omitted.

It is a pleasure to acknowledge the many friends and colleagues who have assisted me in the course of this work. These include Kevin Connell, John Hemingway, Tony Phillips, Roy Ryder, David Sandiford, Harbans Sharma and John Simpson. I am particularly grateful for the many thoughtful comments and helpful suggestions given to me by Bill Phillips and by Paddy Regan, who volunteered to read the entire manuscript.

Finally, I owe a special debt of gratitude to my editor Franz Mandl for his patient guidance throughout and for his critical comments and innumerable detailed suggestions, which contributed so much to the production and quality of the final manuscript.

February 2001   **J. S. Lilley**

PART I
PRINCIPLES

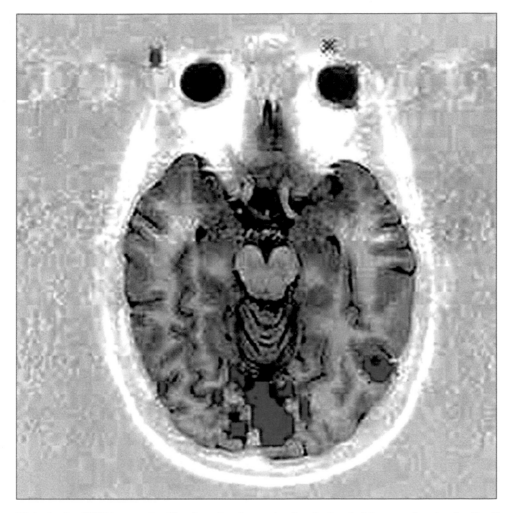

Plate 1 An fMRI image of a slice through a human head at the level of the eyes showing details of the brain cortex. The coloured regions indicate areas of increased activity resulting from a visual stimulus. Courtesy of: Dr. E. J. Burton and Professor A. Jackson, Imaging Science and Bio-medical Engineering, University of Manchester

1

Introduction and Basic Concepts

1.1 INTRODUCTION

It was in 1896 that Becquerel in France detected, by chance, faint traces of the existence of the nucleus in the atom. For many years after that the study of nuclear physics remained a curiosity and intellectual challenge to scientists, but had little practical use outside its own field. The situation changed totally in the 1930s with discoveries that culminated in the cataclysmic demonstrations near the end of the second world war of the immense energy locked up by the force that holds the atomic nucleus together. An unprecedented and irrevocable step had been taken in the degree of power available to humankind with dramatic consequences for good and ill.

Today, nuclear physics has entered into our modern world in a significant way. It influences other branches of science: chemistry, biology, archaeology, geology, engineering, astrophysics and cosmology. It is used widely in society at large – in industry, the environment, medicine, defence, criminology, power production and many other areas. Applications are found even in religion and the arts, where equipment and methods developed originally for nuclear research have found novel application. However, the exploitation of such a powerful force carries with it some danger and is the subject of much debate.

The main aim of this book is to address the broad range and variety of the techniques and applications of nuclear physics used today. The basic physics underlying them is given first in order that the benefits and drawbacks can be properly appreciated. No particular stance is taken on controversial issues. The view taken is that a proper understanding of the subject is important and necessary in order that wise decisions can be taken about how nuclear energy and nuclear radiation should be used.

Essential nuclear physics for understanding the applications is given in this first chapter. Other chapters in Part I give further development of the topics introduced in Chapter 1. The coverage of the applications in Part II is by no means exhaustive. It is intended broadly to inform the reader and provide a suitable preparation for those students who plan to take more advanced courses on any of the separate topics.

Unlike atomic physics, which is underpinned by electromagnetism, there is no fundamental theoretical formalism that completely describes nuclei and nuclear behaviour. For example, there is no formula, analogous to Coulomb's law for the force between two electric charges, which exactly expresses the force between two basic constituents of the nucleus. Progress in understanding the nature of nuclei is made using approximate models, each of which provides insight into the complexity of the real situation, but with a limited range of applicability. Models are drawn from analogy with classical and other branches of physics and are formulated to be consistent with observed properties and behaviour. Conceptual models played a vital role in the first few decades of the twentieth century when the basic framework of the subject was being established. The following section is a short account of this early period.

1.2   EARLY DISCOVERIES

The history of the nucleus dates from the latter years of the nineteenth century with the observation by Becquerel in 1896 of the fogging of photographic plates by an unknown radiation emanating from uranium-bearing rocks. He had encountered *radioactivity*. Detailed studies of this new phenomenon began to be made, notably by Marie and Pierre Curie in France and by Ernest Rutherford, who had come to England from New Zealand earlier in 1895 to work in Cambridge with J. J. Thompson (who discovered the electron in 1897). It was soon revealed that there are three, distinctly different types of radiation emitted by radioactive substances. They were called alpha (α), beta (β) and gamma (γ) rays – terms which have been retained to this day.

The most far-reaching advances in the subject during this early phase were made by Rutherford. He and his co-workers, first in Canada (1898–1907) and later in Manchester, England (1907–1919), began an intensive study of the new radiations. All the laws governing radioactive decay were established. It was shown that α- and β-radioactive decays change the nature of the element and that α particles are helium nuclei. Beta particles were found to be the same as electrons, and γ rays were identified as energetic photons (electromagnetic radiation).

Rutherford used α particles to probe the structure of the atom itself. It was already known that the atom consisted of positively charged and negatively charged components, but there were two very different models for describing how these components might combine to form an atom. The 'planetary' model assumed that light, negatively charged electrons orbit a heavy, positively charged nucleus. The problem with this model was that the electrons would be constantly accelerating and should radiate energy as electromagnetic waves, causing the atom to collapse. In an alternative model, proposed by J. J. Thompson, the electrons are embedded and free to move in an extended region of positive charge filling the entire volume of the atom. Such an atom would not collapse, but Thompson had difficulty in developing his model. For example, he was never able to account successfully for the discrete wavelengths observed in the spectra of light emitted by excited atoms.

The crucial breakthrough came from experiments carried out by Rutherford and his team in Manchester, who were studying the passage of α particles through matter.

It was noted that very thin foils of gold caused α particles to be deflected occasionally through large angles and even in the backward direction. Rutherford realized that this could not be due to the combined effect of a large number of small-angle deflections and could only be explained if the α particle had encountered a tiny, but heavy, charged entity less than 1/1000th of the atom in size. Undaunted by the fact that the planetary model should not exist according to classical theory, he proposed that the atom does consist of a small, heavy positively charged centre surrounded by orbiting electrons which occupy the vast bulk of the atom's volume. The simplest atom, hydrogen, consisted of a proton and a single orbital electron.

Many atomic masses were known to be approximately integer multiples of a basic unit of mass about 1% lighter than the mass of the hydrogen atom. For example, the atomic masses of carbon, nitrogen and oxygen, expressed in these units, are approximately 12.0, 14.0 and 16.0, respectively. However, there are notable exceptions, such as the element chlorine, which has an atomic mass of 35.5 of these units. The idea that an element could consist of differing *isotopes*, which are atoms whose nuclei have different masses but the same charge, was put forward by Soddy in 1911. This explained the existence of anomalous atomic weights, like chlorine, but reinforced an idea, which was current at that time, of nuclei consisting of different numbers of protons and electrons bound together in some way. The proton–electron model persisted for many years until developments in quantum mechanics exposed its shortcomings. No one, for example, could explain why an electron with enough energy to be emitted in β decay was not emitted instantly. Indeed, an estimate of the energy required to keep an electron inside the nucleus (see Problem 1.3) was many times greater than the energies seen in β decay, and an attractive force strong enough to do this would have effects on atomic spectra, which were not observed.

Little progress was made until 1932, when James Chadwick proposed the existence of the *neutron*, an uncharged nuclear constituent whose existence had been anticipated by Rutherford as early as 1920.[1] Bothe and Becker, in 1930, had observed highly penetrating, uncharged radiation from the α-particle bombardment of beryllium. In 1931, F. Joliot and I. Curie measured fast protons emerging from paraffin exposed to this radiation. They surmised that the protons were recoiling from being bombarded by electromagnetic radiation and deduced that, if this were the case, the photon energy would have to be 50 MeV – more energetic than the estimated total energy released in the reaction. Chadwick compared the recoils of protons and nitrogen from different bombarded materials and correctly deduced the mass of the neutral radiation particle to be approximately equal to that of the proton. The neutron was the critical missing ingredient for understanding nuclei. Heisenberg, Majorana and Wigner then took steps to establish the framework that forms the basis of the modern picture of the nucleus consisting of *nucleons* (neutrons and protons) held together by a short-range force whose strength is independent of the type of nucleon.

From the results of subsequent experiments, in which particles were collided at very high energies, it has become apparent that neutrons and protons themselves

---

[1] In a speech given in New York in 1962, Chadwick recalled that Rutherford was considering the possible existence of a neutral particle consisting of a close combination of a proton and an electron.

have an underlying structure composed of three sub-nucleonic particles, which have been called *quarks*. Particle physics deals with the world of the quark and all other particles still thought to be fundamental. These aspects are not dealt with in this text. The structure of neutrons and protons is discerned only at very high energies and, for all practical purposes concerning nuclear structure research and nuclear physics applications in the modern world, the neutron–proton model of the nucleus is entirely adequate.

1.3   BASIC FACTS AND DEFINITIONS

1.3.1   The nucleus and its constituents

An atomic nucleus is the small, heavy, central part of an atom consisting of A nucleons: Z protons and N neutrons; A is referred to as the mass number and Z, the atomic number. Nuclear size is measured in fermis (also called femtometres) where $1\,\text{fm} = 10^{-15}\,\text{m}$.

The basic properties of the atomic constituents can be summarized as follows:

	charge	mass (u)	spin ($\hbar$)	magnetic moment (J T^{-1})
proton	e	1.007276	1/2	1.411×10^{-26}
neutron	0	1.008665	1/2	-9.66×10^{-27}
electron	$-e$	0.000549	1/2	9.28×10^{-24}

Charge: Protons have a positive charge of magnitude $e = 1.6022 \times 10^{-19}\,\text{C}$ (coulombs) equal and opposite to that of the electron. Neutrons are uncharged. Thus, a neutral atom (A, Z) contains Z electrons and can be written symbolically as $^A_Z X_N$.

Mass: Nuclear and atomic masses are expressed in atomic mass units (u), based on the definition that the mass of a neutral atom of $^{12}_6 C_6$ is exactly $12.000\,\text{u}$ ($1\,\text{u} = 1.6605 \times 10^{-27}\,\text{kg}$ – see Appendix A).

Spin: Each of the atomic constituents has a spin $1/2$ in units of $\hbar\, (= h/2\pi)$ and is an example of the class of particles of half-integer spin known as *fermions*. Fermions obey the *exclusion principle*, first enunciated by Wolfgang Pauli in 1925, which determines the way electrons can occupy atomic energy states. The same rule applies to nucleons in nuclei, as we discuss in the next section.

Magnetic moment: Associated with the spin is a magnetic dipole moment. Compared with the magnetic moment of an electron, nuclear moments are very small. However, they play an important role in the theory of nuclear structure. It may be surprising that the uncharged neutron has a magnetic moment. This reflects the fact that it has an underlying quark sub-structure, consisting of charged components. An important application of nuclear moments, based on their behaviour in electromagnetic fields, is the technique of magnetic resonance imaging or nuclear magnetic resonance, which is described in Chapter 9.

1.3.2 Isotopes, isotones and isobars

Isotopes of an element are atoms whose nuclei have the same Z but different N. They
have similar electron structure and, therefore, similar chemical properties. For example, hydrogen has three isotopes: $^1_1H_0$, $^2_1H_1$ and $^3_1H_2$, whose nuclei are, respectively,
the proton p, the deuteron d, and the triton t. Carbon has three, naturally occurring
isotopes: $^{12}_6C_6$, $^{13}_6C_7$ and $^{14}_6C_8$. Nuclei with the same N and different Z are called
isotones, and nuclides with the same mass number A are known as *isobars*. In a
symbolic representation of a nuclear species or *nuclide*, it is usual to omit the N and Z
subscripts and include only the mass number as a superscript, since $A = N + Z$ and
the symbol representing the chemical element uniquely specifies Z.

1.3.3 Nuclear mass and energy

Inside a nucleus, neutrons and protons interact with each other and are bound within
the nuclear volume under the competing influences of attractive nuclear and repulsive electromagnetic forces. This binding energy has a direct effect on the mass of an
atom.

In 1905, Einstein presented the equivalence relationship between mass and energy:
$E = mc^2$. The speed of light c is very large and so even a small mass is equivalent to a
large amount of energy. The complete conversion of 1 g of matter releases about as
much energy as 20 000 tons of TNT. On the atomic scale, 1 u is equivalent to
931.5 MeV/c^2, which is why energy changes in atoms of a few electron volts cause
insignificant changes in the mass of the atom. Nuclear energies, on the other hand,
are millions of electron volts and their effects on atomic masses are easily detectable.
For example, the mass of an atom can be measured to high precision (of the order 1
part in 10^6) in a modern mass spectrometer, and the error in the mass–energy of ^{12}C,
measured to 1 part per million, is about 11 keV. Mass differences can be determined
to even greater precision with these instruments, and many relative masses are known
to an accuracy of a few keV.

Relative masses of nuclei can also be determined from the results of nuclear
reactions or nuclear decay. For example, if a nucleus is radioactive and emits an α
particle, we know from energy conservation that its mass must be greater than that of
the decay products by the amount of energy released in the decay. Therefore, if we
measure the latter, we can determine either of the initial or final nuclear masses if one
of them is unknown. An example of this is presented in Section 1.5.1.

The binding energy of a nucleus B is the energy required to separate it into its
constituent neutrons and protons. The mass of an atom, therefore, is less than the
mass of its constituents by the mass equivalent of B. Symbolically, this is written as

$$m(A, Z) = Zm_H + Nm_n - B/c^2 \qquad (1.1)$$

where $m(A, Z)$, m_H and m_n are the atomic masses of the nuclide (mass number A and
atomic number Z), a hydrogen atom and a neutron, respectively, and B is expressed
in energy units. This equation assumes that differences in the average binding

energies of electrons in different atoms are negligible. All the effects of forces acting on the nucleons inside a nucleus are contained in the binding-energy term. Variations in atomic masses due to variations in binding energy are invariably small compared with an atomic mass unit, which is equivalent to nearly 1 GeV. For this reason, an atomic mass is often presented in the literature as the difference $m(A, Z) - A$ between the mass (in atomic mass units) and the atomic number. This is called the *mass excess* (*me*). Mass excesses of a number of nuclides are listed in Appendix F in units of micro atomic mass units (μu).[2]

Nuclear binding energy increases with the total number of nucleons A and, therefore, it is common to quote the average binding energy per nucleon (B/A). The variation of B/A with A is shown in Figure 1.1 and reveals a number of important features. There is an initial sharp rise with A to a broad maximum of about 8.6 MeV per nucleon near a mass number of 60. This is followed by a gradual decrease to about 7.6 MeV per nucleon for the heaviest nuclei. Nuclei with A greater than 238 are not found in significant quantities in the earth's crust. Several sharp peaks in the light-nuclear region correspond to the nuclei ^{4}He, ^{8}Be, ^{12}C, ^{16}O, ^{20}Ne and ^{24}Mg. The ^{4}He nucleus (α particle) has a particularly stable structure, and the A and Z of the other nuclei are multiples of the α particle. Their extra stability is taken as evidence that they have a structure which bears some resemblance to that of a collection of α particles.

The overall form of the curve in Figure 1.1 is the result of the combined effect of the nuclear and electrostatic (Coulomb) forces and is discussed in some detail in the

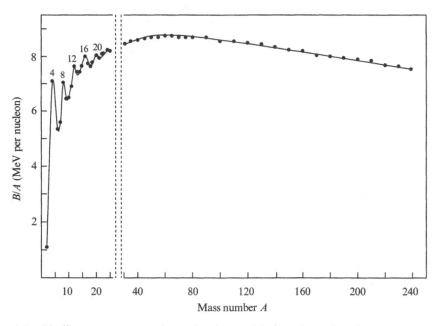

Figure 1.1 Binding energy per nucleon of stable nuclei plotted as a function of mass number A. The horizontal scale below $A = 30$ is twice that above $A = 30$.

---

[2] It is also common to find the mass excess listed in units of energy, which is obtained by multiplying *me* (in u) by $m_u c^2 = 931.494$ MeV (see Appendix A).

next chapter. The nuclear force is very complex and it is not possible to deduce its form precisely. However, the important property we need for the present discussion is that it operates over a very short distance of a few fermis. In a light nucleus, where there are only a few nucleons, any one of them interacts with all the other nucleons and, therefore, its binding energy increases with mass number. However, the size of a nucleus increases with A and when it exceeds that of the range of the internucleon force, the nucleon will interact only with its neighbours which lie within that range. Consequently, the binding energy of a nucleon due to the nuclear force alone will tend to approach a constant value at large A. The Coulomb force, on the other hand, acts over a much larger range, and the electrical potential energy per proton grows steadily as Z increases. It is of opposite sign to the nuclear binding energy and, eventually, this negative term becomes the determining factor causing the fall in B/A beyond $A \sim 60$.

From the form of the binding energy curve, it is clear that when $A \gtrsim 120$, energy could be released by breaking the nucleus into two, roughly equal fragments. This is called *fission* and it occurs spontaneously if A is sufficiently large or if a very heavy nucleus, like ^{235}U or ^{238}U, is excited to a higher energy state. At the other end of the mass scale, energy can be gained by combining two light nuclei together in a *fusion* reaction to form a heavier system with higher B/A. The fission process is already being exploited for commercial power production, and nuclear fusion is a potential source of energy for the future. Some of the underlying physics of fission and fusion is covered in Sections 2.2, 4.5 and 4.6. A fuller discussion of their practical importance is presented in Chapters 10 and 11.

1.4 NUCLEAR POTENTIAL AND ENERGY LEVELS

1.4.1 Nucleon states in a nucleus

It can be shown to a first approximation that a nucleon in a nucleus experiences an average (or mean) attractive energy due to the strong nuclear interaction with its neighbours. This is approximately constant in the nuclear interior and, for the uncharged neutron, it can be represented, as in Figure 1.2(a), by a potential well. This well does not have a sharp edge because of the range of the nuclear force and because the distribution of nucleons in the surface of a nucleus is diffuse. Outside the nucleus, the neutron experiences no force and its potential energy (PE) does not change until it approaches the nuclear surface. There, under the influence of the attractive force, it 'falls' into the nuclear interior with a gain in kinetic energy corresponding to the decrease in PE.

A proton experiences in addition a repulsive Coulomb PE, V_c, due to its positive charge. The general effect of this is shown in Figure 1.2(b). Outside the nucleus, where there is no nuclear force acting on the proton, the form of V_c is that due to a point charge. Inside the nucleus, the Coulomb energy reduces the depth of the total potential for protons compared with that for neutrons by an amount that increases with the nuclear charge Ze. However, compared with the nuclear potential, the difference is always small and can often be neglected to a first approximation. Near the edge of the nucleus, the total PE reaches a maximum B of what is called the *Coulomb barrier*. This energy barrier maintains the stability of the Universe, since

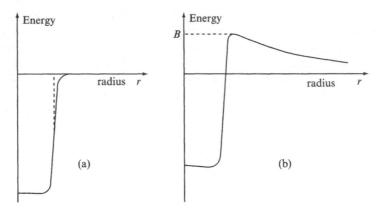

Figure 1.2 Schematic representation of the potential energy (PE) experienced by (a) a neutron and (b) a proton, as a function of distance r from the centre of a nucleus. Outside the nucleus, the proton experiences the electric force only and its PE is given by the Coulomb potential: $V_c = Ze^2/4\pi\varepsilon_0 r$.

it is what prevents low-energy (charged) nuclei coming into contact with each other and initiating nuclear reactions.

A model of a nucleus is to consider that neutrons and protons exist inside nuclei in certain, allowed energy states or levels within the average potential well. Heisenberg's Uncertainty Principle $\Delta x \Delta p \geq \hbar$ helps us to understand why nuclear energies must be large. The small nuclear size Δx implies a large uncertainty in the momentum of a nucleon confined within it. Hence, there is a large uncertainty in its kinetic energy which, typically, is several MeV. We can make the argument more quantitative and, at the same time, illustrate the existence of discrete energy states, by choosing a particular form for the confining potential well.

The simplest example is that of a particle in a one-dimensional box, which we will take to be of size a centred at $x = 0$. Solving the Schrödinger equation for this problem, together with the boundary condition that the wave function must vanish at the boundary of the box, one finds that solutions are only possible for certain values of the energy, known as eigenvalues (see Appendix B):

$$E_{n_x} = \frac{h^2}{8ma^2} n_x^2, \qquad n_x = 1, 2, 3, \cdots . \qquad (1.2)$$

These are the discrete levels of the particle in the box. The corresponding solutions are

$$
\begin{aligned}
\psi_{n_x} &= \text{const.} \times \cos n_x \pi x/a \quad n_x = 1, 3, 5, \cdots \text{ and} \\
\psi_{n_x} &= \text{const.} \times \sin n_x \pi x/a \quad n_x = 2, 4, 6, \cdots
\end{aligned}
\qquad (1.3)
$$

the first few of which are shown in Figure 1.3. Note that they vanish at $x = \pm a/2$ as required.

A very simple model of a nucleon in a nucleus is that of a particle confined inside a cubic box of side a. The above results in one dimension are easily generalized to this

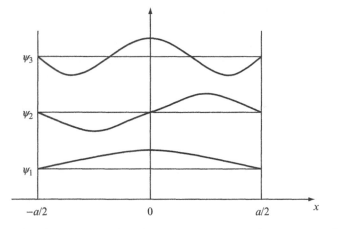

Figure 1.3   Wave functions representing the lowest three states of a particle confined to a one-dimensional box of side a.

case (see Appendix B) and we quote the main result. The energy eigenvalues are now given by

$$E_{n_x n_y n_z} = \frac{h^2}{8ma^2}(n_x^2 + n_y^2 + n_z^2), \quad n_x, n_y, n_z = 1, 2, 3, \cdots. \tag{1.4}$$

The energy of the lowest state a nucleon can occupy is given, with $n_x = n_y = n_z = 1$, as $E_{111} = 3h^2/8ma^2$. Substituting for h^2 (e.g. from Appendix A), with $m = 1$ u, and taking $a = 10$ fm for a typical nuclear size, we obtain $E_{111} \sim 6$ MeV. Other allowed states are given by different combinations of n_x, n_y and n_z (see Problem 1.5).

In general, a particle in a three-dimensional potential may move with orbital angular momentum with respect to the centre of the well. Angular momentum, like energy, is a conserved quantity and states are labelled according to angular momentum as well as energy.

In quantum mechanics, angular momentum occurs in discrete amounts. Orbital angular momentum is specified by a positive, integer quantum number ℓ (≥ 0) and an integer projection quantum number m along an axis, which is normally taken to be the z-axis. The magnitude of the total angular momentum $\mathbf{L}$ and its z projection L_z are given by

$$|\mathbf{L}|^2 = \ell(\ell+1)\hbar^2 \quad \text{and} \quad L_z = m\hbar \quad \text{with } |m| \leq \ell. \tag{1.5}$$

A state with angular momentum quantum number ℓ has $(2\ell + 1)$ independent projections (sub-states) all with different m values between $-\ell$ and $+\ell$. Individual nucleons in nuclei occupy states which can have angular momentum $\ell = 0, 1, 2, 3 \cdots$. These are referred to as $s, p, d, f \cdots$ states, as are the corresponding electron states in atoms.

Angular momentum is also associated with the intrinsic spin property of particles referred to in the previous section. The quantum numbers (s and m_s) for spin angular momentum $\mathbf{S}$ may be either integral or half integral, with equations for $\mathbf{S}$ and its projection S_z similar to Equation (1.5). As stated earlier, a proton, neutron or an electron each has spin quantum number $s = 1/2$ and there are two projections: $m_s = \pm 1/2$ given by the $(2s + 1)$ rule.

Neutrons and protons in a nucleus occupy energy states subject to the exclusion principle rule that no two identical particles can have the same set of quantum numbers. Thus, two protons and two neutrons can exist in the lowest state provided that their spins are anti-parallel to each other, i.e. one has spin 'up' and the other spin 'down'. This forms the ^{4}He nucleus or α particle, which has a total spin of zero. As we have noted, the α particle is particularly stable, requiring over 20 MeV to remove either a neutron or proton. In nuclei heavier than ^{4}He, the additional nucleons must occupy states at progressively higher energies as more nucleons are added.

1.4.2   Energy levels of nuclei

Like atoms, nuclei can exist in different energy levels corresponding to different arrangements of the nucleons in their allowed states. At low excitation, these levels occur at discrete energies, and each nuclide has its own characteristic energy spectrum. That for ^{12}C is shown in Figure 1.4 up to an excitation energy of 10 MeV. The integer labelling each level is a quantum number I specifying its total angular momentum in units of $\hbar$. The total angular momentum of a state $\mathbf{I}$ is a combination of $\mathbf{L}$, the total orbital angular momenta of all the A nucleons in the nucleus, and $\mathbf{S}$, the vector sum of their intrinsic spins. In nuclear physics, the term 'spin' is used to denote this total angular momentum I and, hereafter, this is what the term spin by itself will mean when talking about nuclei. As we shall see in Chapter 2, most of the nucleons in a nucleus combine in pairs, each contributing a net angular momentum or spin of zero, which is why the ground states of all nuclei with even N and even Z have a total spin of zero.

A nucleus in an excited state normally remains there for a very short time. Often, it decays or becomes de-excited by emitting electromagnetic radiation in the form of a γ ray while undergoing a transition to a state lower in energy. However, if the energy of the state is high enough, the nucleus may decay by emitting a particle such as a neutron or α particle. For example, the energy of the second excited state in ^{12}C at 7.654 MeV is above that required for the nucleus to separate into an α particle and

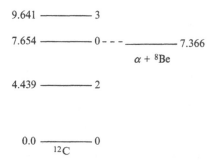

Figure 1.4   Energy-level spectrum of ^{12}C; each level is labelled by its energy in MeV and total angular momentum (or spin) quantum number. Also shown on the same energy scale is the state corresponding to separating ^{12}C into an α particle and a ^{8}Be nucleus.

[8]Be. As we shall see in Chapter 11, this state plays a key role in stellar evolution and the creation of elements inside stars.

The energy of a quantum state that decays is not precisely defined. This is a consequence of the uncertainty principle and it manifests as a spread in energy (or energy width Γ), which increases as the lifetime of the state decreases. As the excitation energy of a nucleus increases, the average width Γ increases. Also, the extra energy means that the levels become more closely spaced because there are many more ways the energy can be shared among the nucleons and each way corresponds to a different energy level. The levels begin to overlap in energy and, eventually, the energy spectrum becomes a continuum in which an individual level cannot be discerned.

1.4.3 Occurrence and stability of nuclei

The occurrence of nuclei and their relative abundance in nature reveal important aspects of nuclear structure which are considered further in Chapter 2. Known stable and unstable (radioactive) nuclides are presented, according to their Z and N numbers, in Figure 1.5. In this diagram, the stable nuclei and very long-lived unstable nuclei (black squares) follow a particular line referred to as the line or valley of stability.

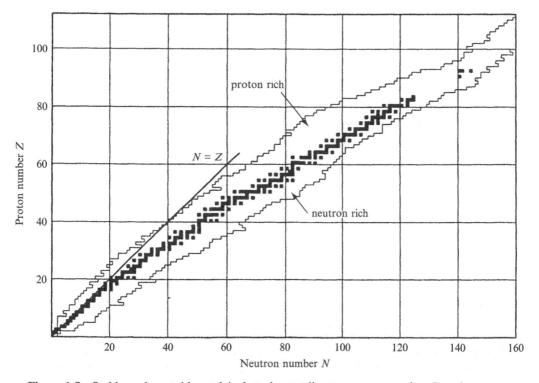

Figure 1.5 Stable and unstable nuclei plotted according to proton number Z and neutron number N. Regions of known proton-rich and neutron-rich unstable nuclei are indicated on either side of the band of stable nuclei (and very long-lived unstable nuclei), which are represented as black squares.

A nucleus lying either above or below this line has a binding energy greater than the minimum value corresponding to its particular mass number A and, therefore, is unstable. Some of the excess energy of such an unstable nucleus may be recovered by converting a neutron into a proton or vice versa. There is a force in nature which allows this to occur, called the weak nuclear force. Its strength is about 10^{-12} times that of the strong force and it operates over an even shorter range. Consequently, it plays no part in holding the nucleus together. As we shall see in the next section, its importance lies in the fact that it can cause a neutron-rich or proton-rich nucleus to change into a more stable form in a process called β radioactivity.

There is an upper mass limit to the line of stability at ^{209}Bi. As A increases beyond this point, it becomes increasingly likely that a nucleus will lose mass by emitting an α particle and be transformed into a more stable product. Naturally occurring thorium and uranium nuclei decay in this way, albeit with very long average lifetimes. Alpha and beta radioactivity and the laws governing these processes are discussed in the next section and also in Chapter 3.

1.5   RADIOACTIVITY AND RADIOACTIVE DECAY

In radioactive decay, an unstable nuclide or 'parent' is transformed into a more stable nuclide called the 'daughter'. If the daughter product is also radioactive, the process continues in a decay chain until a stable product is reached. Radioactivity is a random process. We cannot know exactly when a given unstable nucleus will decay and can only specify a probability per unit time that it will do so. This is normally described by the *half-life* ($t_{1/2}$), which is the time taken for half the nuclei in a sample to decay.

All naturally occurring, and the majority of artificially produced, radioactive nuclei are either α active, β active or (occasionally) both and emit a combination of α, β and γ radiation. Those that continue to exist in the environment, together with cosmic rays from outer space, generate the background radiation to which we all are subjected. Artificially produced unstable nuclei may decay by emitting protons, neutrons or even heavy ions. In spontaneous fission, very heavy fragments are emitted. Some of these decay modes are dealt with at the appropriate places later in the text. This section deals with α, β and γ radioactivity only, describing the changes that occur and the radioactivity decay laws. Further details and insight into the underlying physics are covered in Chapter 3.

1.5.1   Alpha emission

Many naturally occurring, heavy nuclei, with $82 < Z \leq 92$, and artificially produced transuranic elements ($Z > 92$) decay by α emission, in which the parent nucleus loses both mass and charge: $(A, Z) \rightarrow (A - 4, Z - 2)$. The daughter may or may not be stable, but it invariably lies closer to the region of stable nuclei which, eventually, is reached. The α particle is emitted in preference to other light particles (d, t, ^{3}He ...) because energy must be released in order for the decay to take place at all. The α particle has a very stable, tightly bound structure and can be emitted spontaneously with positive energy in α decay, whereas d, t or ^{3}He decay would require an input of energy (see Problem 1.6).

The energy released in α decay (Q_α) is given by the difference in mass energy between the parent nucleus and final products and appears as kinetic energy shared between the outgoing particles. Thus, we can write

$$Q_\alpha = (m_P - m_D - m_\alpha)c^2 = E_D + E_\alpha \tag{1.6}$$

where m_P, m_D and m_α are the masses of the parent, daughter and α particle, respectively, and E_D and E_α are the kinetic energies of the daughter and α particle.

Assuming the decaying nucleus is at rest, the daughter must recoil in the opposite direction to the α particle and with the same momentum, i.e. $m_D v_D = m_\alpha v_\alpha$. The ratio of their kinetic energies, therefore, is

$$\frac{E_D}{E_\alpha} = \frac{\frac{1}{2}m_D v_D^2}{\frac{1}{2}m_\alpha v_\alpha^2} = \frac{(m_D v_D)^2 m_\alpha}{(m_\alpha v_\alpha)^2 m_D} = \frac{m_\alpha}{m_D}. \tag{1.7}$$

From Equations (1.6) and (1.7), we can determine E_α and E_D uniquely in terms of Q_α, m_D and m_α, since

$$Q_\alpha = E_\alpha\left[1 + \frac{E_D}{E_\alpha}\right] = E_\alpha\left[1 + \frac{m_\alpha}{m_D}\right]. \tag{1.8}$$

Alternatively, if we have measured E_α, we can use the equations to determine m_P or m_D if either of them is unknown.

For example, ^{238}U emits an α particle of kinetic energy 4.196 MeV. The daughter is ^{234}Th and its mass is obtained from Equations (1.6) and (1.8) as

$$m_D = m_P - m_\alpha - \frac{Q_\alpha}{c^2} = m_P - m_\alpha - \frac{E_\alpha}{c^2}\left[1 + \frac{m_\alpha}{m_P - m_\alpha}\right] = 234.0436\,\mathrm{u}$$

where we have substituted $E_\alpha/c^2 = 4.196\,(\mathrm{MeV})/931.5\,(\mathrm{MeV\,u^{-1}}) = 0.0045\,\mathrm{u}$, and $238.0508\,\mathrm{u}$ and $4.0026\,\mathrm{u}$ for the masses of ^{238}U and the α particle, respectively. Note that it is accurate enough to use $m_P - m_\alpha$ for m_D (neglecting Q_α/c^2) in the factor multiplying E_α/c^2.

1.5.2 Beta emission and electron capture

Unstable nuclei, which are described as either neutron rich or proton rich (see Figure 1.5), decay towards the line of stability into other isobaric nuclei by positive or negative β-particle emission or by the capture of an atomic electron.

A negative β particle is identical to an electron (e$^-$) and when it is emitted, the charge on the nucleus increases by one unit. Conversely, when a positive β particle is emitted, the nuclear charge decreases by one unit. A positive β particle is called a *positron* (e$^+$). It behaves like a positively charged electron and has the same mass. It is an example of *antimatter*, the existence of which was predicted by Dirac several years before it was first identified by Anderson in 1932.

Unlike α particles, β particles are emitted with a continuous spectrum of energies. An example is shown in Figure 1.6. The maximum energy (end point) in the spectrum is the energy expected according to the mass–energy relationship, but the average electron energy is only about a third of this. Energy, apparently, is missing. Wolfgang Pauli, in 1931, accounted for this apparent violation of energy conservation by postulating the existence of a third, undetected particle in the decay, which Enrico Fermi later called a neutrino (ν) ('little neutral one'). The neutrino is an uncharged fermion (spin 1/2), with zero or negligible rest mass, which interacts extremely weakly with matter. It is difficult to detect the neutrino and it was not until the 1950s, some 25 years after it was postulated, that Reines and Cowan, in a very sophisticated experiment incorporating a nuclear reactor, gave objective proof of its existence.

Beta decay is considered to be the transformation of one of the nucleons in the nucleus from a neutron into a proton or vice versa, creating light particles, called leptons, in the process and leading to a more stable final product. The two processes may be written symbolically as

$$n \rightarrow p + e^- + \bar{\nu} \quad \text{and} \quad p \rightarrow n + e^+ + \nu. \tag{1.9}$$

The leptons (e^-, ν) and anti-leptons $(e^+, \bar{\nu})$ are created by the action of the weak nuclear force, which is why β decay, generally, is a relatively slow process compared with transitions involving the strong force or the electromagnetic interaction. Energy must be conserved and so the transformations given in Equation (1.9) can occur only if the daughter nucleus is lighter (i.e. more stable) than the parent.

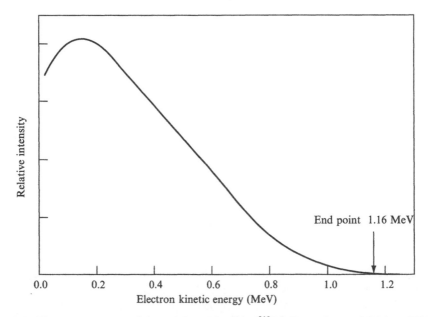

Figure 1.6   Energy spectrum of β particles emitted by ^{210}Bi. From Evans (1955) p. 538.

An alternative to β^+ decay, for a proton-rich nucleus, is *electron capture* (EC) in which a proton and an atomic electron are transformed into a neutron and a neutrino:

$$p + e^- \rightarrow n + \nu. \qquad (1.10)$$

The electron is usually captured from the innermost orbit, or K shell, of the atom, which is why the process is often referred to as K capture. In principle, electrons from less tightly bound atomic shells (L, M, N, etc.) could participate in EC. However, the probability depends on the overlap of the electron wave function with the nuclear volume and this is smaller for L or higher-shell electrons than it is for a K-shell electron.

An example of a β-decay energy-level diagram is shown in Figure 1.7 for ^{22}Na ($t_{1/2} = 2.602$ years). In this case, the main β^+ and EC decay branch is to an excited state of the daughter, which subsequently decays by emitting a 1.275-MeV γ ray. It is common in β decay to have a strong decay branch to an excited state of the daughter. By accurately measuring the γ-ray energy, as described in Chapter 6, it is possible to identify γ-emitting radioactive isotopes uniquely making them much more useful than pure β emitters for applications such as tracing, irradiation, medical imaging and diagnosis (see Chapters 8 and 9).

1.5.3   Gamma emission and internal conversion

As noted earlier, an excited nucleus may lose energy in a transition to a state lower in energy in the same nucleus. When this occurs, most of the transition energy ΔE, which is the energy difference between the initial and final states, may appear in the form of a γ-ray photon.[3] Alternatively, the nucleus may de-excite by ejecting an electron from one of the atomic orbits in a competing process called *internal conversion*. Both these decay modes are due to the action of the electromagnetic force. If the transition energy is sufficiently high, a third type of electromagnetic decay is possible, called *internal pair formation*, in which an electron–positron pair is created. This process is usually weak, except for very high transition energies, and can only occur if the available energy exceeds that needed to create the pair, which is 1.022 MeV.

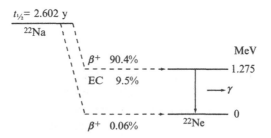

Figure 1.7   Decay scheme for the positron-emitting radioactive nucleus ^{22}Na. The intensities of the β^+ and electron-capture (EC) branches are indicated as percentages.

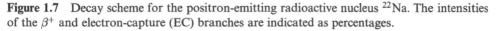

---

[3] A very small amount of energy is taken by the recoiling nucleus.

Gamma-decay and internal-conversion lifetimes are usually very short compared with α- or β-decay lifetimes and are, typically, less than 10^{-9} s. However, there are excited states that last for a much longer time. States with readily measurable half-lives are called metastable or isomeric states (isomers) and the transitions are known as isomeric transitions. The term isomer is sometimes applied to a state with a half-life as short as 10^{-9} s but, more usually, the lifetime is much longer. We shall see an example of a long-lived isomer at the end of Section 1.5.5.

1.5.4   Rate of radioactive decay

The probability per unit time that a given nucleus will decay is called the decay constant λ, and if there are N radioactive nuclei in a sample, the rate of decay is given by

$$dN/dt = -\lambda N \tag{1.11}$$

where the minus sign indicates that N is decreasing with time. If the nucleus decays to several different final states (i.e. has several decay branches), the decay constant λ is the sum of the decay probabilities of all the branches.

The solution to Equation (1.11) is

$$N(t) = N(0)e^{-\lambda t} \tag{1.12}$$

where $N(0)$ is the number of nuclei at $t = 0$.

The mean life τ is defined as the average lifetime of a radioactive nucleus. From Equation (1.11), the number of nuclei which decay between t and $t + dt$ is just $dN = \lambda N dt = \lambda N(0)e^{-\lambda t}dt$. Therefore,

$$\tau = \frac{\int t\, dN}{\int dN} = \frac{\int_0^\infty te^{-\lambda t}dt}{\int_0^\infty e^{-\lambda t}dt} = \frac{1}{\lambda}. \tag{1.13}$$

The half-life, $t_{1/2}$, may be expressed in terms of either λ or τ by substituting $N(t_{1/2}) = N(0)/2$ in Equation (1.12) and solving for $t_{1/2}$:

$$t_{1/2} = \frac{\ln 2}{\lambda} = \tau \ln 2. \tag{1.14}$$

The *activity* $\mathcal{A}$ is the rate of decay of a radioactive sample. It is equal to λN ($= N\ln 2/t_{1/2}$) and, therefore, follows the same time dependence as $N(t)$. The SI unit of activity is the becquerel (Bq), which is one disintegration per second. An older unit, based on the activity of one gram of radium and still in common use today, is the curie (Ci), defined as 3.7×10^{10} Bq. A typical laboratory radioactive source would have a strength of a few tens of kBq or microcuries (μCi). If a sample consists of a mixture of radioactive substances, the activity is the sum of all the activities of the constituents and no longer follows a simple, one-component exponential law.

1.5.5   Radioactive decay chains

In many cases, the daughter product of a nuclear decay is itself radioactive and part of a decay chain.

Consider a chain of decays: A $\rightarrow$ B $\rightarrow$ C $\rightarrow \cdots$ with decay constants λ_A, λ_B, etc. The variation of A with time is given by the solution of a differential equation like Equation (1.11), i.e.

$$N_A(t) = N_A(0)e^{-\lambda_A t}. \tag{1.15}$$

The differential equation for B has an extra term due to the decay of A:

$$dN_B/dt = -\lambda_B N_B + \lambda_A N_A. \tag{1.16}$$

This is solved by using a useful mathematical trick. Multiplying by $e^{\lambda_B t}$ and rearranging gives

$$\frac{dN_B}{dt}e^{\lambda_B t} + \lambda_B N_B e^{\lambda_B t} = \frac{d}{dt}\left(N_B e^{\lambda_B t}\right) = \lambda_A N_A e^{\lambda_B t} = \lambda_A N_A(0)e^{(\lambda_B - \lambda_A)t}$$

after substituting for N_A. Carrying out the integration we obtain

$$N_B e^{\lambda_B t} = \lambda_A N_A(0)\int e^{(\lambda_B - \lambda_A)t}dt = \frac{\lambda_A}{\lambda_B - \lambda_A}N_A(0)e^{(\lambda_B - \lambda_A)t} + K.$$

If, at $t = 0$, $N_B(0) = 0$, the integration constant is given by $K = -\frac{\lambda_A}{\lambda_B - \lambda_A}N_A(0)$, which gives, finally,

$$N_B(t) = \frac{\lambda_A}{\lambda_B - \lambda_A}N_A(0)(e^{-\lambda_A t} - e^{-\lambda_B t}). \tag{1.17}$$

Similar differential equations can be written for C and other daughter products in the decay chain, leading to the so-called Bateman equations for their solution. Figure 1.8 shows the time variation of A, B and C for the case where $\lambda_A = \lambda_B$ and C is a stable nucleus.

If the daughter is relatively long lived compared with the parent (i.e. $\lambda_A \gg \lambda_B$), the decay can be considered to proceed in two stages: first the rapid decay of A $\rightarrow$ B followed by the slower decay of B $\rightarrow$ C for which, after a time long compared with $1/\lambda_A$, Equation (1.17) gives: $N_B(t) \sim N_A(0)e^{-\lambda_B t}$. At the other extreme, if the parent is relatively long lived ($\lambda_A \ll \lambda_B$), N_B initially grows rapidly and, eventually, after a time t such that $e^{-\lambda_A t} \gg e^{-\lambda_B t}$ it approaches an equilibrium ratio with N_A, and both A and B then decay at the same rate governed by λ_A:

$$N_B(t) \rightarrow \frac{\lambda_A}{\lambda_B}N_A(0)e^{-\lambda_A t} = \frac{\lambda_A}{\lambda_B}N_A(t). \tag{1.18}$$

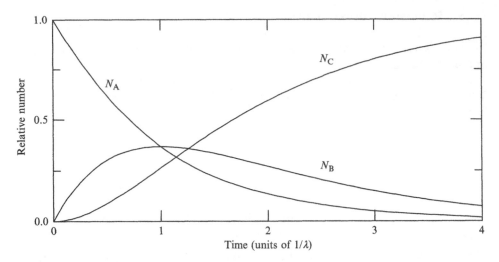

Figure 1.8   Variation with time of the amounts of isotopes A, B and C in a decay chain A → B → C, where the decay probabilities $\lambda_A = \lambda_B = \lambda$ and C is stable. Note that $N_A + N_B + N_C = 1$ at any time.

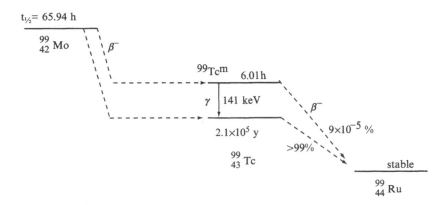

Figure 1.9   Level scheme showing the β^- decay branches of ^{99}Mo to the metastable, excited state (^{99}Tc^m) and ground state of technetium 99. ^{99}Tc^m nearly always decays to its ground state and emits a 141-keV γ ray about 90% of the time.

An example of the latter situation is the production of the important medical radioisotope ^{99}Tc^m. The superscript m denotes that technetium-99 is not in its ground state but in a long-lived (metastable) excited state. The decay chain is shown in Figure 1.9; ^{99}Tc^m is formed from the decay of the parent, ^{99}Mo ($t_{\frac{1}{2}} = 65.94$ h) and decays, with a half-life of 6.01 h, nearly 100% of the time to the ground state. It has only a very weak β branch to ^{99}Ru. The ground state of ^{99}Tc is also radioactive but its half-life is very long compared with the average human lifespan and for all practical medical purposes, it can be considered to be stable. As described in Chapter 9, ^{99}Tc^m is routinely used for diagnostic purposes in nuclear medicine.

1.5.6   Radioactivity in the environment

The material of the earth was created in a series of nuclear processes which, as outlined in Chapter 11, began with the Big Bang origin of the Universe, continued relatively quietly in the cores of burning stars, then violently in stellar explosions (supernovae) which distributed the products into space to be available for planet formation. Many of the nuclides formed were radioactive. Most have decayed away, but a few of them have half-lives longer than or comparable with the age of the earth and remain in varying amounts to this day. They constitute the bulk of the natural radioactivity in the environment, which is a major source of the background radiation we experience throughout our lives. Several of these radioisotopes, such as ^{40}K, decay directly to a stable daughter but many, especially very heavy nuclei ($A > 208$), decay into other active isotopes as part of a decay chain or radioactive series.

There are three naturally occurring radioactive series, which have existed since the earth was formed about 4.5×10^9 years ago. Each is headed by a very long-lived parent (^{238}U, ^{235}U or ^{232}Th), which controls the decays of the active daughters all of which have much shorter half-lives. The nuclides in each chain decay by emitting α and/or β particles until a final (stable) nuclide is reached.

As an example, Figure 1.10 shows the thorium series, headed by ^{232}Th and ending at ^{208}Pb. Since the chain is in equilibrium and the half-life of ^{232}Th (14×10^9 years) is so much longer than any of the daughters, all active nuclides in the chain, except where the chain branches, have essentially equal activities and occur in amounts inversely proportional to the respective decay probabilities according to Equation (1.18). This chain branches at ^{212}Bi, which β decays 64.1% of the time into ^{212}Po and 35.9% via α decay into ^{208}Tl. The activities of ^{212}Po and ^{208}Tl are reduced from that of ^{212}Bi by the corresponding decay branch fractions.

In general, all the naturally occurring radioisotopes, with one exception, are chemically bound to minerals in rocks and pose no biological hazard. The exception

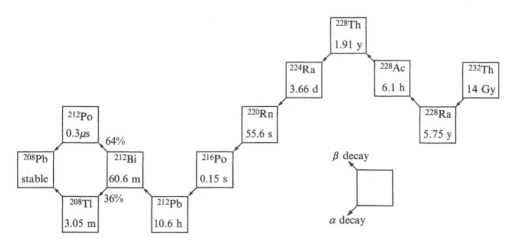

Figure 1.10   The thorium series – a radioactive decay chain headed by the long-lived radioactive nuclide $^{232}_{90}$Th.

is radon (Rn), which has isotopes in each of the radioactive series. Radon is a noble gas and can be dangerous if a sufficient amount is inhaled. Other, natural sources of radioactivity include the products of cosmic-ray nuclear interactions in the upper atmosphere, such as tritium and ^{14}C. In addition, there are man-made sources, mainly from nuclear weapons testing and waste from nuclear power generation. Some of the issues associated with radiation in the environment from these sources are taken up again in Chapters 7 and 10.

1.5.7   Radioactive dating

There are many examples of the use of the radioactive-decay law to determine the ages of materials. Perhaps the best known is that of carbon dating.

Carbon-14 is a radioactive isotope of carbon formed by the action of cosmic rays on nitrogen in the atmosphere. The intensity of cosmic rays is approximately constant over long periods and so the ratio of the isotope ^{14}C to the most abundant isotope of carbon ^{12}C reaches an equilibrium level, which is about $1:10^{12}$. Both of these isotopes are taken up in this proportion by living organisms, but after death there is no more exchange with the environment and the ratio slowly changes with time as the ^{14}C β^- decays to ^{14}N with a half-life of 5730 years. This ratio can be determined by measuring the ^{14}C activity in a known mass of carbon. The activity of a typical sample is rather small. One gram of carbon in living organic material contains about 5×10^{10} atoms of ^{14}C and has an activity of less than 0.25 Bq. Very sensitive, low-background counting techniques have been developed to measure such weak activities, and carbon dating is widely used for dating archaeological objects, which may be up to several tens of thousands of years old. The method has been calibrated over a considerable time by measuring the activities of tree rings, which gives an independent measure of the age of the sample. This has also shown how the flux of cosmic rays has varied with time.

In Section 8.6, we describe another technique for measuring small amounts of radioactive material, which also is widely used for dating purposes.

1.6   NUCLEAR COLLISIONS

Apart from a few that are radioactive, naturally occurring nuclei are inert and in order to unlock their secrets, scientists actively disturb them and note the consequences. They usually do this by firing a stream of projectiles at a target. From time to time, there is a collision between a projectile and a target nucleus during which a reaction takes place.

In general, the result of a nuclear collision may be one of a number of possible reactions, each one of which sheds light on a particular aspect of nuclear structure or nuclear behaviour. Research scientists measure different reactions in order to study the nuclear force, synthesize new nuclei, determine nuclear sizes and shapes and investigate the properties of excited nuclei and the dynamical behaviour of nuclear matter during different types of collision. In addition, as we shall discuss in Part II of this book, there are many applications of nuclear reactions in other fields. The fission reaction has been harnessed for commercial power generation, and a variety of reactions are employed in techniques such as trace element detection and materials analysis and in the production of radioisotopes for use in industry, chemistry, biology, medicine and the arts. In this section, we outline only the basic elements of nuclear reactions needed for Part II. A more extended, but still largely descriptive introduction to the subject is given in Chapter 4.

The outcome of a nuclear collision depends on the nature of the interacting particles and very much on the amount of energy available and how much of it goes into nuclear excitation. At high energies, violent collisions can take place resulting in many outgoing particles. At low energies, there are generally only two final products. Such a reaction is called a binary reaction and is the type we will mainly be concerned with in this book.

1.6.1   Nomenclature

Consider a binary reaction in which a particle a is incident on a target nucleus A leading to final products b and B.

The reaction is written in the following forms

$$a + A \rightarrow B + b \quad \text{or} \quad A(a, b)B \tag{1.19}$$

where a + A is called the initial (entrance) channel and b + B is the final (exit) channel.

The reaction energy, or Q value, is given by energy conservation as the difference in the masses of the initial and final particles multiplied by c^2:

$$Q = (m_a + m_A - m_b - m_B)c^2. \tag{1.20}$$

The reaction is endothermic or exothermic, depending on whether the Q value is negative or positive.

1.6.2   Probes

Nuclear reactions can be initiated by any type of projectile. Protons, deuterons, α particles and other nuclei are positively charged and so require enough energy to overcome the Coulomb barrier. This barrier can be considerable; for example, it is about 20 MeV for an α particle incident on a uranium target.

There is no barrier opposing neutrons, however, and because neutrons typically are bound in stable nuclei by 7 or 8 MeV, they can trigger exothermic reactions at very low energies. This important property is discussed further below and is the basis of all current commercial nuclear power generation (see Chapter 10).

Electrons and γ-ray photons also experience no barrier as they approach a nucleus. They interact via the electromagnetic force and, when they do so in a nuclear reaction, give detailed information about the distribution of charges and currents inside the target nucleus.

1.6.3   Cross section, differential cross section and reaction rate

We now introduce a framework for describing reaction rates between colliding nuclei.

Consider, first, the situation shown in Figure 1.11 of a stream of projectiles all moving in the same direction and incident on a stationary target. We can define a flux Φ as the number of particles crossing unit area perpendicular to the direction of motion per unit time. If all particles have the same speed v, the flux is given by

$$\Phi = n_p v \tag{1.21}$$

where n_p is the density of projectiles in the beam. In general, there is a distribution of particle speeds and if $n_p(v)dv$ is the density of projectiles with speeds between v and $v + dv$, the flux is given by the integral $\Phi = \int v n_p(v)dv$.

Suppose a given reaction A (a,b) B is occurring at a certain rate. If the nuclei in the target act independently, the event rate (or reaction rate) per nucleus exposed to the beam is proportional to the incident flux. The constant of proportionality is called the *cross section* (σ), which can be written as

$$\sigma = \frac{\text{event rate per nucleus}}{\text{incident flux}}$$

and, if N target nuclei are exposed to the beam, we have a reaction rate:

$$R = N\sigma\Phi. \tag{1.22}$$

This rate was derived for a unidirectional beam incident on a target. However, we know that the rate does not depend on the direction of the incident particles and, for the case in which particles are moving in different directions, we can define flux as the total path length travelled by all particles in a unit volume per unit time. This

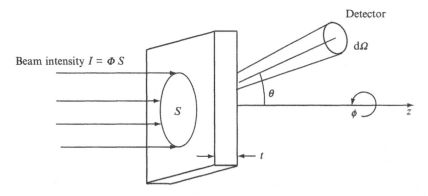

Figure 1.11   Schematic diagram showing a beam of intensity I and cross-sectional area S incident on a (thin) target of thickness t. For a beam of particles of the same speed v and of density n_p, the flux $\Phi = n_p v$. A detector, subtending a solid angle $d\Omega$ at the target, receives outgoing particles emerging at polar angles θ and ϕ. The distance to the detector is normally very large compared with the diameter of the beam.

definition is consistent with the one given above for a unidirectional beam but it expresses better the true nature of flux, which is a scalar quantity.

In the geometry of Figure 1.11, the beam intensity is $I = \Phi S$ particles per unit time, where S is the cross-sectional area of the beam, and we can write down an alternative expression for the reaction rate in terms of I and target thickness t:

$$R = N\sigma I/S = I\sigma n_t t \qquad (1.23)$$

where n_t is the number of target nuclei per unit volume. If the target consists of a certain isotopic species of atomic mass M_A (in atomic mass units), we know that $n_t = \rho N_A/M_A$ where ρ is the density of the target and N_A is Avogadro's number. Thus, we can rewrite Equation (1.23) as

$$R = I(\rho t)\sigma N_A/M_A \qquad (1.24)$$

where the quantity (ρt) is a measure of the amount of material in the target, expressed in units of mass per unit area.

In a typical experiment, the products of a reaction emerge from the target at different polar angles (θ, ϕ), as shown in Figure 1.11, and enter a detector, which subtends a solid angle of $d\Omega$ at the target. They enter at a rate $dR(\theta, \phi)$, which is proportional to $d\Omega$ as well as to the number of irradiated nuclei N and the incident flux Φ. The constant of proportionality is the *differential cross section* $d\sigma/d\Omega$, which sometimes is written as $d\sigma(\theta, \phi)/d\Omega$ or, more concisely, as $\sigma(\theta, \phi)$ to emphasize the dependence on the polar angles. Thus, we have

$$dR(\theta, \phi) = \frac{d\sigma}{d\Omega} N\Phi d\Omega = \sigma(\theta, \phi)N\Phi d\Omega. \qquad (1.25)$$

The cross section σ defined above is equal to the differential cross section integrated over the whole sphere, i.e. over all angles θ and ϕ:

$$\sigma = \int \sigma(\theta, \phi)d\Omega. \qquad (1.26)$$

For a given set of initial conditions, many different reactions may occur with different probabilities. Each has its own partial cross section σ_i (and partial differential cross section) and the total interaction probability is measured by the total cross section, which is the sum of the partial cross sections: $\sigma_T = \sum_i \sigma_i$.

The cross section has the dimensions of area and is expressed in several different units in the literature. The most common unit is the *barn* ($1\text{b} = 10^{-28}\text{m}^2$), which is approximately equal to the geometric cross-sectional area of a nucleus of mass number $A = 100$.

1.6.4 Isotope production

When a target is bombarded by a beam of particles, it is common for nuclear reactions to take place leading to radioactive residual nuclei, and, if the cross section

σ is known, the number of these radioactive nuclei can be determined as a function of time during the bombardment. Target nuclei are lost during the bombardment and so, eventually, the number of target nuclei becomes depleted. However, in most practical situations, the fraction of nuclei converted is very small (see Problem 1.11), in which case we can consider the number of target nuclei and, hence, the rate of production P to be independent of time.

The net rate of increase of the radioactive nuclei (decay constant λ) is the difference between the rate of production and the rate of decay:

$$\frac{dN}{dt} = P - \lambda N. \tag{1.27}$$

Taking P to be constant, this equation is easily solved (Problem 1.10), using the same trick that was used in the previous section to solve Equation (1.16), and yields the result

$$N(t) = \frac{P}{\lambda}(1 - e^{-\lambda t}) \tag{1.28}$$

where it is assumed that $N(0) = 0$. The production rate P is given by one of the forms for the reaction rate in Equation (1.22), (1.23) or (1.24).

The buildup of N with time is shown in Figure 1.12. The initial rate of increase $(dN/dt)_{t=0} = P$ and, after a long time (several half-lives), N approaches an equilibrium value of P/λ, when the left-hand side of Equation (1.27) is zero and the rates of production and decay are equal.

The activity of the sample is defined as

$$\mathcal{A} = \lambda N = P(1 - e^{-\lambda t}) \tag{1.29}$$

and follows the same form as $N(t)$.

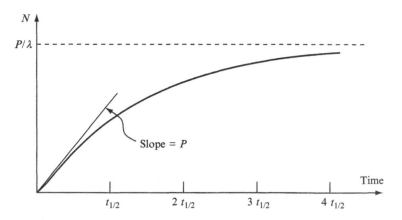

Figure 1.12   Variation with time of the number of radioactive nuclei (decay constant λ), which are being produced at a constant rate P.

1.6.5   Examples of nuclear reactions

In the remainder of this section, we present a broad survey of the principal types of nuclear reaction processes that may occur during a nuclear collision. In most cases, it is not possible to calculate the cross section from first principles and recourse has to be made to nuclear models for approximate treatments. However, for applications where an accurate knowledge of the reaction cross section is needed, extensive tabulations of measured cross sections are often available.

Elastic scattering

This is the simplest type of interaction in which the projectile and target nuclei scatter off each other with no loss of total kinetic energy. The most famous example of this is Coulomb scattering, or Rutherford scattering, in which charged particles interact via the Coulomb force obtained by treating them as point charges.

The elastic scattering of a charged particle in a Coulomb field is important in physics. Nuclei and many fundamental particles are charged and so Coulomb scattering is commonly encountered in general collision phenomena. It is also one of the few reaction processes for which there is an exact formula for the differential cross section. If we assume the scatterer is infinitely massive and remains at rest, the expression takes the form:

$$\frac{d\sigma_{Ruth}}{d\Omega} = \sigma_{Ruth}(\theta) = \left(\frac{Z_1 Z_2 e^2}{16\pi\varepsilon_0 E}\right)^2 \mathrm{cosec}^4\left(\frac{\theta}{2}\right) \tag{1.30}$$

where Z_1 and Z_2 are the atomic numbers of the colliding nuclei, θ is the scattering angle, and E is the initial kinetic energy of the projectile. A slightly more complicated version of the equation is used if the effects of target recoil are to be taken into account. Rutherford obtained this formula using non-relativistic, classical mechanics. Derivations are presented in many books and one, also based on classical physics, is given in Appendix E. Note that if the force is attractive (i.e. one of the charges is negative – as in the scattering of an electron by a nucleus), the projectile particle is deflected *towards* the scattering centre, but the expression for the differential cross section is unchanged.

In Rutherford's experiments, using α particles to bombard heavy target nuclei, the energy of the α particles was insufficient to allow them to reach the surface of the target nuclei and be affected by the nuclear force. However, if the energy is gradually increased, a point is reached at which the α particles can begin to touch the nuclear region, the orbits become distorted and the cross section deviates from Rutherford's formula. An example is shown in Figure 1.13 for the scattering of α particles by lead. The distance of closest approach, or apsidal distance, of a Coulomb trajectory to the scattering centre can easily be calculated and, therefore, knowing the energy and angle at which $\sigma(\theta)$ deviates from $\sigma_{Ruth}(\theta)$, one can obtain a value for the interaction radius R at which the nuclear force comes into play. An analysis of experimental scattering data for many different targets gave the result

$$R = 1.41A^{1/3} + 2.11\,\text{fm} \tag{1.31}$$

where A is the mass number. It suggests that the effective radii of the target nucleus and α particle are $1.41\,A^{1/3}$ fm and 2.11 fm, respectively. This result was one of the early indications that the radius of a nucleus varies as $A^{1/3}$ and, therefore, since its volume is proportional to A, the density of a nucleus is approximately constant, like a liquid drop.

We must remark here that the effective nuclear radius obtained from Equation (1.31), is not the same as the radius marking the edge of the nuclear matter distribution $\rho(r)$ representing the spatial distribution of the A nucleons in the nucleus. This is partly because $\rho(r)$ does not have a sharp edge and partly because the interaction radius includes the finite range of the nuclear force. We can define a half-way radius $R_{1/2}$, where the matter density is half its central value, but the scattering data indicate that the α particle begins to experience the nuclear force due to the target at radii larger than $R_{1/2}$ and where the matter density is quite low.

When the bombarding energy is well above the Coulomb barrier, elastic scattering angular distributions exhibit diffraction-like features, which can be analysed to reveal

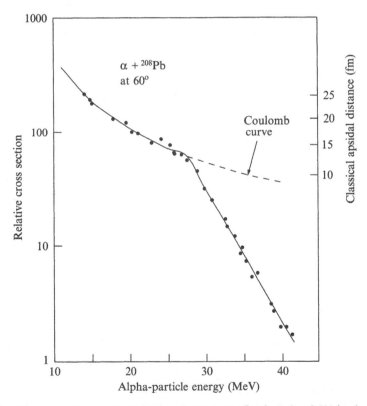

Figure 1.13   Elastic scattering of α particles by lead at a fixed angle of 60° in the laboratory system plotted as a function of incident α energy. The dashed curve is given by Coulomb scattering. The classical distance of closest approach for pure Coulomb scattering is given on the right-hand ordinate, using the dashed curve. Source: Farwell *et al.* (1954).

details about $\rho(r)$. In Chapter 4, we outline briefly, with simple examples, how some of this information is obtained. Important practical applications, described in Chapters 8 and 9, are the use of elastic scattering as a tool for analysing thin films and material surfaces and as the mechanism by which neutrons are slowed down in nuclear reactors to energies where they are most effective in inducing nuclear fission for energy production.

Direct reactions

A direct nuclear reaction occurs when the projectile interacts primarily in the surface of the target nucleus. Energy and material transfer are small and, therefore, the masses and energies of the outgoing products are related to those of the initial particles. The time scale of the reaction is rapid ($\sim 10^{-22}$ s), given by the time it takes for the projectile to cross the nuclear diameter and, since the projectile-like outgoing particle retains most of the momentum of the initial channel (a + A), its angular distribution is generally peaked in the forward direction.

In *inelastic scattering* (as in elastic scattering), the initial and final products are the same. Part of the initial kinetic energy is used to raise one or both of the final particles to excited states, which subsequently de-excite by γ emission. It is an important reaction for determining the properties of nuclear excited states, which reveal a wealth of information about the internal structure of nuclei.

A *transfer* reaction is one in which one or more nucleons are exchanged as the target and projectile pass each other. The final nuclei may or may not be left in excited states. Conventionally, if the transfer is from the projectile to the target, it is called a *stripping* reaction. A *pickup* reaction is when the reverse takes place. Common examples of the former reaction are: (d,p), (α,d) and (^{16}O, ^{12}C); and of the latter: (p,d), (p,t) and (^{16}O, ^{17}O). Transfer reactions are used to identify and measure the properties of nuclei that are close in mass to the initial ones. They also are widely used to produce radioactive isotopes for which, as we describe in Chapters 8 and 9, there are many applications in industry and in medicine.

Compound nucleus reactions

In a central or near-central collision, we can have compound-nucleus or non-direct reactions, which are considered to proceed in two independent stages. First, the projectile and target fuse together and then, by successive nucleon–nucleon collisions within this combined system, the reaction energy becomes shared among many nucleons. Once this equilibration process has taken place, the average energy of a nucleon is below its binding energy and the compound nucleus can exist in an excited state for a long time (10^{-16} to 10^{-18} s) compared with the collision time for a direct reaction. Eventually, de-excitation occurs when, by chance, a single nucleon or group of nucleons acquires enough energy to escape. In this way, the compound nucleus loses its energy rather like a heated liquid drop, which cools down by evaporating molecules. The energy and nature of the outgoing products are determined by the properties of the heated compound nucleus and not by those of the colliding particles from which it is formed. This is known as the independence hypothesis of Niels Bohr. A combination of γ rays, nucleons and nuclear fragments are emitted by an excited compound nucleus, depending on the amount of internal excitation energy available. However, if this excitation energy is close to the threshold for evaporating a particle, the compound nucleus will often decay by emitting only γ rays and internal-conversion electrons in what is called a *capture* reaction.

We have already noted in reference to Figure 1.1 that the fusing together of two light nuclei will liberate a great deal of energy, but these reactions are inhibited by the Coulomb barrier at low energies, which is why it is proving difficult to harness the fusion reaction for power generation. However, inside stars, thermal energies are high enough that nuclear fusion does occur and, in Chapter 11, we describe how fusion reactions determine the life cycle of a star and the creation of the elements we find on earth.

Neutrons, on the other hand, do not experience a Coulomb barrier and can fuse at very low energies. In fact, the probability increases as the energy is reduced, varying inversely as the neutron velocity. The effect is known as the '$1/v$' law, which was discovered by Fermi when he found that neutron-induced reaction rates increased dramatically when he slowed the neutrons down. Figure 1.14 is a plot of the energy dependence (or excitation function) of the cross section for the ^{6}Li(n, α)t reaction. As can be seen, the slope of this (log–log) graph over a large energy range is close to minus one half and, thus, illustrates the $1/v$ dependence very well.

Resonance

The smooth $1/v$ dependence is not all that is shown in Figure 1.14. There is a peak at the upper end of the excitation function, which is due to a resonance in the initial

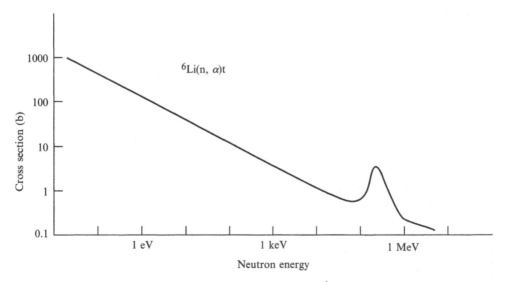

Figure 1.14   Excitation function of the cross section for the ^{6}Li(n, α)t reaction. Source: Stehn *et al.* (1958).

fusion process of the neutron with the ^{6}Li target. At about 250 keV, the neutron energy is such that the compound nucleus ^{7}Li is formed at an excitation which corresponds exactly to one of its higher energy states or natural frequencies. When this happens, energy is drawn from the entrance (exciting) channel into the reaction channel in the same way that a resonating mechanical or electrical oscillator draws energy from its driving force.

The cross section across the resonance has a Lorentzian or Breit–Wigner form:

$$\sigma(E) \propto \frac{1}{(E - E_r)^2 + (\Gamma/2)^2}. \tag{1.32}$$

This resonance shape is sketched in Figure 1.15. It peaks when E is equal to the resonance energy E_r and has a full width at half maximum of Γ. A striking example of a resonance, shown in Figure 1.16, is the low-energy, neutron-capture (n, γ) reaction by cadmium, which is dominated by a strong resonance at about 0.17 eV superimposed on the $1/\upsilon$ trend. Here, the cross section reaches a huge value of nearly 10 kb!

Resonance is very important in nuclear research studies and in many applications, such as optimizing production yields of radioisotopes and in certain analytical techniques. It will be shown in Chapter 10 how it has a major influence on the lifetime of a neutron in a reactor and, in Chapter 11, it will be seen that resonance effects are crucial in allowing the creation in stars of nuclides beyond $A = 4$.

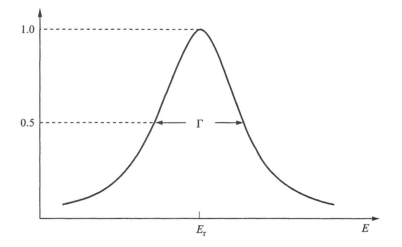

Figure 1.15   Sketch illustrating the Breit–Wigner resonance shape of Equation (1.32).

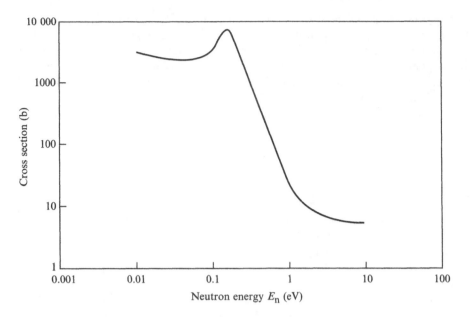

Figure 1.16   Excitation function of the total cross section for neutrons incident on cadmium, showing a strong resonance at a neutron energy of about 0.17 eV. Source: Goldsmith *et al.* (1947).

PROBLEMS 1

1.1 Assuming that a nucleus is a sphere of nuclear matter of radius $1.2 \times A^{1/3}$ fm, express the average nuclear density in SI units.

1.2 Calculate the wavelengths of 1 MeV, 10 eV and thermal (0.025 eV) γ rays, electrons, neutrons and fission fragments ($A = 100$).
    What is the ratio of the momenta of a 10-MeV ^{13}C ion and a 10-MeV photon?

1.3 Use the uncertainty principle to estimate the minimum kinetic energy of an electron confined within a nucleus of size 10 fm. Hint: Assume the electron is fully relativistic.

1.4 Before the discovery of the neutron, it was proposed that the penetrating radiation produced when beryllium was bombarded with α particles consisted of high-energy γ rays (up to 50 MeV) produced in reactions such as $\alpha + {}^9\text{Be} \rightarrow {}^{13}\text{C} + \gamma$.
    (a) Calculate the Q value for this reaction.
    (b) If 5-MeV α particles are incident on ^{9}Be, calculate the energy of the ^{13}C nucleus and, hence, determine the energy of γ radiation assuming it is emitted as a single photon. Hint: You may neglect the momentum of the γ ray relative to the ^{13}C nucleus (see Problem 1.2). Masses: $m(^4\text{He}) = 4.0026\,\text{u}$, $m(^9\text{Be}) = 9.0122\,\text{u}$, $m(^{13}\text{C}) = 13.0034\,\text{u}$.

1.5 (a) If the kinetic energy of a neutron, confined inside a cubic box, is 10 MeV in its ground state, calculate the size of the box.
    (b) What are the energies of the next three excited states?

1.6 Calculate the Q value for the α decay of ^{234}U. Compare this with the Q values for ^{234}U to decay by emitting a deuteron, triton or ^{3}He particle. Use atomic mass data given in Appendix F.

1.7 The α-active nucleus ^{212}Po in its ground state decays to the ground state of ^{208}Pb by emitting an 8.784-MeV α particle. Calculate the energy of the recoiling ^{208}Pb nucleus and, hence, the mass (in atomic mass units) of ^{212}Po. Use mass data from Appendix F.

1.8 ^{36}Cl decays into ^{36}S (35.967081 u) and ^{36}Ar. If the energy release is 1.142 MeV to ^{36}S and 0.709 MeV to ^{36}Ar, calculate the masses of ^{36}Cl and ^{36}Ar. Describe the modes of decay.

1.9 An initial number $N_A(0)$ of nuclei A decay into daughter nuclei B, which are also radioactive. The respective decay probabilities are λ_A and λ_B. If $\lambda_B = 2\lambda_A$, calculate the time (in terms of λ_A) when N_B is at its maximum. Calculate N_B (max) in terms of $N_A(0)$.

1.10 Derive the formula $N(t) = P(1 - e^{-\lambda t})/\lambda$ for the production of a radioactive nuclide (decay constant λ) as a function of time, given that the production rate is constant at P nuclei per second.

   Estimate the time it will take to produce a 100 μCi source of ^{36}Cl by irradiating 1 g of natural nickel chloride (molecular weight 129.6) in a neutronflux of 10^{14} cm^{-2} s^{-1}. The cross section for the neutron capture reaction ^{35}Cl$(n, \gamma)^{36}$Cl is 43 b and the half-life of ^{36}Cl is long (3×10^5 years). 75.8% of natural chlorine consists of ^{35}Cl.

1.11 Using the information given in Problem 1.10, calculate the fraction of ^{35}Cl which is transformed if 1 g of nickel chloride is irradiated for 1 day.

1.12 A thin (1 mg/cm^2) target of ^{48}Ca is bombarded with a 10-nA beam of α particles. A detector, subtending a solid angle of 2×10^{-3} steradians, records 15 protons per second. If the angular distribution is measured to be isotropic, determine the total cross section (in mb) for the ^{48}Ca (α,p) reaction. Take the atomic mass of ^{48}Ca to be 48 u.

2

Nuclear Structure

2.1 INTRODUCTION

No complete theory exists which fully describes the structure and behaviour of complex nuclei based solely on a knowledge of the force acting between nucleons. However, great progress has been and is being made with the aid of conceptual models designed to give insight into the underlying physics of the inherently complex situation. A model embodies certain aspects of our knowledge and, almost invariably, incorporates simplifying assumptions which enable calculations to be made. A successful model should be able to give a reasonable account of the properties it was designed to address and also make predictions of other properties which can be checked by experiment.

One of the simplest nuclear models is one in which the nucleus is regarded as a collection of neutrons and protons forming a droplet of incompressible fluid which behaves in some ways like a classical liquid drop. In the 1930s, this *liquid-drop model* was shown to account very well for the systematic behaviour of the nucleon binding energy with mass number, and confidence in its validity grew. However, as more information about nuclei became available, certain discrepancies with the liquid-drop model were noted and efforts to understand them led to the development of the *nuclear shell model*. The basis of this model is that there is an ordered structure within the nucleus in which the neutrons and protons are arranged in stable quantum states in a potential well that is common to all of them. Indeed, many nuclei behave as if most of the nucleons form an inert core and low-energy excited states are determined by a few nucleons 'outside' the core. The picture is similar to that of an atom in which electrons are arranged in shells and any chemical activity is determined by the most weakly bound, valence electrons.

The nuclear shell model has proved to be very successful in accounting for the ground-state properties and low-lying excited states of very many nuclei. However, there are certain modes of excitation, found in most nuclei, which are better described as collective vibrations and rotations of the nucleus represented as a liquid drop. Conceptually, the liquid-drop and shell models appear to be very different and the idea of nucleons moving in stable orbits and interacting weakly is not obviously consistent with that of a collection of nucleons forming a drop of nuclear fluid. Reconciling these two apparently contradictory pictures of nuclear behaviour has been a central goal of scientists developing unified theories of nuclear structure.

In the next section, we outline some of the properties of the force acting between nucleons in nuclei and describe how the liquid-drop and simple, independent-particle models led to the development of a semi-empirical formula, which gives a remarkably good representation of nuclear masses across the whole range of the periodic table. Deviations from this formula, and the development of the nuclear shell model are described in Section 2.3. In Section 2.4 we present examples of nuclear properties that can be described in terms of a single, active neutron or proton in the nucleus, and in Section 2.5 we discuss collective nuclear behaviour and present evidence for it, including sequences of excited states which imply that certain nuclei are non-spherical even in their ground states.

2.2   NUCLEAR MASS

2.2.1   The nuclear force

The properties of a nucleus are determined by the forces acting between its constituent neutrons and protons. The nuclear part of this force is extremely complicated and difficult to derive from first principles and express in an explicit form. Nuclear-structure calculations generally employ empirical forms derived from experimental studies of nucleon–nucleon scattering and from the properties of the deuteron (^{2}H nucleus), which is the simplest example of a nucleus with more than one nucleon. In this text, a detailed knowledge of the nucleon–nucleon force is not essential and we summarize here only those properties which we will need to account for some general features of nuclear structure.

- *Short range*: The nuclear force between nucleons has a strongly attractive component, which acts only over a short range. The clearest evidence for this was found in early scattering experiments, which showed that deviations from the Rutherford scattering cross section appear only when the surfaces of the colliding nuclei were within a few fermis of each other. The short-range property of the force was confirmed later by detailed studies of nucleon–nucleon scattering.

- *Repulsive core*: Nucleon–nucleon scattering at very high energies revealed that the effective internucleon force also has a repulsive component, which dominates at very short distances (less than about 0.5 fm). The nucleon–nucleon potential energy as a function of distance is shown schematically in Figure 2.1. It has a minimum at the point where the attractive and repulsive components are equal and opposite. This is a position of equilibrium and, therefore, a nucleon inside a nucleus will tend to maintain an average separation with its neighbours. This leads to the concept of the *saturation of the nuclear force*, which is that the nuclear binding energy of a nucleon due to the interactions with its neighbours approaches a saturation value independent of mass number A. Evidence for this is the approximately constant density of nuclear matter in nuclei (see Section 1.6.5) and the relatively small variation of the binding energy per nucleon for A greater

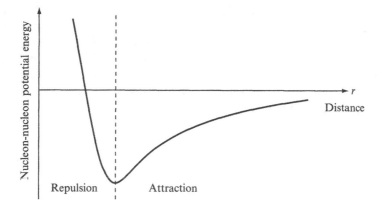

Figure 2.1   Schematic representation of the potential energy (PE) as a function of distance between two nucleons. As the nucleons approach each other, they experience an attractive force, which leads to a decrease in PE. At shorter distances ($\lesssim 0.5$ fm), the force becomes repulsive and the PE increases. A minimum in PE occurs where the attractive and repulsive forces are equal and opposite.

than about 20 (see Figure 1.1). The situation is similar to that of a liquid where the separation between molecules is roughly constant and is the reason why the density of the liquid and the binding energy of a molecule in it are both independent of the size of a drop.

- *Charge dependence*: It is found that the nuclear force is charge symmetric, i.e. the force between two neutrons in given states is the same as that between two protons in the same states. An indication of this is the fact that, in the case of light nuclei ($A \lesssim 40$), for which the energy due to the Coulomb force is small compared with the nuclear binding energy, the stable ones are those for which $N \simeq Z$. This would not be the case if the n–n and p–p forces were significantly different.

  It is also observed that the force is almost *charge independent*, i.e. it is the same for n–p, n–n and p–p. However, we must be careful since nucleons are spin 1/2 particles and must obey the Pauli exclusion principle. This means that, for two neutrons or two protons to occupy the same spatial state and be physically close to each other, they must be in opposite spin states. One must be 'spin up' and the other 'spin down,' which gives a net spin of zero (quantum number $S = 0$) for the lowest state of the diproton or dineutron. A neutron-plus-proton combination is not restricted in this way, and their spins can be arranged to be either parallel ($S = 1$) or anti-parallel ($S = 0$). In the $S = 0$ state, the force is the same as it is for either the diproton or dineutron. However, the n–p force is stronger in the $S = 1$ state, which accounts for the fact that a neutron and proton can form a bound system (the deuteron), but neither the diproton nor the dineutron has a bound state. Effectively, this means that the average force between a neutron and a proton inside the nucleus is greater than the force between two identical nucleons. The difference is about a factor of two.

- *Spin dependence*: We have just noted that the nucleon–nucleon force is different for two nucleons in the $S = 0$ or $S = 1$ state and, therefore, is spin dependent.

There is a further contribution to the internucleon force which depends on the orientation of the total spin angular momentum **S** of the two nucleons and any orbital angular momentum **L** in their relative motion. This is called a *spin–orbit force* and it varies according to the scalar product **L·S**. The force is attractive if **S** and **L** are parallel and repulsive if they are anti-parallel. If a nucleon is well inside the nucleus, it encounters as many other nucleons with spin up as with spin down and so some forms of spin dependence will average out. However, as we shall see in Section 2.3.3, the spin–orbit force does not average out everywhere and leads to an important contribution to the force experienced by a nucleon in a nucleus.

2.2.2   Semi-empirical mass formula

The semi-empirical mass formula (SEMF) was originally devised in 1935 by C.F. Von Weizsäcker to represent known nuclear masses in terms of a few parameters and enable useful estimates to be made of the masses of unknown nuclei. The mass of an electrically neutral atom $m(A, Z)$ is given in Equation (1.1) as the sum of the masses of the constituent N neutrons, Z protons and Z electrons reduced by B/c^2, the mass equivalent of the binding energy of the nucleus. In the SEMF, the total binding energy is expressed as a sum of five terms:

$$B = a_v A - a_s A^{2/3} - a_c \frac{Z^2}{A^{1/3}} - a_a \frac{(N-Z)^2}{A} \pm \Delta \qquad (2.1)$$

where a_v, a_s, a_c, a_a and Δ are constants obtained by fitting the formula to experimentally determined values. The form of each term, as outlined below, is based on physics principles and general properties of the nuclear force noted above. In deriving these terms, the nucleus is treated as an assembly of particles, which behave rather like molecules in a liquid drop. Quantum effects, including the Pauli principle, are taken into account using a simple model of spin $1/2$ particles confined to a box. The five terms in Equation (2.1) from left to right are due to volume, surface, Coulomb, symmetry and pairing effects, respectively, as will now be shown.

- *Volume energy*: The mass dependence of this term follows directly from the saturation of the nuclear force, described in the last subsection. A nucleon in the nuclear interior interacts, on average, with a fixed number of neighbouring nucleons within the short range of the nuclear force. The binding energy per nucleon, therefore, will be constant and the total contribution to the nuclear binding energy will be proportional to A and, therefore, proportional to the nuclear volume.

- *Surface correction*: In a finite nucleus, a nucleon near the surface interacts with fewer nucleons and, therefore, is less tightly bound than if it is in the interior. Thus, a term proportional to the surface area (or the square of the nuclear radius R) must be subtracted from the volume term to correct for this and, since $R \propto A^{1/3}$, the surface term is proportional to $A^{2/3}$.

- *Coulomb energy*: A second negative contribution to the pure liquid-drop binding energy arises from the long-range, repulsive electric force acting between protons. The electrical potential energy of a uniformly charged sphere, radius R and total charge Q, is equal to $\frac{3}{5}(Q^2/4\pi\varepsilon_0 R)$ (see Problem 2.1) and, since the nuclear charge is Ze, we obtain a term of the form given in Equation (2.1).

- *Symmetry term*: This term expresses the charge-symmetric nature of the nucleon–nucleon force, which has the consequence, noted above, that, in the absence of the Coulomb force, the most stable nuclei would have equal numbers of neutrons and protons. The form of the symmetry term follows from the Pauli principle and the fact that the effective force in the nucleus is stronger between unlike nucleons (n–p) than between identical nucleons (n–n or p–p). There are two contributions to the symmetry term: one from the nuclear potential energy (PE) of the nucleons and the other from their kinetic energies. In the simple derivation presented below, we assume that both N and Z are large compared with unity.

  The PE of each nucleon (in the nuclear volume) arises from its interaction with a fixed amount of nuclear matter of which a fraction $f_n = N/A$ consists of neutrons and $f_p = Z/A$ consists of protons. Let v_{nn}, v_{pp} and v_{np} represent the average PE arising from the interaction between two neutrons, two protons, and a neutron and a proton, respectively. The binding energy due to N neutrons is proportional to $N(v_{nn} f_n + v_{np} f_p)$ and, for Z protons, it is proportional to $Z(v_{np} f_n + v_{pp} f_p)$. Since $v_{nn} = v_{pp} = v$ and $v_{np} \approx 2v$, we can write the total PE contribution to the binding energy as being proportional to

$$N(vf_n + 2vf_p) + Z(2vf_n + vf_p) = \frac{v}{A}(N^2 + 4NZ + Z^2) = \frac{v}{2}\left(3A - \frac{(N-Z)^2}{A}\right)$$

where we have substituted for f_n and f_p. Thus, there are two components to the PE: one which depends on A and is included in the volume term and another one, the symmetry term, which depends on $-(N-Z)^2/A$.

The kinetic energy contribution to the symmetry term is obtained by using a result from a simple, quantum-mechanical model of particles occupying states in a three-dimensional box with rigid walls. This model, called the Fermi-gas model, is widely used for obtaining results for the behaviour of fermions in different situations.

Calculating the energies of the allowed states given by the model is straight forward and details are given in Appendices B and C. For a nucleus in its ground state, the A nucleons will occupy the states according to the exclusion principle. Since neutrons and protons are distinguishable, a given state can hold up to two neutrons and two protons. For a nucleus with the same number of neutrons and protons ($N = Z$), the arrangement (configuration) with the lowest energy is the one in which neutrons and protons fill the lowest available spatial states in pairs up to the same energy, which is known as the Fermi energy ε_F. This configuration is shown schematically in Figure 2.2(a). If now $N - Z$ protons are changed into neutrons [see Figure 2.2(b)] or vice versa, the total kinetic energy must increase

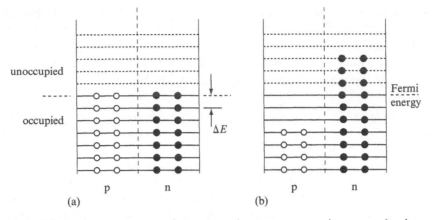

Figure 2.2   Schematic arrangement of protons and neutrons occupying energy levels near the Fermi energy ε_F. In (a), proton and neutron states are filled to the same level, whereas, in (b), a number of protons have been changed into neutrons. The average energy between adjacent levels is ΔE.

because of the exclusion principle. The $N - Z$ nucleons come from $(N - Z)/2$ states, which are at an average energy of $(N - Z)\Delta E/4$ below ε_F and transfer to states at an average energy $(N - Z)\Delta E/4$ above ε_F, where ΔE is the average energy spacing between states close to ε_F. Therefore, the average energy change per nucleon is $\frac{1}{2}(N - Z)\Delta E$. It is shown in Appendix C that, if the nucleon density is constant, as it is in nuclear matter, the Fermi energy is independent of A and the spacing of states near ε_F is proportional to $1/A$. Therefore, the kinetic energy contribution to the change in binding energy is proportional to $-(N - Z)^2/A$, which is the same form as the PE contribution.

- *Pairing*: The final term in the SEMF reflects the tendency for like nucleons to form spin-zero pairs in the same spatial state. When coupled like this, extra binding comes from the strong overlap of their spatial wave functions, which means that they will spend more time being closer together within range of the nuclear force than when they occupy different orbitals. The pairing contribution is positive, i.e adds to the binding energy if N and Z are both even and the nucleons are all coupled to form spin-zero pairs, and is negative if both N and Z are odd. It is zero if either N or Z is odd (A odd). The need for this term can be justified by noting that all the other terms in the mass formula vary smoothly with A whenever N or Z changes and the pairing term corrects for the dependence on whether the change causes a pair to be formed or broken. Different forms for the A dependence of the pairing term have been published. A common version gives $\Delta = 12/A^{1/2}\,\text{MeV}$.

The parameters in the SEMF are adjusted to give the best overall agreement with empirical measurements. As shown in Figure 2.3, a typical parameter set reproduces B/A values reasonably well over a wide range of A.

We can now use the formula to examine the relative stability of nuclei.

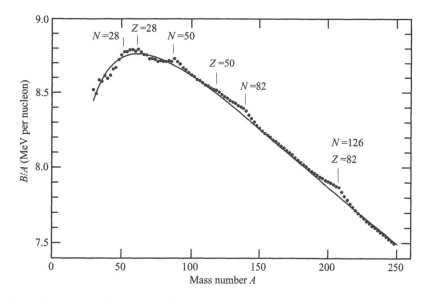

Figure 2.3 Experimental values of binding energy per nucleon B/A plotted as a function of mass number A. The smooth curve represents the semi-empirical mass formula with $a_v = 15.56$ MeV, $a_s = 17.23$ MeV, $a_a = 23.28$ MeV and $a_c = 0.7$ MeV. Each point represents an odd–even nucleus or an average of neighbouring nuclei (for A even) so that there is no effect due to the pairing term. Significant differences between experimental values and the SEMF occur near indicated values of N and Z.

2.2.3 Nuclear stability

Beta stability

If we substitute Equation (2.1) for the binding energy B into Equation (1.1), we see that for isobaric nuclei (A constant), the atomic mass $m(A, Z)$ is quadratic in Z. There will be a particular value of Z that corresponds to the most tightly bound nucleus, i.e. the one with the smallest mass. If A is odd, a plot of m versus Z will be a single parabola because the pairing term Δ is zero. If A is even, there will be two parabolas, separated in energy by 2Δ, because the sign of Δ depends on whether N and Z are both even or odd. Pairing favours N and Z as both even and so the parabola for even–even nuclei lies below that for odd–odd nuclei. Examples are shown in Figure 2.4 for $A = 121$ and 122.

The mass–energy difference between two adjacent isobars is the energy available for a radioactive transition ($\beta^\pm$ decay or electron capture) from the heavier to the lighter one. For A odd, since there is a single parabola, only one isobar is expected to be stable and to exist in nature unless the energy difference is so small that the half-life of the heavier one is very long ($> 10^9$ years). Examples of the latter are: ^{113}Cd (7.7×10^{15} years), ^{115}In (4.4×10^{14} years) and ^{123}Te ($> 6 \times 10^{14}$ years). For A even, there are two parabolas and it is common to have two stable, even–even isobars, as is

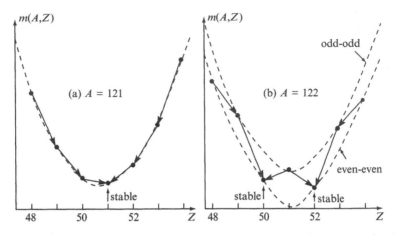

Figure 2.4   Stability of nuclei in isobaric mass sequences. The nuclear mass $m(A, Z)$, calculated using the SEMF, is plotted as a function of atomic number for (a) $A = 121$ and (b) $A = 122$. Note that when A is even, points for the different isobars fall on two parabolas because of the pairing energy term in the SEMF. Beta-decay transitions are indicated by arrows.

the case in Figure 2.4(b), because the intervening odd–odd nuclide is more massive than either of them. For the same reason, it is also common to find an odd–odd nucleus which can decay in either direction, as shown in the figure, releasing energy by converting one of its protons into a neutron or vice versa.

   The existence of the pairing energy means that the SEMF does not predict any stable odd–odd nuclei. However, there are four stable odd–odd nuclei: ^{2}H, ^{6}Li, ^{10}B and ^{14}N, but these are very light and the formula is not reliable for such low values of A. There are also a few others: ^{40}K, ^{50}V, ^{138}La, ^{176}Lu and ^{180}Ta, which are not strictly stable but have very long half-lives because of certain factors which hinder their decay (see Section 3.3).

Instability of heavy nuclei

As A increases, the relative importance of the Coulomb term increases and this results in B/A decreasing with A from a maximum at $A \sim 60$ (Figure 2.3). Eventually, it becomes energetically possible for the nucleus to divide into two smaller fragments (fission). For example, if ^{238}U were to split into two mass 119 nuclei, about 200 MeV of energy would be liberated because B/A for the final nuclei is about 1 MeV greater than it is for ^{238}U. This process does not happen spontaneously because, in order for fission to occur, a nucleus must first become deformed from its equilibrium shape and this increases the surface area, which requires energy according to the second term in the SEMF. The increase in surface energy is offset to some extent by an energy gain from the Coulomb term, which decreases with deformation. The stability of a heavy nucleus, therefore, depends on a sensitive balance between these two tendencies.

For all naturally occurring nuclei, the surface-energy term dominates and the nucleus exists in a state of stable equilibrium, as shown in Figure 2.5, which is a schematic plot of the potential energy of a heavy nucleus as a function of deformation. Initially, energy is required to deform the nucleus from equilibrium and there is a strong restoring force that acts to return the nucleus to its equilibrium shape. However, at a certain point, the relative strengths of the surface and Coulomb terms reverse. Coulomb repulsion begins to dominate, driving the deformation irreversibly on to larger and larger values until the nucleus breaks in two. Then, as the mutual electrical repulsion continues to drive them apart, potential energy is converted into kinetic energy of the fragments.

The energy required to overcome the barrier to fission is called the *activation energy* or *fission barrier* and is about 6 MeV for $A \sim 240$. It is found that the activation energy decreases as A increases. Eventually, a point is reached where the activation energy disappears altogether, the nucleus has no position of stable equilibrium and, even if it could be created, it would undergo very rapid spontaneous fission. The SEMF enables us to obtain an estimate for this condition.

Consider a spherical nucleus, radius R being deformed into an ellipsoid, which has a volume $\frac{4}{3}\pi ab^2$ where $a = R(1 + \varepsilon)$ is the semi-major axis, $b = R(1 + \varepsilon)^{-1/2}$ is the semi-minor axis, and ε is the eccentricity of the elliptical shape. Note that the volume is unchanged by the deformation as it should be if we regard nuclear matter to be incompressible. As the nucleus deforms, it can be shown that the surface area increases as $4\pi R^2(1 + \frac{2}{5}\varepsilon^2 + \cdots)$ and, therefore, there will be an increase, to first order, in the surface energy term of $\frac{2}{5}a_s A^{2/3}\varepsilon^2$. It can also be shown that the deformation will change the Coulomb energy term by a factor $(1 - \frac{1}{5}\varepsilon^2 + \cdots)$, leading to a decrease of $\frac{1}{5}a_c Z^2\varepsilon^2/A^{1/3}$ compared with the spherical state. The *change* in binding energy is given by the sum of the two changes:

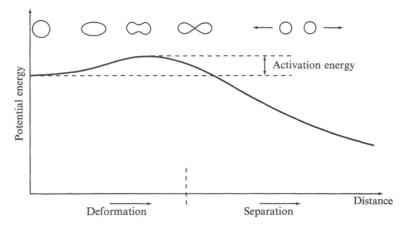

Figure 2.5 Schematic plot of the potential energy of a nucleus, first, as a function of deformation from a spherical shape and then as a function of separation after fission has occurred. The activation energy (fission barrier) is the input energy required to overcome the barrier preventing fission from taking place spontaneously.

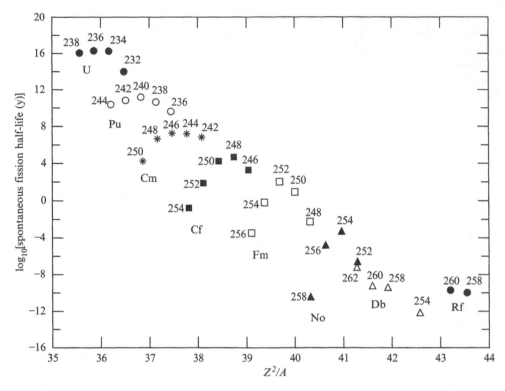

Figure 2.6   Logarithm of spontaneous fission half-lives (in years) for several sequences of isotopes plotted as a function of the parameter Z^2/A.

$$\Delta B = -\left(\frac{2}{5}a_s A^{2/3}\varepsilon^2 - \frac{1}{5}a_c Z^2\varepsilon^2/A^{1/3}\right). \tag{2.2}$$

If the magnitude of the second term exceeds the first, any change in deformation from equilibrium will lead to a *gain* in binding energy. The nucleus will be shape unstable and undergo very rapid spontaneous fission. The limiting condition for this to occur is when $\Delta B = 0$ and, therefore,

$$\frac{Z^2}{A} = \frac{2a_s}{a_c} \sim 49 \tag{2.3}$$

where we have substituted typical values for a_s and a_c. This estimate is approximate because of quantum-mechanical effects and because many of the heaviest nuclei are non-spherical even in their groundstates (see Section 2.5). Spontaneous fission is observed in nuclei with Z^2/A as low as 35, as is shown in Figure 2.6. However, if we extrapolate the trend in half-life to a very short time ($\sim 10^{-20}$ s) corresponding to zero activation energy, we obtain a Z^2/A value reasonably close to the SEMF estimate.

2.3   NUCLEAR SHELL MODEL

2.3.1   Evidence for shell structure

There is evidence which suggests that nuclei with certain numbers of neutrons and protons are particularly stable.

- The number of stable isotopes for $Z = 20$ and 50 is larger than average, as is the number of stable isotones with $N = 50$ and 82.

- Experimental values of binding energy per nucleon deviate most notably from the SEMF curve for certain values of N and Z, as indicated in Figure 2.3. There is a particularly sharp discontinuity in the trend at ^{208}Pb for which $N = 126$ and $Z = 82$.

- The energy required to remove a neutron or a proton from a nucleus (the single-nucleon separation energy) shows variations from a smooth trend at values of N or Z equal to 2, 8, 20, 28, 50, 82 and 126. This is illustrated in Figure 2.7 in which the difference between the energy required to remove two neutrons from a nucleus and

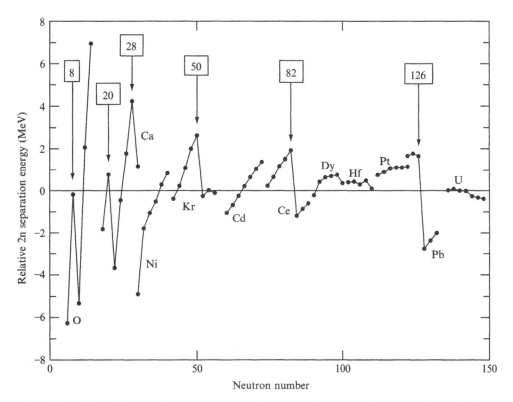

Figure 2.7   The difference between measured two-neutron separation energies and those predicted by the SEMF, plotted against neutron number. The lines connect sequences of isotopes. Discontinuities in the otherwise smooth trend are evident.

that predicted by the SEMF is plotted as a function of N. The sharp discontinuities at particular values of N are evident.

The SEMF based on the liquid-drop model does not predict any of these effects, and the special values of N and Z noted above are referred to as the nuclear magic numbers.

The effect of magic numbers has a familar consequence in atomic physics, namely, the existence of the inert (noble) gases. Within an atomic shell, electron states are closely spaced in energy and there is a sizeable energy gap between one shell and the next. Electrons in an unexcited atom occupy states in order of increasing energy with the Pauli principle restricting the number allowed in each state to two. The numbers of electrons required to complete successive shells are the atomic 'magic numbers'. They correspond to the stable, inert gases: He ($Z = 2$), Ne ($Z = 10$), Ar ($Z = 18$), Kr ($Z = 36$), Xe ($Z = 54$), etc. The closed-shell structure is particularly stable because a considerable amount of energy is required to break an electron away from a filled shell and promote it across the energy gap to the next shell of states.

The existence of shells implies that electrons occupy well-defined states and, to a first approximation, move in these states independently of each other. It is this concept of independent-particle motion that is crucial to understanding the shell model of the nucleus.

2.3.2   Independent particle motion and the shell model

In the early days of nuclear physics, it was not expected that the independent-particle approximation would be valid for nuclei because, unlike an electron in an atom, a nucleon is not very small compared with a nucleus. It was thought that nucleons would collide with each other and that the picture of them moving in stable quantum-mechanical states in an average nuclear potential would not be useful. However, this is a view based on classical expectation. In reality, a nucleus is a quantum-mechanical object and an important consequence of this is that the number of possible nucleon–nucleon collisions is greatly restricted by the exclusion principle.

As we noted in the previous section, nucleons will occupy allowed energy states up to a certain energy – the Fermi energy ε_F (see Figure 2.2). In such a situation, there can be no transfer of energy or momentum between two colliding nucleons in filled states because all states lower in energy are occupied and, provided the nucleus is close in energy to its ground state, there is insufficient energy available to move either or both of them into unoccupied states. Thus, collisions cannot occur and the nucleons move within the nucleus as if they were transparent to each other.

The potential energy V experienced by a nucleon in a nucleus is generated by its interactions with the other nucleons. Its spatial form, therefore, will be similar to that of the density of nuclear matter, which is roughly constant in the nuclear interior and decreases in the surface over a distance approximately equal to the range of the nuclear force. The nuclear potential energy experienced by a nucleon in a nucleus is often parametrized in the following way:

$$V(r) = \frac{-V_o}{1 + \exp[(r - R)/a]} \tag{2.4}$$

which is known as the Woods–Saxon or Fermi form of the potential. This form is plotted in Figure 2.8; R is the radius where the potential is half its central value (for $R \gg a$) and a is a measure of the diffuseness of the nuclear surface. A typical set of parameters is: $V_0 = 50$ MeV, $R = 1.25 \times A^{1/3}$ fm and $a = 0.6$ fm. The Schrödinger equation can be solved in three dimensions by using this potential, and the allowed energy states are shown as the central set of states in Figure 2.9. Also shown on the left-hand side of this figure are levels predicted for an infinite, spherical well with a sharp edge. This shows that the ordering of the states is affected by changing the shape of the potential, but the number of them stays the same.

The states in Figure 2.9 are labelled s, p, d, f, g, etc., according to the orbital angular momentum quantum number $\ell = 0, 1, 2, 3, 4$, etc. Successive ℓ states in a nucleus are labelled sequentially in order of increasing energy: e.g. 2p and 3s indicate the second $\ell = 1$ and third $\ell = 0$ states, respectively. For each value of ℓ, there are $2\ell + 1$ substates corresponding to allowed orientations of the angular momentum along a given direction. For spherical nuclei, the potential wells are spherically symmetric and in this case, substates of a given ℓ all have the same energy, and each ℓ state is said to consist of $2\ell + 1$ degenerate states.

The A nucleons in a nucleus occupy states in ascending order according to the exclusion principle. Thus, each ℓ state can contain up to $2(2\ell + 1)$ nucleons of each type. A magic number occurs when there is a large energy gap between the last filled level and the next unoccupied one. The first magic number is 2, which is when the 1s state is filled with either neutrons or protons. The next one is 8 when both the 1s and 1p states are filled. The magic numbers predicted by the different wells are indicated in Figure 2.9. The first three correspond to experimental values, but neither of these wells accounts for the higher ones. Many attempts to solve the discrepancy were made in the 1940s using a variety of different well shapes but it was not until 1949 that it was shown by Mayer and by Haxel, Jensen and Suess that the proper spacing of levels with the observed shell closings could be obtained by adding a *spin–orbit* term to the nuclear potential.

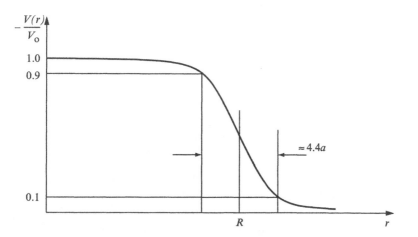

Figure 2.8   The Woods–Saxon or Fermi form of a realistic nuclear potential. The distance over which $-V(r)/V_0$ falls from 90 to 10% of its central value is 4 ln3 times the diffuseness parameter a.

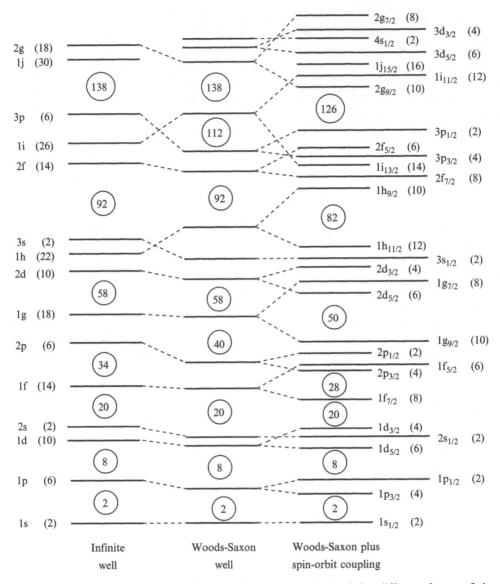

| Infinite | Woods-Saxon | Woods-Saxon plus |
| well | well | spin-orbit coupling |

Figure 2.9 Sequences of bound single-particle states calculated for different forms of the nuclear shell-model potential. The number of protons (and neutrons) allowed in each state is indicated in parentheses and the numbers enclosed in circles indicate magic numbers corresponding to closed shells.

2.3.3   The spin–orbit potential

The spin–orbit (SO) potential has the form $-V_{so}(r)\mathbf{l}\cdot\mathbf{s}$ where $\mathbf{l}$ and $\mathbf{s}$ are the orbital and spin angular momenta, respectively of a nucleon moving in the nuclear well. It increases the binding energy if $\mathbf{l}$ and $\mathbf{s}$ are parallel and reduces it if they are anti-parallel. Thus, nucleon states with different values of total angular momentum

$\mathbf{j} = \mathbf{l} + \mathbf{s}$ will have different energies. A nucleon has a spin quantum number $s = 1/2$, which means that each $\mathbf{l}$ state can have a total angular momentum quantum number $j = \ell \pm 1/2$ (except $\ell = 0$ for which only $j = 1/2$ is allowed). The particular form used for $V_{so}(r)$ is not important. The interesting features stem from the $\mathbf{l}\cdot\mathbf{s}$ term, as we show below.

The SO force for nucleons in nuclei arises from the SO force between pairs of nucleons, noted at the end of Section 2.2.1. This force exists if the spins of the two nucleons are parallel ($S = 1$) and it is attractive or repulsive, depending on whether the nucleon spins point in the same or in the opposite direction to the relative angular momentum between the nucleons. There is no SO force if the nucleon spins oppose each other ($S = 0$).

In order to understand the effect of the nucleon–nucleon SO force inside a nucleus, we will use a crude, semi-classical picture of a nucleon (x) orbiting as shown in Figure 2.10 and passing between two other nucleons labelled a and b. We need only consider the case where the spins of a and b are parallel to the spin of x, since there is no SO force between any pair of nucleons if their spins are anti-parallel. Assume, first, that the spins of all three nucleons point out of the page. In this case, the spin of x is parallel to its orbital angular momentum $\mathbf{l}$. However, the relative angular momenta between x and nucleons a and b are in opposite directions and so the nucleon–nucleon SO force between x and a is attractive and between x and b is repulsive. In the interior of the nucleus, the effects tend to cancel out because the nucleon x will pass equal numbers of nucleons with spins parallel to its own on either side. However, in the surface, there will be more nucleons at smaller radii (like a) and fewer at larger radii (like b). Therefore, there will be a net nuclear SO attraction which will be concentrated in the nuclear surface. If the nucleon moves in the opposite direction, the direction of $\mathbf{l}$ is reversed and the SO force is repulsive.

The qualitative effects of the SO potential on the splitting and ordering of the energy levels is easily obtained. Since,

$$\mathbf{j}^2 = (\mathbf{l} + \mathbf{s})^2 = \mathbf{l}^2 + \mathbf{s}^2 + 2\mathbf{l}\cdot\mathbf{s} \qquad (2.5)$$

we have, for a nucleon in an orbital with angular momentum quantum numbers ℓ and j:

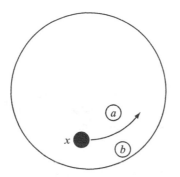

Figure 2.10 Origin of the spin–orbit force. The diagram is a classical representation of a nucleon x orbiting inside a nucleus and passing between two other nucleons a and b.

$$\langle \mathbf{l \cdot s} \rangle = \frac{1}{2}[j(j+1) - \ell(\ell+1) - s(s+1)]\hbar^2. \tag{2.6}$$

Since $s = \frac{1}{2}$, this gives $\langle \mathbf{l \cdot s} \rangle = \frac{1}{2}\ell\hbar^2$ for $j = \ell + \frac{1}{2}$, and $\langle \mathbf{l \cdot s} \rangle = -\frac{1}{2}(\ell+1)\hbar^2$ for $j = \ell - \frac{1}{2}(\ell \neq 0)$. Thus, each ℓ state is split by an energy proportional to $2\ell + 1$.

The effect of adding the SO term to the central potential is shown by the set of states on the right-hand side in Figure 2.9. Here, the label on each state includes the j quantum number as a subscript and the occupation of each state is equal to $2j + 1$. Note that labelling the states with j includes the effect of spin and so the number of nucleons that can be accommodated is simply equal to the number of substates of j.

The addition of the spin–orbit potential significantly alters the way the states group together in shells. The 1f ($\ell = 3$) state has $j = 7/2$ and $5/2$ and, with the appropriate value of V_{so}, the splitting of these $f_{7/2}$ and $f_{5/2}$ states is sufficient to give a magic number at 28. The g state ($j = 9/2$ and $7/2$) is split even further between the $g_{9/2}$ and $g_{7/2}$ states, resulting in a magic number at 50. The remaining magic numbers also occur at the observed values. The success of the independent-particle shell model proved to be the crucial step in understanding the properties and behaviour of nuclei in terms of their structure.

2.4   SINGLE-PARTICLE FEATURES

We have discussed the independent-particle shell model in which a nucleon is assumed to experience only an average field due to the other nucleons in the nucleus. Although it is a poor approximation in general, this extreme version of the shell model does reproduce many features of nuclei, which consist of closed shells of nucleons with either a single extra nucleon (single particle) or with one nucleon short of a closed shell (single hole). We shall illustrate this by comparing predicted and observed properties of energy levels for several examples of such nuclei.

Nuclear states are characterized by two quantum numbers associated with angular momentum and parity. We reviewed some basic quantum aspects of angular momentum in Chapter 1. Here, we introduce the concept of *parity*, which is important in understanding many aspects of nuclear structure and nuclear transitions.

2.4.1   Parity

Parity is a property of a wave function that depends on the result of the operation of inversion in which all co-ordinates are reflected through the origin, i.e. $\mathbf{r} \rightarrow -\mathbf{r}$. In Cartesian co-ordinates, inversion results in: $x \rightarrow -x, y \rightarrow -y, z \rightarrow -z$ and in spherical co-ordinates: $r \rightarrow r, \theta \rightarrow \pi - \theta, \phi \rightarrow \pi + \phi$.

If the system is invariant under this inversion, [i.e. if $V(\mathbf{r}) = V(-\mathbf{r})$ in the Schrödinger equation] the wave function describing the system has the property that $\psi(\mathbf{r}) = \pm\psi(-\mathbf{r})$. This means that there are two classes of wave function, namely, those with even (+) parity, which are unchanged by inversion, and those with odd

(−) parity, which change sign.[1] Accordingly, we shall assign a parity quantum number $\pi(=\pm)$ to wave functions (states) of even and odd parity, respectively.

As a simple example, consider the allowed wave functions for a particle confined to a one-dimensional box (see Figure 1.3). The box is centred at $x = 0$ and so the potential is symmetric under inversion, i.e. $V(x) = V(-x)$. The allowed wave functions are either cosine or sine functions of x, which have even or odd parity as required. In a spherically symmetric three-dimensional well, the potential is independent of angle, and any solution to the wave equation in spherical co-ordinates is a product of a radial part and an angular part and has the form: $R(r) Y_\ell^m(\theta, \phi)$, where ℓ and m are the quantum numbers for the orbital angular momentum and its projection, respectively. The function $Y_\ell^m(\theta, \phi)$ is called a *spherical harmonic* (see Appendix D). A few of them are listed in Table D.1. The radial part of the total wave function is unchanged by inversion, but the angular part does change and it can be shown (see Appendix D) that

$$Y_\ell^m(\pi - \theta, \pi + \phi) = (-1)^\ell Y_\ell^m(\theta, \phi). \tag{2.7}$$

Thus, all states of a particle in a central potential have a definite parity, which is odd if ℓ is odd and even if ℓ is even. If all the forces acting on a system of particles are invariant under inversion, it can be shown that parity is conserved in any time development of the system that may take place. There is strong evidence that electromagnetic and strong nuclear forces conserve parity and, therefore, wave functions of nuclear states, which are products of wave functions of nucleons interacting under the influence of these forces, will have definite parity. The parity of a wave function representing a collection of particles will be odd if there is an odd number of particles with odd parity and even otherwise. We shall see in the next chapter that parity is an important factor in determining what nuclear transitions can take place.

2.4.2 Spectra of single-particle or single-hole nuclei

We can now examine spins and parities of some energy states of nuclei near closed shells and compare them with predictions of the simple shell model. Energy spectra of the ground and low-lying energy levels of a number of nuclei with single-particle or single-hole configurations are shown in Figure 2.11. For the listed states, the energy is given in MeV and the spin and parity quantum numbers (I^π) are also shown.

The shell model would predict the ground state of ^{17}O to consist of protons and neutrons completely filling the $1s_{1/2}$, $1p_{3/2}$ and $1p_{1/2}$ shell-model states with an extra neutron in the $1d_{5/2}$ state in agreement with the ground-state I^π assignment of $5/2^+$. This state is said to have a neutron configuration of $(1d_{5/2})^1$. The remaining part of the configuration representing the closed shells, in this case $(1s_{1/2})^2 (1p_{3/2})^4 (1p_{1/2})^2$, is normally omitted since these nucleons are considered to form an inert core with

---

[1] See, for example, Mandl (1992), Section 4.1.

(a)

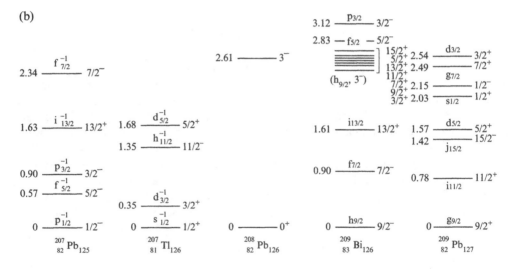

(b)

Figure 2.11   Energy-level diagrams for nuclei near the doubly closed-shell nuclei (a) ^{16}O and (b) ^{208}Pb. The excitation energy is given in MeV on the left-hand side of each level and the spin and parity quantum numbers I^π on the right. For the nuclei neighbouring the closed-shell nuclei, single-particle and single-hole configurations are indicated on some levels.

$I^\pi = 0^+$. The first excited state is also correctly predicted and corresponds to promoting the extra-core (valence) neutron to the $2s_{1/2}$ state.

The nucleus of ^{17}F should be similar to that of ^{17}O, the only difference being that the extra particle is a proton rather than a neutron. Indeed, the level structures of the two nuclei are similar and this is expected for light nuclei because the change in Coulomb energy due to changing a neutron into a proton should not significantly affect the relative spacing of the neutron and proton shell-model states.

The nucleus of ^{15}O is an example of a single-hole nucleus with a configuration $(1p_{1/2})^{-1}$, indicating a vacancy in the $1p_{1/2}$ shell-model state. The spin and parity assignments of this state are easily obtained, since the $1p_{1/2}$ shell-model state with one vacancy contains just one $1p_{1/2}$ nucleon. Hence, the spin and parity assignments of the ground state are $1/2^-$, in agreement with experiment.[2] The ground state of the corresponding proton-hole nucleus ^{15}N is the same and is also in accord with the shell model.

$^{17}_{8}O_9$, $^{17}_{9}F_8$ and $^{15}_{8}O_7$, $^{15}_{7}N_8$ are examples of pairs of so-called mirror nuclei, which are isobaric and differ only to the extent that the Z value of one nucleus is the N value of the other. According to the SEMF, the difference in the binding energies of mirror nuclei arises from the Coulomb-energy term and can be related to nuclear size (see Problem 2.1). Other examples of mirror nuclei are: $(t, ^3He)$, $(^{13}C, ^{13}N)$ and $(^{39}K, ^{39}Ca)$.

At the other end of the periodic table are the nuclei ^{207}Tl, ^{207}Pb, ^{209}Bi and ^{209}Pb neighbouring the doubly closed-shell nucleus ^{208}Pb. The ground states and several excited states of ^{209}Bi and ^{209}Pb are given by promoting the single proton or neutron within the shells above $Z = 82$ or $N = 126$, respectively. States of ^{207}Tl and ^{207}Pb are similarly given by the shell model by moving the proton or neutron hole down the otherwise-filled shells below $Z = 82$ or $N = 126$, respectively. Similar correspondence with the shell model is found in regions near other closed shells, and this level of agreement in accounting for spins and parities of low-lying nuclear states has been a great success for the shell model. However, even these examples show that the approximation of an inert, closed-shell core soon begins to break down.

In each of the spectra of the single-particle and single-hole nuclei neighbouring ^{208}Pb, there are several states with excitation energies near 2.6 MeV. These states have been identified in ^{209}Bi and are shown in Figure 2.11. They are not single-particle or single-hole states, but states in which the ^{208}Pb core has been excited into a 3^- state at 2.61 MeV (see Figure 2.11). This is an example of a collective state, which we discuss in the next section. Spins of states in the neighbouring nuclei are given by combining spin 3 of the core state with the spin of the valence particle (or hole) in the ground-state configuration. The rules of quantum mechanics determine that the allowed spins of the combination have quantum numbers, which differ by at least unity and lie in the range:

$$|I_1 - I_2| \leq I \leq |I_1 + I_2| \tag{2.8}$$

where I_1 and I_2 are the quantum numbers of the spins being added together. So, for example, in ^{209}Bi, we find a group of seven states varying in spin from 15/2 to 3/2, all with positive parity, formed by coupling a proton in a $h_{9/2}$ state to the 3^- core state. We shall see in Chapter 4 how nuclear reactions excite the different states in nuclei in different ways and can be used to reveal their character.

In the level spectra of both ^{17}F and ^{17}O, negative-parity states $1/2^-, 5/2^-$ and $3/2^-$ are found at excitation energies below the $3/2^+$ state, which corresponds to the valence particle occupying the $1d_{3/2}$ shell-model state. Promoting a proton or neutron to the next higher ($1f_{7/2}$) shell and above would produce negative-parity states, but they should occur well above the energies shown in Figure 2.11. These low-lying,

---

[2] It is easy to generalize this result to any shell-model state containing only one vacancy. (We are all the time considering nucleons of one kind only.) Consider the $n\ell_j$ state containing z_j nucleons, i.e. with one vacancy. Its spin and parity assignments (I^π) must be $I = j$ and $\pi = (-)^\ell$. Only with these assignments does the addition of the $(2j + 1)$th nucleon, with spin j and parity $(-)^\ell$ give the resultant angular momentum 0 and positive parity for full state.

negative-parity states can be explained by breaking a particle from the 1p shell of nucleons forming the ^{16}O core, and performing a shell-model calculation that treats the resulting two particles and a single hole explicitly.

The independent-particle shell model is even more limited if there are several valence nucleons outside closed shells, because it cannot be assumed that identical pairs of nucleons will always couple together to form spin-zero units even if the nucleus is in its ground state. In general, the possible resulting spin values of two nucleons are given by Equation (2.8). However, if identical nucleons occupy the *same* shell-model state, the Pauli exclusion principle additionally restricts the resulting spin to be even and with even parity. For example, two neutrons or two protons in a $d_{5/2}$ state can be coupled together to a net spin and parity of 0^+, 2^+ or 4^+, but not 1^+, 3^+ or 5^+.

Shell-model calculations can be done for these more complex situations, but the independent-particle principle has to be abandoned. All possible configurations, in which the extra-core nucleons can couple together to form different spins and parities, have to be included. Also, the calculations must contain a residual inter-action to account for the difference between the true effective internucleon inter-action and that part given by the average shell-model potential. When all this is done, it is found that the shell model can successfully reproduce the low-lying, energy spectra and other properties, such as magnetic moments, of a very wide range of nuclei.

Unfortunately, there is a practical limit to this approach. As the number n of extra-core particles and holes grows, the calculations require an increasing amount of computer power because the number of configurations increases as a high power of n. Also, the microscopic wave functions representing the states become increasingly complex and difficult to interpret. However, in the midst of this complexity, an interesting feature emerges, which is the appearance of states, often low down in energy, of a particular character in which the nucleons appear to move in a coherent or co-ordinated way. These are the collective states, which we will now discuss.

2.5   COLLECTIVE STATES

Low-lying states of many nuclei exhibit regular features that vary rather smoothly over wide regions of A and do not depend significantly on the number of valence nucleons or the particular shell-model states they occupy. For example, the lowest excited state of nearly all even–even nuclei has $I^\pi = 2^+$. The excitation energies $E(2^+)$ of a number of them are plotted as a function of A in Figure 2.12 and, except near closed shells and for very light nuclei, $E(2^+)$ generally varies slowly with A, especially across a given set of isotopes. For stable nuclei with A ranging from about 30 to 150, $E(2^+)$ is of the order of 1 MeV. In the mass regions 150 to 190 and above 230, the 2^+ energy is significantly lower and changes remarkably little with A. As we shall see, the character of these states is that of co-ordinated motion of not just the valence nucleons, but of a large fraction of the entire nucleus. Below $A \sim 150$, the motion is mainly vibrational in character and, in the two higher mass regions, it is rotational. For many of these nuclei, the number of extra-core (valence) nucleons is such that obtaining a microscopic representation of a state in terms of the spherical shell model

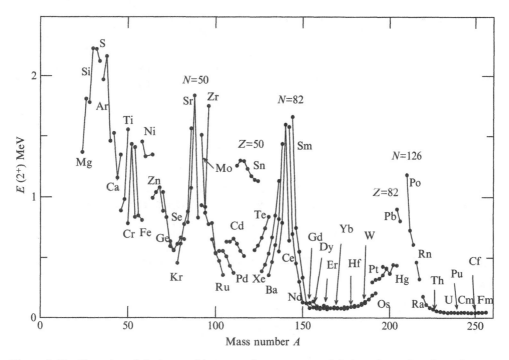

Figure 2.12 Energies of the lowest 2^+ states of even–even nuclei plotted as a function of mass number. The solid lines connect sequences of isotopes.

would be virtually impossible but, even if it could be done, it would be of little value in describing the character of a collective state. There would be a huge number of terms (one for each possible shell-model configuration) and writing down the complete wave function for the state would not be the most transparent way to describe its collectivity. In an analogous situation in classical physics, one would not describe a vibration of a liquid droplet in terms of the equations of motion of all the contributing particles. Insight into the nature of collective nuclear states is obtained by using a model that describes them as quantum states of a vibrating or rotating liquid drop. We shall consider vibrational states first.

2.5.1 Vibrational states

The liquid-drop model predicts that a nucleus will be spherical in its ground state. Any deformation of the shape from equilibrium increases the surface-energy term and, provided that Z is not too high, there will be a minimum in the graph of potential energy versus deformation (see Figure 2.5). For small deformations near the minimum, the shape is parabolic, like the harmonic-oscillator potential. It is possible for the nucleus to vibrate about its equilibrium shape and exist in quantum states that can be obtained by solving the Schrödinger equation, using a suitable harmonic-oscillator potential. These states can be classified by considering the different vibrational modes of a liquid drop.

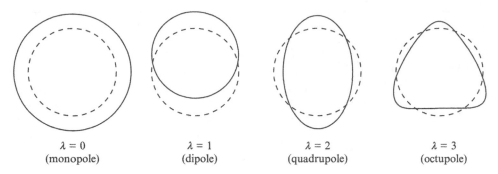

$\lambda = 0$
(monopole)

$\lambda = 1$
(dipole)

$\lambda = 2$
(quadrupole)

$\lambda = 3$
(octupole)

Figure 2.13   The lowest vibrational modes of a spherical system. In each case, the dashed line is the equilibrium shape and the solid line indicates an instantaneous view of the vibrating surface.

Any three-dimensional surface can be described by expressing the radius as a sum of spherical harmonics:

$$R = \sum_{\lambda=0}^{\infty} \sum_{m=-\lambda}^{\lambda} \alpha_{\lambda m}(t)\, Y_{\lambda}^{m}(\theta, \phi) \tag{2.9}$$

where θ and ϕ denote the angular co-ordinates of a point on the surface. The $\alpha_{\lambda m}(t)$ are amplitudes, which may be constant to describe a fixed shape or time dependent to describe a time-varying distribution such as a vibration. The expansion is in terms of *multipoles*, each represented by a spherical harmonic $Y_{\lambda}^{m}(\theta, \phi)$. Figure 2.13 shows sketches of the first four axially symmetric, vibrational modes about a spherical shape for the multipoles $\lambda = 0$ to 3. Each mode has a characteristic frequency ω_{λ}, which depends on the amount of material in motion and the strength of the restoring force. The energy is quantized in units of $\hbar\omega_{\lambda}$. A single quantum of vibrational energy is called a phonon and its spin and parity are given by the spin and parity of the spherical harmonic, which are λ and $(-1)^{\lambda}$ (see Section 2.4.1 and Appendix D). We shall briefly describe the first few modes as they occur in nuclei.

- *Monopole* ($\lambda = 0$): The spherical harmonic for $\lambda = 0$ is spherically symmetric and the vibration is an expansion and contraction of R. If the nucleus is incompressible, $\alpha_{00}(t)$ is constant and this mode is forbidden. Actually, nuclear matter is not completely incompressible but the restoring force is very large, which means that the frequency and energy of this 0^{+} state will be high. A 0^{+} collective state is found in nuclei at an excitation of several hundred MeV and has been attributed to excitation of the monopole vibrational mode.

- *Dipole* ($\lambda = 1$): For this mode, the spherical harmonic gives a shift in the centre of mass and, therefore, a dipole vibration cannot occur for a nucleus as a whole in the absence of external forces. However, a highly collective dipole (1^{-}) state is observed in nuclei at an excitation energy between 10 and 20 MeV. It is interpreted as a vibration in which protons and neutrons move in opposition to each other. The centres of mass of the neutrons and protons move (as they must in a dipole oscillation), but in such a way that the centre of mass of the nucleus as a whole remains fixed. This is not a low-energy state because the separation of protons and neutrons is resisted by a strong restoring force, which implies a high frequency and, therefore, high energy.

- *Quadrupole* ($\lambda = 2$): This is the lowest vibrational mode, which requires no compression or separation of neutrons and protons. In one type of quadrupole vibration, the entire nucleus participates in the motion, rather like a rubber ball vibrating in this mode. Such modes are observed in nuclei, generally at energies greater than 10 MeV, and are referred to as giant quadrupole resonances (GQRs). However, a quadrupole mode is also possible which involves only the movement of nucleons in the nuclear surface. It occurs much lower in energy than the GQR, and low-energy, collective 2^+ states in spherical nuclei, which we noted in reference to Figure 2.12, are considered to be examples of the excitation of surface vibrational phonons. One quadrupole phonon added to the 0^+ ground state of an even–even nucleus puts the nucleus in a 2^+ excited state.

  Suppose we consider a state of two quadrupole phonons. Adding their spins ($I_1 = I_2 = 2$) can give states with the resulting angular momentum quantum number I taking on values of 0, 1, 2, 3 and 4. However, not all these states are possible because phonons have integral spin and must satisfy Bose–Einstein statistics. This imposes a symmetry restriction on their wave functions that they must be symmetric under interchange of the phonons. A detailed analysis shows that only the states with I equal to 0, 2 or 4 have symmetric wave functions (the remaining ones are anti-symmetric) and so these are the only possible ones for our two-phonon system. Since the parity of each phonon is even, the parity of the combination will also be even. Therefore, a nucleus with two quadrupole phonons of vibrational excitation could be in any one of a triplet of even parity states at about twice the energy of the single-phonon state. Figure 2.14 shows an example for the nucleus ^{118}Cd. The energy of the 0^+, 2^+, 4^+ triplet is about twice the energy of the first excited 2^+ state as predicted. They are not all at exactly the same energy because this simple picture neglects many effects that depend on the detailed structure of the nucleus. Nevertheless, the collective model is a useful representation, which is further supported by the fact that these vibrational states decay rapidly by γ emission (see Chapter 3) as would be expected if a significant number of the nucleons in the nucleus were moving in phase with each other and radiating electromagnetic energy.

- *Octupole* ($\lambda = 3$): An octupole phonon carries three units of angular momentum and has negative parity, which gives a 3^- state when coupled with a 0^+ ground state. The first-excited state in ^{208}Pb at 2.61 MeV is an example of a collective octupole-phonon state (see Figure 2.11). More usually, however, octupole vibrational states are found at energies somewhat higher than the 2^+ vibrational state. Two-phonon octupole states are not clearly discernible experimentally since they occur at energies high enough to break a pair of nucleons in the ground-state configuration, and a proper treatment of the situation is complicated.

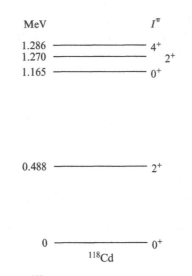

Figure 2.14   Low-lying levels of ^{118}Cd. On the collective model, the first excited state corresponds to the single 2^+ phonon vibration. A two-phonon triplet of states occurs at approximately twice the energy of the single-phonon state, as would be expected if the nucleus were a simple vibrating system.

2.5.2   Deformed nuclei

It is well established that many nuclei with N and Z values between magic numbers are permanently deformed in their ground states. Analyses of hyperfine splitting in optical spectra indicate that these nuclei have electric quadrupole moments that can only be due to a non-spherical distribution of nuclear charge.

The deformation arises because of the way valence nucleons arrange themselves in an unfilled shell. Nucleons filling a shell-model state with a given ℓ value will tend to group into substates with similar values of projection quantum number m because this maximizes the wave function overlap and, therefore, the binding energy from the nucleon–nucleon attraction. However, the spatial distribution of a given substate is non-spherical, becoming increasingly so the larger the value of ℓ. For example, a nucleon with the maximum projection of its angular momentum parallel to the z-axis ($m = \pm\ell$) will have its distribution concentrated in the xy plane. Additional nucleons entering these substates will tend to reinforce the non-sphericity, and the net effect will be a non-spherical contribution to the shell-model potential. Once these substates are full, further nucleons will go into other substates which have the greatest overlap with the filled ones. Also, certain substates of higher levels may be favoured by the deformed potential and fill preferentially, thus adding to the deformation. The tendency to drive the nucleus into a non-spherical shape is greatest when the shell is about half filled. Beyond this point, additional nucleons are constrained to enter the remaining, unfilled substates in the shell. Gradually, the tendency to deformation is reversed and sphericity is restored when shell closure is complete. A closed shell resists becoming deformed and it is only when *both* proton and neutron shells are partially filled that we find permanently deformed nuclei.

The most common non-spherical shape is an ellipsoid where two of the three principal axes are equal. One, like a rugby ball, where the third axis is larger than the others, is called prolate, and a flattened sphere, where the third axis is shorter, is called oblate. In deformed nuclei, the deviation from sphericity is, typically, about 20%. Deformed stable nuclei are found throughout the periodic table and are most common in the mass regions $150 < A < 190$ and $A > 230$. In the first group, protons and neutrons are filling the shells above $Z = 50$ and $N = 82$ and, in the heavier group, shells above $Z = 82$ and $N = 126$ are being filled. Even–even nuclei in these regions are characterized by having very low-lying, first excited 2^+ states (see Figure 2.12). They are examples of rotational states, which we now discuss.

2.5.3 Rotational states

Collective rotational motion can only be observed in nuclei with a non-spherical shape. This is related to the fact that, in quantum mechanics, a wave function representing a perfectly spherically symmetric system has no preferred direction in space and a rotation does not lead to any observable change. Only if there is a deviation from spherical symmetry can a rotation be detected. For this reason, a nucleus shaped like an ellipsoid can rotate about one of its equal axes, but not about its third (symmetry) axis.

If the nucleons in a nucleus couple to give a net spin of zero in the ground state, as is the case for an even–even nucleus, the rotational angular momentum is the same as the total nuclear angular momentum $\mathbf{I}$ (quantum number I) and the angular part of the wave function, which describes the rotation of the state, is a spherical harmonic $Y_I^M(\theta, \phi)$, where M is the z projection of I. The parity of this wave function, therefore, is $(-1)^I$. It can be shown that, as a consequence of the mirror symmetry of an ellipsoid about an axis perpendicular to its symmetry axis, the sequence of allowed rotational states contains only even or odd values of I and so, for an even–even nucleus (with a 0^+ ground state), this gives a set of levels with $I = 0, 2, 4, 6 \cdots$ all with even parity.

The magnitude of the rotational angular momentum is given by $\mathcal{I}\omega$ where $\mathcal{I}$ is the effective moment of inertia and ω is the angular frequency of the rotation. The expression for rotational energy is obtained from the classical relation by substituting the quantum mechanical result $I(I + 1)\hbar^2$ for the expectation value of the square of the rotational angular momentum. Thus, we find that

$$E(I) = \frac{1}{2}\mathcal{I}\omega^2 = \frac{(\mathcal{I}\omega)^2}{2\mathcal{I}} = \frac{I(I + 1)\hbar^2}{2\mathcal{I}}. \tag{2.10}$$

An example of a ground-state rotational band of levels is shown in Figure 2.15 for the even–even nucleus ^{180}Hf. The spins and parities have the predicted values, and the energies in the band follow the $I(I + 1)$ trend remarkably well for such a simple model. There are many other examples that exhibit this pattern of states, and Figure 2.16 compares energy ratios in the ground-state rotational bands for even–even nuclei in the mass regions $150 < A < 190$ and $A > 230$ with expectation based on Equation (2.10). The experimental ratios follow the $I(I + 1)$ rule quite closely for the

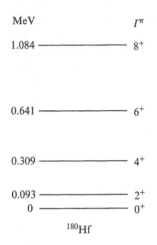

Figure 2.15   Low-lying level spectrum of the nucleus ^{180}Hf showing a band of states typical of a rotating, non-spherical nucleus with an axially symmetric quadrupole deformation.

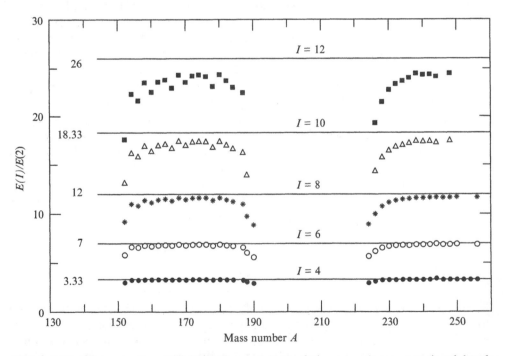

Figure 2.16   Energy ratios $E(I)/E(2)$ for members of the ground-state rotational bands in even–even nuclei in the mass regions $150 \leq A \leq 190$ and $224 \leq A$. The points are experimental values and the horizontal lines give the theoretical energy ratios predicted by Equation (2.10).

lower states (up to $I = 8$) but there is a general fall below expectation for higher spins. Equation (2.10) assumes that the moment of inertia is independent of rotational frequency and the trend shown in Figure 2.16 can be interpreted as a tendency for $\mathcal{I}$ to increase as the frequency of rotation increases. Thus, the nucleus is 'soft' in the sense that its internal structure can change in response to the internal stresses that arise due to the high rate of rotation. Further insight into the rigidity of the nucleus is obtained by considering the magnitude of its moment of inertia. This can be calculated by substituting experimental energies into Equation (2.10) and, when this is done, it is found that the effective moment of inertia is about one-third of that obtained if the nucleus is assumed to be a rigid body (see Problem 2.8). Evidently, the nucleus does not rotate like a rigid body and only part of its nucleons can be considered to be involved in the collective motion.

2.5.4 Superdeformation

We close this section with a few brief comments on a particularly striking example of nuclear collective motion. In certain circumstances, collective rotation occurs where the shape of the nucleus is highly deformed and the nuclear moment of inertia is close to the rigid-body value. The phenomenon is known as *superdeformation*.

In a spherical potential well, magic numbers occur because shell-model states group together to form shells with large energy intervals between them. A filled shell forms a stable structure because considerable energy is required to break a nucleon away from a closed shell and move it across the energy gap to the next one. If the Schrödinger equation is solved for a non-spherical well, all states occur at energies that are different to those given by a spherical well and, in general, the grouping of levels into well-defined shells is lost. However, if the deformation is increased to a value which turns out to be close to that corresponding to an ellipsoid with an axis ratio of 2:1, it is found that the states calculated in such a well tend to group together again and we regain the concepts of shell closure and magic numbers – this time for a deformed nucleus. In the deformed well, energy gaps occur at N and Z values that are quite different to the magic numbers for a spherical well. Indeed, some of the magic numbers in the 2:1 deformed well occur at N, Z values that correspond to nuclei with partially filled spherical shells and which may be deformed in their ground states. Such nuclei will be expected to exhibit a quasi-stable structure at the deformation corresponding to the 2:1 axis ratio, which is known as the super-deformed state.

Consider how the potential energy (PE) of such a nucleus changes with deformation. There will be a local potential minimum at the deformation corresponding to the ground state. Any change away from the ground-state deformation shape will lead to an increase in PE as shown in Figure 2.17. However, when the deformation approaches the superdeformed (SD) value, the extra stability due to the new shell closures will give rise to a second local minimum. The effect is known as shell stabilization.

If such a nucleus is rotated, the influence of the centrifugal force on the internal structure of the nucleus favours an increase in deformation. In this case, if the nucleus is made to rotate sufficiently rapidly, it becomes increasingly likely that it

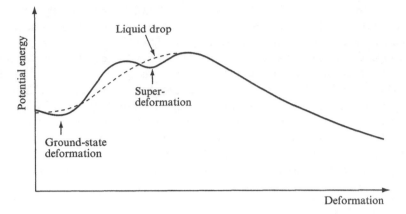

Figure 2.17   Schematic representation of the double-humped barrier of a nucleus which is deformed in its ground state and for which shell structure gives extra stability at an axis ratio corresponding to a superdeformed shape. The dashed line gives the variation predicted by the liquid-drop model, which contains no effects due to shell structure.

will be in an energy state which is one of those of a rotational band based on the SD shape.

Nuclei can be formed with high angular momentum and in the SD state following a collision between two heavy ions that fuse together (see Section 4.6.2). If the electromagnetic decays of the resulting highly excited and rapidly spinning nuclei are measured, a spectrum of γ rays is obtained that is characteristic of a rotating nucleus with a high moment of inertia close to the rigid-body value. The γ rays are emitted in a series of transitions that proceed sequentially down the rotational band of states. From Equation (2.10), the γ-ray energy for a transition between states with spins $I + 2$ and I is $E_\gamma(I + 2 \rightarrow I) = (2I + 3)\hbar^2/\mathcal{I}$ and, hence, the difference between two adjacent γ transitions is

$$\Delta E_\gamma = 4\hbar^2/\mathcal{I}. \tag{2.11}$$

This is independent of spin I and so peaks in the γ-ray spectrum should appear at regular energy intervals. An example is shown in Figure 2.18 for the nucleus ^{152}Dy. In this case, the energy difference between sequential γ transitions is approximately constant at about 47 keV. The corresponding figure for transitions within the ground-state rotational band of a typical deformed nucleus in this mass region is considerably greater at about 120 keV. This suggests a large increase in the effective moment of inertia for the nucleus in the SD state compared with the ground state. Analyses of γ-ray energies for a number of SD transitions and measurements of the rates at which the SD states decay by γ radiation are consistent with the nucleus having entered into a stable SD configuration with an effective moment of inertia considerably higher than that of a rotational band built on the ground state. Further evidence of the stability of the shape is the observation of the high degree of regularity in the spacing of the γ transition energies. The SD rotational band follows the $I(I + 1)$ rule quite closely, even at high-spin values, and this implies that the

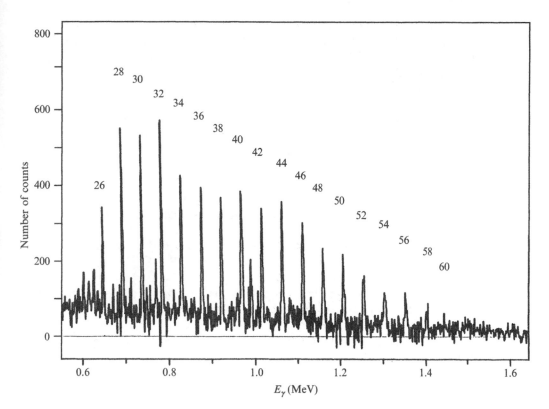

Figure 2.18   Spectrum of γ rays emitted in successive radiative transitions down a super-deformed rotational band in the nucleus ^{152}Dy. The numbers marking the peaks indicate the proposed spin values I for the values of $E_\gamma(I + 2 \rightarrow I)$. Below $I = 24$, the nucleus is no longer in the superdeformed state and the sequence of transitions terminates. From Twin *et al.* (1986).

effective moment of inertia changes little with rotational frequency, which is consistent with that of a stable (rigid) object.

PROBLEMS 2

2.1 Derive an expression for the Coulomb energy of a uniformly charged sphere, total charge Q and radius R. If the binding energies of the mirror nuclei ^{41}Sc and ^{41}Ca are 343.143 and 350.420 MeV, respectively, estimate the radii of the two nuclei by using the semi-empirical mass formula (SEMF).

2.2 Calculate (relative) nuclear masses to determine which $A = 46$ nuclei are stable. Use the SEMF coefficients given in the caption to Figure 2.3 and $\Delta = 2$ MeV.

2.3 Use the SEMF to obtain an expression for the Z value of the isobar which will have the lowest mass for a given value of A. Hence, determine which nuclide with $A = 101$ is predicted to be the most stable.

2.4 Use the SEMF and the values of the coefficients given in the caption to Figure 2.3 (with $\Delta = 2$ MeV) to calculate the energy release if a ^{238}U nucleus were to split into two equal fragments ($^{119}_{46}$Pd). Compare this with an estimate of the symmetric fission of a nucleus with $A = 238$, based on Figure 2.3 and comment on any difference you may find.

2.5 The spin–orbit force splits each shell-model state with a given ℓ (other than $\ell = 0$) into two states with $j = \ell \pm 1/2$, as shown in Figure 2.9. Show that the average energy of this spin–orbit pair of states, weighted according to occupancy, is independent of the strength of the spin–orbit force.

2.6 (a) A certain odd-parity, shell-model state has total and orbital angular momentum quantum numbers j and ℓ, respectively. If the state can hold up to 16 nucleons, what are the values of ℓ and j?
(b) Give the expected ground-state spins and parities, based on the independent-particle shell model, for the following nuclei: ^{7}Li, ^{11}B, ^{31}P, ^{39}K, ^{59}Co and ^{127}I.

2.7 Suggest some shell-model configurations (out of many which are possible) for each of the $1/2^-$, $5/2^-$ and $3/2^-$ excited states in ^{17}O (see Figure 2.11).

2.8 (a) Using the information in Figure 2.15, compare the energies of the low-lying levels in ^{180}Hf with the $I(I + 1)$ rule.
(b) Show that the effective moment of inertia of the ^{180}Hf nucleus is about 40% of its rigid-body value, calculated assuming the nucleus to be a uniform sphere of radius $1.25 \times A^{1/3}$ fm.

2.9 Estimate the classical rotational period of the ^{180}Hf nucleus in its 2^+ state and compare it to the time it takes for a nucleon to cross the nucleus with a kinetic energy equal to the depth of the shell-model potential: 50 MeV.

2.10 Show that the excited ^{152}Dy nucleus, emitting the spectrum of γ rays shown in Figure 2.18, exhibits rotational behaviour with an effective moment of inertia (MI) comparable with that of a rigid superdeformed (SD) nucleus. Note that the MI of a rigid SD nucleus (2:1 axis ratio) is about 1.3 times that of the nucleus in the spherical shape. Use $R = 1.25A^{1/3}$ fm to calculate the spherical MI.

3

Nuclear Instability

3.1 INTRODUCTION

In Chapter 1, we noted several ways whereby an unstable or excited nucleus may change into a more stable system and emit excess energy in the form of radiation. The various nuclear decay modes are classified according to the force causing the change. The strong nuclear force is involved when the nucleus loses nuclear material, as in α decay or fission, the action of the weak force changes the ratio of neutrons to protons via β decay, and the electromagnetic force comes into play when a nucleus de-excites by γ emission.

In the following three sections, we outline the main physics underlying the emission of α, β and γ radiation during nuclear decay and present expressions which enable the term containing the nuclear structure information of interest to be extracted from a measurement of the decay rate. We consider γ emission first because the other types of decay are often accompanied by γ decay. Also, γ spectroscopy provides more information about nuclear structure than either α or β spectroscopy because γ emission can be measured between many more pairs of nuclear states and for many more nuclei than are accessible in either α or β decay. Furthermore, unlike α and β particles, γ rays experience negligible absorption and scattering in air and can be detected easily with good energy resolution, which is often crucially important in identifying which states are involved in a particular transition.

3.2 GAMMA EMISSION

An excited nucleus may lose energy by emitting a γ-ray photon, whose energy E_γ is equal to the energy difference or transition energy ΔE between the initial and final nuclear states, apart from a small correction due to recoil (see Problem 3.1). Gamma-ray energies cover a wide range up to tens of MeV but, typically, are of the order of 1 MeV. For reference, the wavelength of a 1-MeV γ ray is about 1240 fm, which is much smaller than an atom but large compared with the size of a nucleus. In principle, the initial state could decay to any state at a lower energy but, in practice, the probability of a particular transition is very dependent upon the quantum numbers of the states and the transition energy.

As we noted in Section 1.5.3, there are other electromagnetic processes competing with γ emission, namely, internal conversion, in which an electron is ejected from the atom, and internal pair creation (for high-energy γ rays). In this section, we will be dealing mainly with γ emission, but there are situations where internal conversion and, possibly, internal pair creation can occur and γ emission cannot.

3.2.1   General features and selection rules

Gamma radiation is the same as any other type of electromagnetic radiation in which a changing electric field induces a magnetic field and vice versa. Radiation can be generated either by an oscillating charge, which causes an oscillation in the external electric field, or by a varying current or magnetic moment, which sets up a varying magnetic field. Radiation emitted by the former mechanism is called electric (E) radiation and the latter is said to give rise to magnetic (M) radiation. The classification of the different processes leading to the emission of photons is based on conservation of angular momentum and parity between the radiating system and the radiation field. A proper analysis is beyond the scope of this book and we refer the interested reader to the bibliography. Here, we only state the results.

A photon carries away angular momentum of magnitude given by a quantum number L, which can have an integer value greater than zero. The value $L = 0$ is excluded as photons with zero angular momentum do not exist. Photons are referred to as having multipolarity L and we talk about multipole radiation and transitions as dipole $(L = 1)$, quadrupole $(L = 2)$, octupole $(L = 3)$ and so on. The angular momentum of the photon is related by angular-momentum conservation to the angular momenta (spins) I_i and I_f of the initial and final nuclear states. Thus, we have the vector equation:

$$\mathbf{I_i} = \mathbf{I_f} + \mathbf{L} \qquad (3.1)$$

which leads to a selection rule that L can have any integer value between the sum and difference of the quantum numbers of the initial and final states:

$$|I_i - I_f| \leq L \leq |I_i + I_f|. \qquad (3.2)$$

We have already noted that $L = 0$ γ emission is not allowed. Thus, single-photon emission is forbidden by angular-momentum conservation for a transition between two spin–zero states. In principle, two photons could be emitted, but this is extremely rare and can be neglected in the context of this text. There are a few cases of even–even nuclei (e.g. ^{16}O, ^{68}Ni and ^{90}Zr) where both the ground and first excited states have zero spin. In these cases, the excited state decays by internal conversion (see below) which is allowed.

Parity is conserved to a high degree of accuracy in electromagnetic transitions and a parity selection rule must also be obeyed. The radiation field can have even or odd parity for a given L, depending on whether the radiation type is electric or magnetic. Electric multipole radiation, denoted by $E1, E2, \cdots EL$, has parity $(-1)^L$ and magnetic multipole radiation $(M1, M2, \cdots ML)$ has parity $(-1)^{L+1}$.

Figure 3.1   (a) Electric dipole moment formed by positive and negative charges separated by a distance **r**. The dipole moment is equal to $q\mathbf{r}$. (b) Magnetic dipole moment created by a charge q moving with speed v in a circular loop of radius r. The dipole moment $\boldsymbol{\mu}$ is proportional to $q\mathbf{r} \times \mathbf{v}$.

It is not obvious that the parity change for a given L should be different for electric and magnetic types of radiation. The difference is related to the fact that electric and magnetic multipoles of the same order L have opposite parity. Although this is not simple to show in general, we can see that it is the case for electric and magnetic dipole moments in classical physics. An electric dipole, shown in Figure 3.1(a), is a displacement of charge q and is of the form $q\,\mathbf{r}$, which, under the parity operation of inversion, transforms into $-q\mathbf{r}$ and, therefore, has odd (negative) parity. A magnetic dipole is like a circulating charge which forms a current loop as shown in Figure 3.1(b). It has a magnetic moment proportional to $q\mathbf{r} \times \mathbf{v}$, which, under inversion, transforms into $q(-\mathbf{r}) \times (-\mathbf{v}) = q\,\bar{\mathbf{r}} \times \bar{\mathbf{v}}$ i.e. it does not change sign and so has even parity.

The general results described above are presented in Table 3.1. These, together with Equation (3.2), summarize the selection rules that determine the type and multipole order of photons emitted in a given radiative transition. In general, several multipolaritics may be allowed. For example, in a $3^- \rightarrow 2^+$ transition, we can have any L value between 1 and 5 given by Equation (3.2). There is a parity change and so, according to the rules, we could have any of the following types of emitted radiation: $E1, M2, E3, M4, E5$. In practice, only $E1$ radiation is likely because there is a very strong dependence of the transition rate on multipolarity as we shall see in the next subsection.

Table 3.1   Selection rules for γ emission.

Multipolarity	Dipole		Quadrupole		Octupole		$\cdots$
Type of radiation	$E1$	$M1$	$E2$	$M2$	$E3$	$M3$	$\cdots$
Parity change	Yes	No	No	Yes	Yes	No	$\cdots$

3.2.2   Transition rate

In the classical theory of electromagnetic radiation, the power emitted by an antenna is given by

$$P(\sigma L) = \frac{2(L+1)c}{\varepsilon_0 L[(2L+1)!!]^2} \left(\frac{\omega}{c}\right)^{2L+2} (\mathscr{M}(\sigma L))^2 \tag{3.3}$$

where σ denotes the type of radiation: E (electric) or M (magnetic), ω is the angular frequency of the radiation, and the double factorial

$(2L + 1)!! = (2L + 1) \times (2L - 1) \times \cdots \times 3 \times 1$; $\mathcal{M}(\sigma L)$ is the amplitude of the oscillating electric or magnetic multipole moment.

Quantum mechanics requires that we recognize that radiation is emitted as discrete photons, and the multipole moment amplitude is replaced by a matrix element, which has the form:

$$\mathcal{M}_{fi}(\sigma L) = \int \psi_f^* m(\sigma L)\psi_i dv \tag{3.4}$$

where the integral is carried out over the volume of the nucleus. The wave functions ψ_i and ψ_f refer to the initial and final states of the nucleus, respectively. The multipole operator $m(\sigma L)$ is related to the multipole moment. Its function is to change the nucleus from its initial to its final state and create a photon of the appropriate type (E or M) and multipolarity L.

Emitted photons have energy $\hbar\omega$ and we obtain their rate of emission or emission probability per unit time directly from Equation (3.3) as

$$T(\sigma L) = \frac{P(\sigma L)}{\hbar\omega} = \frac{2(L + 1)}{\varepsilon_0 L[(2L + 1)!!]^2 \hbar} \left(\frac{\omega}{c}\right)^{2L+1} B(\sigma L) \tag{3.5}$$

where $B(\sigma L)$ is called the reduced transition probability and is the square of the modulus of the matrix element $\mathcal{M}_{fi}(\sigma L)$.

Calculation of the reduced transition probability, i.e. of the matrix element in Equation (3.4), requires a knowledge of the nuclear wave functions involved in the transition and can be very complicated. An estimate, referred to as a *single-particle* estimate, can be obtained by assuming that the transition is due to a single proton making a transition between two shell-model states. The reduced transition probabilities depend not only on the multipolarity L of the emitted radiation, but also on the angular momentum quantum numbers of the shell-model states involved. However, these dependencies are not large, and Weisskopf[1] has shown that, making various simplifying assumptions, a reasonable approximation for the single-particle reduced transition probability for an electric transition is given by

$$B_{sp}(EL) = \frac{e^2}{4\pi} \left(\frac{3R^L}{L+3}\right)^2 \tag{3.6}$$

and for a magnetic transition by

$$B_{sp}(ML) = 10\left(\frac{\hbar}{m_p cR}\right)^2 B_{sp}(EL) \tag{3.7}$$

where m_p is the mass of a proton. These are the Weisskopf single-particle estimates.

Substituting $R = R_0 \times A^{1/3}$ for the nuclear radius, we obtain the following expressions for the electric and magnetic, single-particle reduced transition rates:

---

[1] For a derivation and discussion of these results, see Blatt and Weisskopf (1952), pp. 623–627.

$$B_{sp}(EL) = \frac{e^2}{4\pi} \left(\frac{3}{L+3}\right)^2 (R_o)^{2L} A^{2L/3} \tag{3.8}$$

$$B_{sp}(ML) = \frac{10}{\pi} \left(\frac{e\hbar}{2m_p c}\right)^2 \left(\frac{3}{L+3}\right)^2 (R_o)^{2L-2} A^{(2L-2)/3}. \tag{3.9}$$

Taking $R_o = 1.2$ fm for the nuclear radius parameter, we find that the ratio $B_{sp}(ML)/B_{sp}(EL)$ is numerically equal to $0.31 \times A^{-2/3}$, which is always less than unity.

Inspection of Equations (3.5) and (3.6) shows that the main dependence of $T(\sigma L)$ on multipolarity L appears as $(\omega R/c)^{2L}$ or $(kR)^{2L}$, where k is the wavenumber of the radiation. Since $k = E_\gamma (\text{MeV})/197\,\text{fm}^{-1}$, we obtain $kR = E_\gamma$ (MeV) $\times$ (1.2/197) $\times$ $A^{1/3}$, which is much smaller than unity and occurs to the power $2L$ in the transition probability. For a typical nuclear transition, kR is small – approximately 10^{-2} for a few hundred keV γ ray emitted by a medium-mass nucleus. Therefore, since other factors in Equation (3.5) vary slowly with L, we can see that the γ-ray emission probability will decrease by several orders of magnitude for each successive increase in L.

The expressions for the single-particle, reduced transition rate estimates can be substituted into Equation (3.5) to give single-particle estimates for radiative decay probabilities (s^{-1}) in terms of L, E_γ (MeV) and A. Results for multipole orders up to 5 (with $R_o = 1.2$ fm) are

$$
\begin{array}{ll}
T(E1) = 1.0 \times 10^{14} A^{2/3} E_\gamma^3 & T(M1) = 3.1 \times 10^{13} E_\gamma^3 \\
T(E2) = 7.3 \times 10^7 A^{4/3} E_\gamma^5 & T(M2) = 2.2 \times 10^7 A^{2/3} E_\gamma^5 \\
T(E3) = 34 A^2 E_\gamma^7 & T(M3) = 10 A^{4/3} E_\gamma^7 \\
T(E4) = 1.1 \times 10^{-5} A^{8/3} E_\gamma^9 & T(M4) = 3.3 \times 10^{-6} A^2 E_\gamma^9 \\
T(E5) = 2.4 \times 10^{-12} A^{10/3} E_\gamma^{11} & T(M5) = 7.4 \times 10^{-13} A^{8/3} E_\gamma^{11}.
\end{array}
\tag{3.10}
$$

Estimates for a medium-mass nucleus (up to $L = 5$) are plotted in Figure 3.2, which graphically illustrates the very strong dependence of emission rate on photon energy as well as multipolarity.

This strong dependence on L means that, normally, radiation will be emitted with the lowest multipolarity allowed by the selection rules. For example, consider the γ emission by ^{60}Ni following the β decay of ^{60}Co. The decay scheme is shown in Figure 3.3(a). The main β-decay branch is to a 4^+ excited state of ^{60}Ni, which decays by γ emission almost 100% of the time to the first-excited (2^+) state. The selection rules allow $E2, M3, E4, M5$ and $E6$. However, $E2$ radiation is strongly favoured and, indeed, dominates this particular transition. Note that the $4^+ \rightarrow 0^+$ transition probability, relative to the $4^+ \rightarrow 2^+$ transition, is extremely weak at about 2×10^{-8}. Even though the $4^+ \rightarrow 0^+$ γ-ray energy is higher than that of the $4^+ \rightarrow 2^+$ transition, the former requires the emission of an $E4$ photon and the multipolarity dependence effectively suppresses it relative to the $4^+ \rightarrow 2^+$ $E2$ transition.

A high multipolarity is seen in the decay of the $11/2^-$ state in ^{137}Ba, which is populated in the β decay of the radioactive nuclide ^{137}Cs [see Figure 3.3(b)]. The

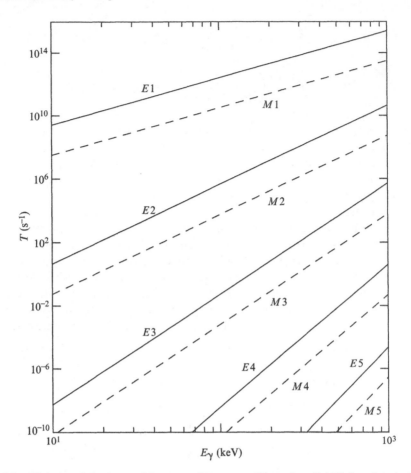

Figure 3.2   Weisskopf single-particle γ transition rates [Equation (3.10)] for electric (E) and magnetic (M) transitions of different multipolarities calculated as a function of photon energy E_γ for a nucleus of mass number 125.

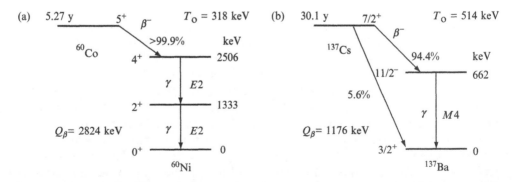

Figure 3.3   Radioactive decay schemes for the nuclei (a) ^{60}Co and (b) ^{137}Cs. The end-point energies T_o of the main β transitions are indicated, given by $Q_\beta - E^*$, where E^* is the excitation energy of the final (daughter) nucleus.

lowest multipolarity in the transition to the $3/2^+$ ground state is 4 and, since there is a parity change, $M4$ radiation is emitted. The low emission probability for $M4$ radiation accounts for the long lifetime of > 2.5 min for this state and it is an example of an isomeric transition.

We introduced the concept of the isomeric or metastable state in Section 1.5.3 and we have already seen an example in Figure 1.9, namely, the medical radioisotope ^{99}Tc^m. Most isomeric transitions are of high multipolarity and there are striking cases where the lifetime may be days or even years. An extreme example is a metastable (16^+) state of hafnium ^{178}Hf^m, which has a half-life of about 31 years. It decays by an $M4$ radiative transition to a 12^- state. However, the high multipolarity accounts for only part of the long lifetime. The probability of this particular transition is further weakened because the wave functions of the initial and final nuclear states have very different structures and so the matrix element, which contains their overlap, will be small. Isomerism is usually found in odd-N nuclei, just below closed shells, where shell-model states differing in spin by a large value can occur close together in energy.

Not all transitions are dominated by a single radiation type and we can have what are called mixed transitions. For example, the γ-emission selection rules for a $2^+ \rightarrow 2^+$ transition allow $M1, E2, M3$ and $E4$, and Figure 3.2 suggests that $M1$ should dominate.[2] However, it would not be uncommon for this to be a mixed $M1/E2$ transition because 2^+ states often exhibit a degree of collectivity, which increases their probability for emitting $E2$ radiation. As described in Section 2.5, a collective state is characterized by the coherent (collective) motion of many nucleons. A consequence of this is that there are many amplitudes contributing to the matrix element of Equation (3.4), each one corresponding to the transition of a single nucleon from a shell-model state it occupies in the initial state (ψ_i) to a different one in the final state (ψ_f). Also, in a collective state, these amplitudes are all approximately in phase with each other, corresponding to the co-ordinated motion of the contributing nucleons. Therefore, when they are added together coherently, they tend to reinforce each other, resulting in a transition amplitude larger than that due to the transition of a single nucleon and a $B(E2)$ value well in excess of the single-particle estimate.

The first excited state of most even–even nuclei is a low-lying, 2^+ collective state which decays strongly to the 0^+ ground state. We can see evidence for this in Figure 3.4, which plots reduced transition probabilities (in single-particle units), derived from experiment, of $2^+ \rightarrow 0^+$ transitions for a range of even–even nuclei as a function of mass number A. As we reported in Section 2.5, below $A \sim 150$ these 2^+ states are mainly vibrational in character and they clearly exhibit enhanced $E2$ transition probabilities consistent with their collective character. For nuclei between $A = 150$ and 190 and above 230, there are regions of permanent nuclear deformation, and the 2^+ state is a member of a rotational band of energy levels. The strengths of these $E2$ transitions are very large and provide striking evidence for the collective nature of the deformation in these nuclei.

---

[2] Note, we are not considering internal conversion here, which would allow an $E0$ transition to occur.

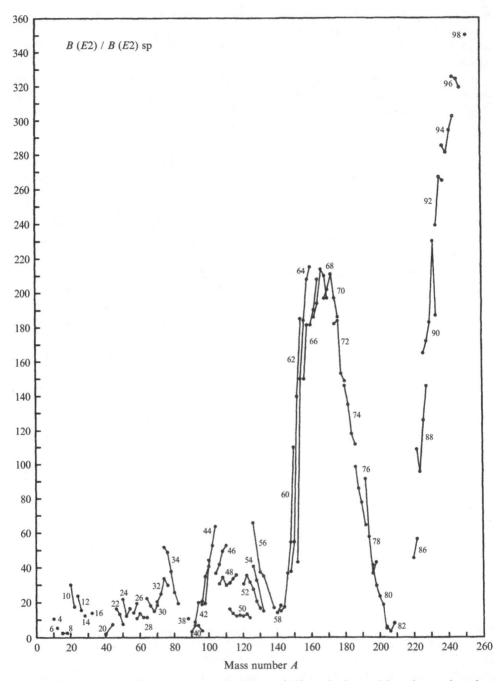

Figure 3.4   Ratios of reduced decay probabilities $B(E2)$ to single-particle estimates based on Equation (3.8) for 2^+ states in even–even nuclei, which are stable below $A = 210$. Numbers on the figure indicate Z values for isotope sequences connected by solid lines. The large increases correspond to quadrupole-deformed nuclei in the lanthanide ($150 < A < 190$) and actinide ($A > 230$) regions. From Bohr and Mottleson (1975). Reproduced by permission of World Scientific Publishing Co. Pte Ltd.

3.2.3 Internal conversion

As noted at the beginning of this section, a transition between two states in the same nucleus can be induced by electric or magnetic interaction with the atomic electrons. This is the process of internal conversion, and the transition results in the emission of an atomic electron, which, if we neglect nuclear recoil, emerges with kinetic energy:

$$E_{el} = \Delta E - B \qquad (3.11)$$

where ΔE is the transition energy and B is the binding energy of the electron in the atom. Electrons emitted from different atomic shells will appear at different energies. They can be measured accurately and, in a heavy nucleus, it is possible even to distinguish between electrons from different orbits within a given shell. These are labelled accordingly. For example, electrons emitted from the L shell (which is specified by the principal quantum number $n = 2$), consisting of $s_{1/2}$, $p_{3/2}$ and $p_{1/2}$ states, are labelled L_I, L_{II} and L_{III}.

The effects on the nucleus are the same in internal conversion as they are in radiative emission. Therefore, the purely nuclear part of the matrix element in Equation (3.4) is the same in both cases. Other factors determining the internal conversion transition rate can be evaluated and so the ratio of internal conversion to γ emission is accurately calculable and is independent of the structure of the initial and final nuclear states. The ratio of the measured rates of emitted electrons and γ rays defines the internal conversion coefficient α. It is written as a sum of partial coefficients each of which corresponds to the particular atomic shell of the emitted electron (K, L, M $\cdots$). Thus,

$$\alpha = \alpha_K + \alpha_L + \cdots. \qquad (3.12)$$

The L conversion coefficient can be further partitioned as $\alpha_L = \alpha_{L_I} + \alpha_{L_{II}} + \alpha_{L_{III}}$, if the electrons from the different orbits in the L shell can be identified experimentally. Internal conversion is accompanied by the emission of X-rays and, possibly, Auger electrons (see Section 5.4.1) as the vacancy in the atomic shell is filled and the atom reverts to its initial state.

Extensive tabulations of calculated conversion coefficients are available. Internal conversion coefficients increase with atomic number and, generally, decrease with increasing transition energy ΔE. Figure 3.5 shows plots of α_K versus ΔE for $Z = 60$. The important feature to note here is that α_K depends on the type (EL or ML) of the transition. Other internal conversion coefficients also depend on the type of transition, but in a different way. Thus, from measurements of different internal conversion coefficients, it is often possible to determine the type of transition, which is a vital piece of information for identifying spins and parities of unknown nuclear states. It is also possible to obtain this information by measuring the ratios of different conversion electrons, for example, the ratio of K to L conversion electrons or ratios of L_I, L_{II} and L_{III} conversion electrons. This method is especially useful if the transition is highly converted and γ emission is weak or negligible.

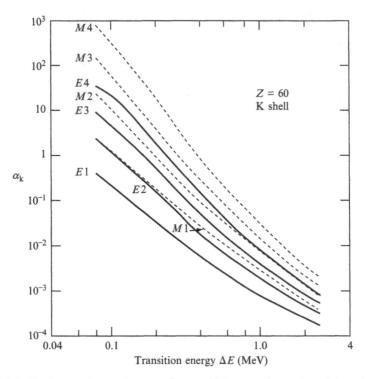

Figure 3.5   The K-electron internal conversion coefficient α_K for various $E\lambda$ and $M\lambda$ transitions for $Z = 60$. From Lederer and Shirley (1978).

3.3   BETA DECAY

As described in Section 1.5.2, β decay is the mechanism by which an unstable nucleus changes to optimize the N/Z ratio for a given mass number. The transformation is brought about by the action of the weak nuclear force, which changes a neutron into a proton, or vice versa, and either creates a lepton–antilepton pair (electron plus antineutrino or neutrino plus positron), which is emitted, or converts an electron from one of the atomic shells into a neutrino, which is then emitted. The basic mechanisms are listed in Eqs. (1.9) and (1.10) and are reproduced again here:

$$\beta^- \text{decay}: \qquad n \rightarrow p + e^- + \bar{\nu}$$

$$\beta^+ \text{decay}: \qquad p \rightarrow n + e^+ + \nu$$

$$\text{Electron capture}: \quad e^- + p \rightarrow n + \nu.$$

It is worth noting that the above transitions satisfy a strict conservation law, which is that the number of leptons minus the number of antileptons is unchanged in a reaction. The lepton–antilepton pair is created at the moment of decay, as is the photon in γ emission.

The β-decay probability depends on a number of factors, such as the energy released in the decay, angular momentum of the emitted particles, the nuclear charge and the nature of the initial and final nuclear states. For our analysis, we will use a fundamental theorem of quantum mechanics known as Fermi's Golden Rule.[3] This states that the transition rate between an initial state (i) and a final state (f) is given by

$$\lambda = \frac{2\pi}{\hbar} |M|^2 \frac{dn}{dE} \qquad (3.13)$$

where E is the energy available to the final-state products, and dn/dE, often referred to as the phase-space factor, is the density of final states, i.e. dn is the number of final states with energy in the interval E to $E + dE$. M is the matrix element of the transition and has the form:

$$M = \int \psi_f^* H \psi_i d\tau \qquad (3.14)$$

where H is the interaction causing the transition between the initial and final state wave functions ψ_i and ψ_f. Equation (3.13) separates the matrix element, containing details of the transition and nuclear structure, from the phase-space factor, which influences the way the particles are emitted. In the remainder of this section, we show how phase space determines the form of the β-particle energy spectrum and then comment on the way β transitions are classified according to the angular momentum carried away by the radiation. In this brief introduction to the subject, we are unable to discuss details of the matrix element or the different ways the weak interaction can operate between nuclear states. For these, the reader is referred to more advanced texts, some of which are listed in the bibliography.

3.3.1 Beta-particle energy spectrum

The most striking difference between β decay and α or γ emission is the continuous energy spectrum of the detected β particles (see Figure 1.6). This observation caused great concern when the process was first being investigated and was explained only when it was proposed by Pauli that an uncharged and undetected particle is emitted at the same time as the β particle. We now know that different particles are emitted in β^+ and β^- decays, called neutrinos and antineutrinos, respectively. As the analysis is the same for β^+ and β^- decays, we shall treat them together and use the suffix v to stand for neutrino or antineutrino, whichever is appropriate. The observed upper limit to a β-particle energy spectrum corresponds to a value calculated assuming that the neutrino (or antineutrino) mass m_v is very small. Attempts to measure m_v have determined an upper limit of a few eV and it is usually taken to be zero.

Clearly, if there is an additional particle involved in the transition, the continuous β spectrum is immediately understandable because the decay energy must be shared between three outgoing particles and there are many different ways this can be done. Note that the residual nucleus is massive compared with the leptons. It will absorb

---

[3] See, for example, Mandl (1992), Section 9.4.

recoil momentum, but its kinetic energy will be very small and we neglect it later on in this discussion. The way the energy released in the transition is converted into the kinetic energy of the leptons depends mainly on the phase-space factor. However, it is also influenced by the Coulomb interaction between the emitted β particle and the daughter nucleus. For simplicity, and to illustrate the general principle, we shall assume first that the leptons are emitted in the absence of the Coulomb field.

If we neglect the kinetic energy of the residual heavy nucleus, the total nuclear energy released in the transition is shared between the electron and the neutrino, i.e. $E_o = E_e + E_\nu$. Since the total electron energy E_e includes its rest mass ($m_e c^2 = 511$ keV), the kinetic energy of the electron is $T_e = E_e - m_e c^2$. The maximum electron kinetic energy T_o is the end point of a β spectrum (see Figure 1.6), and $T_o = E_o - m_e c^2$, taking the neutrino rest mass to be zero.

The number of states available to a β particle with momentum between p_e and $p_e + dp_e$ and constrained to a volume V is $4\pi p_e^2 dp_e V/h^3$. The general result for this is derived in Appendix C. A similar formula applies to the neutrino. Assuming that the phase space available for either particle is independent of the other, i.e. there is no angular correlation between their directions of emission, the total number of states available for the two particles is

$$dn = \frac{(4\pi)^2 V^2 p_e^2 dp_e p_\nu^2 dp_\nu}{h^6}.$$
(3.15)

The factor V^2 drops out from the results below as it is compensated by a factor V^{-2}, which stems from the normalization of the lepton wave functions in the volume.

There is a constraint imposed by energy conservation. If we neglect the kinetic energy of the residual nucleus, we can write the neutrino energy as

$$E_\nu = p_\nu c = E_o - E_e$$
(3.16)

where we have again assumed the neutrino rest mass to be zero. It follows that, for a fixed electron energy E_e,

$$dp_\nu = \frac{dE_o}{c}.$$
(3.17)

Substituting the results of Equations (3.16) and (3.17) into Equation (3.15) gives an expression for the phase-space factor in terms of the energy and momentum of the β particle as

$$\frac{dn}{dE_o} \propto p_e^2 (E_o - E_e)^2 dp_e.$$
(3.18)

The above treatment of phase space does not take into account the distorting effect of the Coulomb field on the emitted electrons or positrons. The correction factor for this is called the Fermi function $F(Z, E_e)$, where Z is the atomic number of the final (daughter) nucleus. It has been evaluated for many different situations and tabulated values are available. Substituting the result of Equation (3.18), including the Coulomb correction factor, into Equation (3.13) we obtain, finally, an expression for the differential β-decay probability per unit time for β-particle momenta in the range p_e to $p_e + dp_e$ as

$$d\lambda(p_e) = C|M|^2 F(Z, E_e) p_e^2 (E_o - E_e)^2 dp_e \tag{3.19}$$

where C is a constant.

3.3.2 Allowed transitions

The matrix element M, Equation (3.14), consists of an integral over the nuclear volume and includes, as part of ψ_f, the wave functions of the emitted leptons. If we represent these wave functions as plane waves, the integral will contain a factor:

$$\exp(-i\mathbf{P}\cdot\mathbf{r}\hbar) \tag{3.20}$$

where $\mathbf{P} = \mathbf{p}_e + \mathbf{p}_\nu$ is the vector sum of the electron and antineutrino momenta and $\mathbf{r}$ is the co-ordinate of the nucleon making the transition. We can estimate the magnitude of the exponent by taking as typical values: $P = m_e c$ (giving $P/\hbar = 1/386 \text{ fm}^{-1}$) and $R = 4 \text{ fm}$. Hence, $PR/\hbar$ is approximately 0.01. This is small and the exponential can be replaced by 1 for $r \lesssim R$. We shall not study the β-decay interaction H in detail. For our purposes it suffices to quote the property that H does not depend on the lepton momenta. This has the important consequences that the dependence of the matrix element M, Equation (3.14), on the lepton momenta arises entirely from the lepton wave functions (3.20) and that the matrix element M is independent of the lepton momenta if one approximates the exponential, (3.20), by unity. Transitions for which we can make this approximation are called *allowed* transitions.

Schematic β-particle spectra for allowed transitions, illustrating the effects of phase space and Coulomb distortion, are shown in Figure 3.6. The solid curve is without the Coulomb correction term $F(Z, E_e)$. Its general shape is determined by

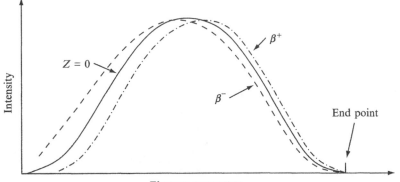

Figure 3.6 Schematic momentum spectra of β particles. The form of the solid curve (labelled $Z = 0$) is that given by Equation (3.18), which contains no Coulomb correction. The dashed curves include the distorting effects due to the Coulomb field, which are different for electrons (β^-) and positrons (β^+). The end points are the same in all cases.

the form of Equation (3.18). For small momenta, it is proportional to p_e^2 and for large momenta, near the end point, it varies as $(E_o - E_e)^2$. It has a maximum in between these two limits. The effect of the Coulomb correction term is indicated by the dashed lines in the figure. Positron spectra are shifted to higher energies and electron spectra to lower energies.

The point at which a curve in Figure 3.6 reaches the end point is difficult to determine because the curve is tangential to the horizontal axis. A more accurate measurement of T_o is obtained by dividing the measured intensity, which is proportional to $d\lambda(p_e)$, by $p_e^2 F(Z, E_e)$ and plotting the square root of the result as a function of T_e. According to Equation (3.19), the graph should be a straight line whose intercept on the horizontal axis is easy to determine. It is called a Fermi–Kurie plot and it has become the conventional way of presenting β spectra. An example is shown in Figure 3.7(a) for the β decay of ^{32}P. The linearity of this plot confirms the validity of the theory for this particular transition in which there is no orbital angular momentum carried by the leptons. As we shall see in the next subsection, Fermi–Kurie plots of decays in which orbital angular momentum is carried away by the leptons do not exhibit the simple linear dependence. The Fermi–Kurie plot also reveals cases where there are β-decay branches to more than one final state. In this case, the graph will be a superposition of several straight lines with different end points. An example is shown in Figure 3.7(b).

The total decay rate is obtained by integrating Equation (3.19):

$$\lambda = C|M|^2 \int F(Z, E_e) p_e^2 (E_o - E_e)^2 dp_e = C|M|^2 f(Z, E_o). \qquad (3.21)$$

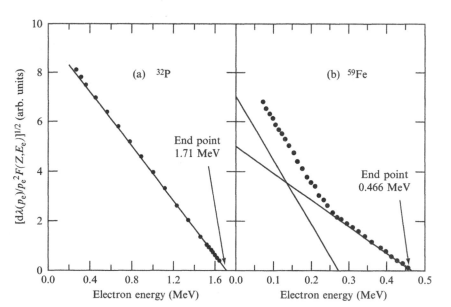

Figure 3.7   Fermi–Kurie plots of β spectra: (a) ^{32}P, which decays with a single transition and shows a single end point; from Porter *et al.* (1957). (b) ^{59}Fe, which has two main β-decay branches to two final states and, therefore, is a composition of two lines with two end points; from Metzger (1952).

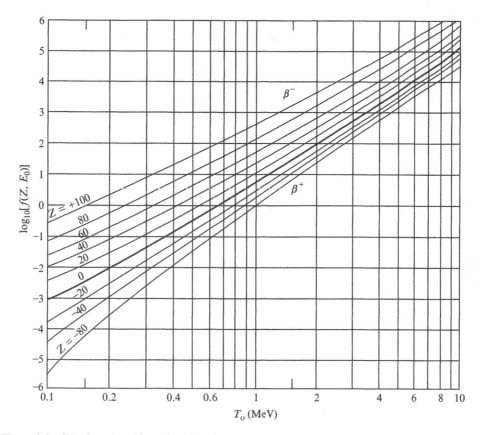

Figure 3.8 The function $f(Z,E_o)$, defined in Equation (3.21), plotted against end-point energy T_o [from Evans (1955), p. 560]. Graphs are shown for different values of atomic number Z of the final nucleus. Positive values are for β^- decays and negative values are for β^+ decays. Note that T_o is equal to the Q value if the parent and daughter nuclei are both in their ground states (neglecting the small amount of energy taken by the final nucleus). Otherwise, $T_o = Q + E_p^* - E_d^*$, where E_p^* and E_d^* are the excitation energies of the parent and daughter nuclei, respectively. Reproduced by permission of The McGraw-Hill Companies.

The function $f(Z, E_o)$ is also a calculable quantity and its logarithm is plotted in Figure 3.8 as a function of end-point energy for a range of values of Z. It is very sensitive to T_o, varying approximately as its fourth power.

Since $f(Z, E_o)$ can be accurately evaluated for any transition, β-decay results are conveniently expressed in terms of the product of $f(Z, E_o)$ and the transition half-life $t_{1/2}$, which is equal to $\ln 2/\lambda$ [Equation (1.14)]. It is important to note that λ refers here to the decay probability of a particular transition between two states. If, as is often the case, a β-active nuclide has several decay branches, λ is equal to the total decay probability (decay constant) multiplied by the fractional decay probability for the particular branch. Thus, $t_{1/2}$ for the branch is obtained by dividing the half-life of the radionuclide by the fractional decay probability. The product of f and $t_{1/2}$ is called the comparative half-life. As we can see from Equation (3.21), $ft_{1/2}$ is proportional to $1/|M|^2$ and so contains the relevant information about the nuclear physics of the transition separated from phase-space effects. As we shall see below, $ft_{1/2}$ varies over many orders of magnitude for different transitions and, therefore, it is customary to quote values of $\log_{10} ft_{1/2}$, where $t_{1/2}$ is in seconds.

So far, we have ignored the fact that the leptons have spin. Both the electron and neutrino (and their antiparticles) have spin $\frac{1}{2}\hbar$, which combine to give a total spin $S = 0$ or $1\hbar$. Each combination corresponds to a different type of transition. The former ($S = 0$) are called Fermi type (F) and the latter ($S = 1$) Gamow–Teller (G-T). Each involves a different interaction with the weak nuclear field. Both types occur and a single transition can be a mixture of the two provided that certain selection rules, which are described below, are satisfied.

Conservation of angular momentum and parity and properties of the initial and final nuclear states determine the type of β decay that can occur (F and/or G-T). For an allowed transition, the lepton wave function is replaced by a constant. This has no angular dependence and, therefore, corresponds to zero orbital angular momentum ($L = 0$). As we have seen in Section 2.4, the parity associated with orbital angular momentum is $(-1)^L$; therefore, there will be no parity change between the initial and final states in an allowed transition. For an allowed Fermi transition (with $S = 0$), no angular momentum of any kind is carried away and, therefore, $I_i = I_f$. For an allowed G-T transition, the leptons carry away a spin angular momentum of one unit and this leads to a different selection rule, which allows a spin change $\Delta I = |I_i - I_f| = (0)$ or 1. The parentheses indicate that $I_i = I_f = 0$ is excluded for G-T decays because, in a $0^+ \rightarrow 0^+$ transition, there can be no angular momentum carried away.

3.3.3   Forbidden transitions

In dealing with allowed transitions, we replaced the exponential of (3.20) by unity. This is justified because the exponent is small (typically, ~ 0.01). However, if the transition does not satisfy the selection rules for an allowed transition, the matrix element is zero. In this case, we must retain the full exponential form, which can be expanded as a power series in $\mathbf{P} \cdot \mathbf{r}/\hbar$. All terms in this series depend on the polar angles (θ, ϕ) of the vector $\mathbf{r}$ and, hence, correspond to orbital angular momenta greater than zero. They are responsible for transitions in which the lepton pair carries away orbital angular momentum with quantum number $L = 1, 2, \cdots$.[4] The corresponding transitions are called first forbidden, second forbidden and so on.

The exponential occurs as a factor in the integral of the transition matrix element M. Since this is integrated over the nuclear volume (i.e. $r \lesssim R$), it follows, from the smallness of the exponent $PR/\hbar$, that successive terms in the expansion of the exponential will make contributions of rapidly decreasing magnitude to $|M|^2$ and, hence, to the transition rate. In general, therefore, one need retain in the expansion only the term of lowest power for which $M \neq 0$, i.e. for which selection rules do not forbid the transition altogether. Thus, a first-forbidden transition, with $L = 1$, will be expected to be hindered relative to an allowed transition, a second-forbidden transition, with $L = 2$, will be even more hindered, and so on.

---

[4]   Since the plane wave contains the total momentum $\mathbf{P}$ of the two leptons, it follows that L is the angular momentum quantum number of their total orbital angular momentum.

In order to see the connection between the terms in the expansion and angular momentum, consider, first, the second term, which is $-i\mathbf{P}\cdot\mathbf{r}/\hbar$. Taking the direction of $\mathbf{P}$ to lie along the z-axis, this term becomes $(-iPr/\hbar)\cos\theta$, which has the angular dependence of the spherical harmonic $Y_1^0(\theta, \phi) = \sqrt{3/4\pi}\cos\theta$ (see Appendix D). Thus, this term affects first-forbidden transitions in which the lepton pair carries away one unit of angular momentum, $L = 1$.

The same procedure can be applied to higher-power terms in the expansion. The quadratic term contains a factor $(\mathbf{P}\cdot\mathbf{r})^2$, which has the angular dependence $\cos^2\theta$ with $\mathbf{P}$ aligned with the z-axis, as before. Since $Y_0^0(\theta, \phi) = \sqrt{1/4\pi}$ and $Y_2^0(\theta, \phi) = \sqrt{5/16\pi}(3\cos^2\theta - 1)$ (see Appendix D), we can write

$$\cos^2\theta = \frac{4}{3}\sqrt{\frac{\pi}{5}}Y_2^0 + \frac{2}{3}\sqrt{\pi}Y_0^0,$$

i.e. the quadratic term contains the spherical harmonic Y_2^0 and, hence, gives rise to second-forbidden ($L = 2$) transitions. In general, one can show that the term $(\mathbf{P}\cdot\mathbf{r})^L$ leads to an angular dependence containing Y_L^M and, therefore, corresponds to Lth-order forbidden transitions in which the lepton pair carries away L units of angular momentum.

In a forbidden transition, we cannot assume that the matrix element M is independent of momentum, as we could for an allowed transition. A consequence of this is that there is an effect on the shape of the β-particle energy spectrum when L is non-zero. A Fermi–Kurie plot will reflect this dependence on L. Indeed, a non-linear plot is an experimental indication that the transition is forbidden.

The selection rules for forbidden transitions are less simple than the rules for allowed transitions because both spin and orbital angular momenta must be taken into account. In a Fermi transition ($S = 0$), the initial and final nuclear spins must satisfy the relation:

$$\mathbf{I}_i = \mathbf{I}_f + \mathbf{L} \tag{3.22}$$

where $\mathbf{L}$ is the orbital angular momentum carried away by the leptons. For a G-T transition, the leptons carry away an additional spin angular momentum of one unit and so, the vector relation becomes

$$\mathbf{I}_i = \mathbf{I}_f + \mathbf{L} + 1. \tag{3.23}$$

Any parity change between the initial and final nuclear states is given by the parity associated with the orbital angular momentum. Thus, there will be a parity change if the orbital angular-momentum quantum number L is odd and no change if L is even. For a first-forbidden transition, therefore, there is a parity change, since $L = 1$. According to the above equations, a Fermi transition is possible for $\Delta I = (0)$ or 1, where the parentheses exclude a $0 \rightarrow 0$ transition, and a G-T transition is possible for $\Delta I = 0, 1$ or 2.

A similar analysis can be done for higher-order forbidden transitions, and selection rules up to the fourth level of forbiddenness are summarized in Table 3.2. Certain

Table 3.2   Selection rules for β decay. Transitions which are not possible if either I_i or I_f is zero are in parentheses; $\Delta\pi$ indicates whether the parities of the initial and final states are different.

Transition Type	L	Fermi		Gamow–Teller	
		ΔI	$\Delta\pi$	ΔI	$\Delta\pi$
Allowed	0	0	No	(0), 1	No
First forbidden	1	(0), 1	Yes	0, 1, 2	Yes
Second forbidden	2	(1), 2	No	2, 3	No
Third forbidden	3	(2), 3	Yes	3, 4	Yes
Fourth forbidden	4	(3), 4	No	4, 5	No

transitions must be of a single type. For example, the decay between two spin-zero states with the same parity must be a Fermi transition. Also, if $\Delta I = L + 1$, the decay is pure Gamow-Teller. Otherwise, mixed transitions are possible.

3.3.4   Comparison of β-decay rates

A large number of β decays have been measured for cases where the spins and parities of the states are known and so the transitions can be classified according to the selection rules. The results are summarized in Table 3.3. Here, an approximate, average value of $\log_{10} ft_{1/2}$ is given for each type of transition with an indication of the spread in values for allowed and first-forbidden decays. Differences of several orders of magnitude in the average $ft_{1/2}$ values for the different types are consistent with calculated estimates.

The category 'superallowed' refers to a special set of allowed β decays in which there is a strong overlap between the initial and final nuclear states. Examples are the so-called mirror transitions, such as n $\rightarrow$ p, t $\rightarrow ^3$He and ^{17}F $\rightarrow$ ^{17}O, in which the only difference between the initial and final states is that an unpaired neutron has been changed into a proton or vice versa. There is a considerable spread in values for a given type of transition, due partly to variations in $|M|^2$. Indeed, there are extreme cases, well outside the limits shown, where the range of $ft_{1/2}$ values for the different categories overlap to some considerable extent. Nevertheless, the classification is useful and a measurement of $ft_{1/2}$ can often help to determine an unknown nuclear spin by identifying the type of transition.

We will consider two examples of allowed decays. The first is the decay of ^{60}Co, which is illustrated in Figure 3.3. The transition is from a 5^+ state to a 4^+ state. Therefore, $\Delta I = 1$ (or greater) and, since there is no parity change, this should be an allowed G-T transition, according to the selection rules. The Q value, which assumes that the parent and daughter nuclei are both in their ground states, is 2.824 MeV, but the decay proceeds almost 100% of the time to the 4^+ second-excited state of ^{60}Ni at 2.506 MeV. Therefore, the end-point energy T_o is $2.824 - 2.506 = 0.318$ MeV and we obtain a value of $\log_{10} f$ of about -0.8 from Figure 3.8. The half-life of 5.27 years gives $\log_{10} t_{1/2} = 8.2$, whence $\log_{10} ft_{1/2} = \log_{10} f + \log_{10} t_{1/2} = 7.4$, which is only just consistent with this being an allowed decay.

Table 3.3 Approximate values of $\log_{10} ft_{1/2}$ for different types of β-decay transition.

Type of transition	$\log_{10} ft_{1/2}$
Superallowed	~ 3.5
Allowed	5.5 ± 1.5
First forbidden	7.5 ± 1.5
Second forbidden	~ 12
Third forbidden	~ 16
Fourth forbidden	~ 21

The second example is the β^+ decay of the 0^+ ground state of ^{14}O to the 0^+ state in ^{14}N at an excitation energy of 2.312 MeV. According to the selection rules, this can only be a pure Fermi transition, which is allowed. The half-life of 70.6 s gives $\log_{10} t_{1/2} = 1.8$. The end-point energy (Q value minus the excitation energy of ^{14}N) is $4.123 - 2.312 = 1.811$ MeV, which leads to a value of about 1.7 for $\log_{10} f$ (from Figure 3.8), whence $\log_{10} ft_{1/2} \approx 3.5$. This suggests that the transition is superallowed and, indeed, it is. The wave functions of the ^{14}O and ^{14}N states are known to be simply related to each other by the operation of transforming a proton in ^{14}O to a neutron and leaving the shell-model structure unchanged.

3.3.5 Electron capture

We have already noted that the weak interaction can induce a process whereby an atomic electron is captured and a proton is changed into a neutron. It is an alternative to β^+ decay and competes with it when both decay modes are possible. The electron-capture (EC) decay rate depends on the overlap of the atomic electron's wave function with that of the nucleus and on the phase-space factor. It also depends on the matrix element M. However, the operation of the weak interaction on the nucleus is the same in both processes; therefore, $|M|^2$ is the same for EC as for β^+ emission.

The electron is most likely to be captured from the K shell of the atom, since the probability of being close to the nucleus is greater for K-shell electrons than it is for electrons in other (outer) shells. The size of the K-electron orbit varies roughly inversely with the atomic number Z. Therefore, the volume it occupies will scale approximately as $1/Z^3$ and its overlap with the nucleus as Z^3. The relative importance of EC increases rapidly with Z because of this factor and also because positron emission is inhibited by the Coulomb barrier for positive particles near the nucleus (see Section 1.4.1) and, therefore, decreases with Z (see Figure 3.8).

The phase-space factor is simpler in EC than in β^+ decay, because only a single particle, the neutrino, is emitted. There is no Coulomb correction factor and so dn/dE_0 is proportional to p_ν^2, which is equal to E_0^2/c^2 if we neglect the electron's binding energy. This energy dependence is weaker than that for positron emission, which varies approximately as the fourth power of the end-point energy (see Figure 3.8). Thus, the ratio of EC to β^+ decay increases with decreasing E_0. Note, also, that the energy released in EC is greater than it is in β^+ emission by $2mc^2$ (1.022 MeV) because, in EC, the mass of an electron is converted into energy whereas, in β^+

emission, energy is required to create the positron (see Problem 3.5). This means that, for decay energies near the threshold for β^+ decay ($T_0 \approx 0$), the relative importance of EC will become very large, because the phase-space factor for it will remain finite whereas that for β^+ emission will be close to zero.

3.4   ALPHA DECAY

Alpha decay is a common form of radioactivity in heavy nuclei and was one of the earliest nuclear physics phenomena to be studied. However, it proved to be one of the most puzzling because the α particles appear at energies which, classically, should be inaccessible. An α particle, incident on a heavy nucleus like uranium, experiences a Coulomb barrier in excess of 20 MeV. Yet α particles are emitted from uranium with energies of less than 5 MeV, apparently violating energy conservation in the vicinity of the barrier.

Another important aspect of α decay, which was noticed as early as 1911 by Geiger and Nuttall, is a very strong correlation of half-life with the decay energy Q value. This is plotted in Figure 3.9 for a representative number of even–even, α-emitting nuclei from which it can be seen that, for nuclei with the same atomic number, there is an approximately linear relationship between $\log t_{1/2}$ and $1/\sqrt{Q}$. The most striking feature of this plot, however, is the degree of sensitivity – a change of a factor of two in Q gives rise to a change of 20 orders of magnitude in the lifetime!

3.4.1   Semi-classical theory of α decay

The explanation of the above general features of α decay was given independently and almost simultaneously in 1928 by G. Gamow and by R. Gurney and E. Condon. In a simple version of the theory, the α particle is assumed to be trapped inside the nucleus by the Coulomb barrier, but has a finite chance of escaping to the outside world by a process known as quantum-mechanical barrier penetration. The situation is illustrated in Figure 3.10 for a spherical nucleus, where $V(r)$, the potential energy between the α particle and the daughter nucleus, is plotted as a function of the distance r between their centres. Clearly, the decay Q value must be positive, otherwise the α is bound and cannot escape at all. Outside the nucleus, $V(r)$ is given by the Coulomb potential. This is inversely proportional to r and, asymptotically, the kinetic energy of the α particle is equal to Q, if we neglect the recoil energy of the daughter nucleus. Close to the nuclear surface, the α particle experiences the attractive, short-range nuclear force and, once inside the nucleus, acquires a kinetic energy that depends on U, the depth of the well. In Figure 3.10, the nuclear potential is taken to be a square well, radius R, with a sharp edge. In practice, the surface will not be sharp and the peak of the Coulomb barrier will have a rounded shape, as in Figure 1.2(b). For simplicity, we shall continue to use the form shown in Figure 3.10.

An important part of the α-decay problem is the probability that the α particle exists as a recognizable entity inside the nucleus before its emission. We shall refer to this as the preformation probability P. It is very dependent on the structure of the states of the parent and daughter nuclei and is a measure of the degree to which the

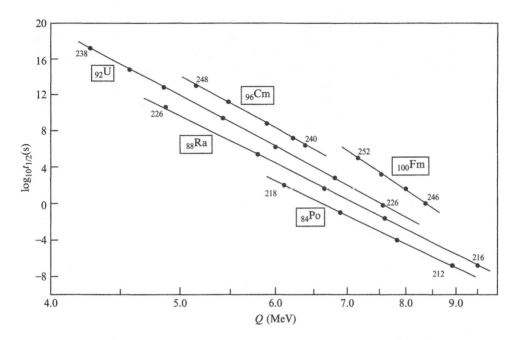

Figure 3.9 Logarithm of the half-life (in seconds) plotted against Q value for various even–even α-emitting nuclei. The lines connect sequences of isotopes. The horizontal scale is linear in $1/Q^{1/2}$.

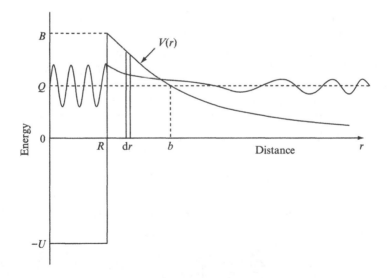

Figure 3.10 Schematic view of barrier penetration in α decay. The α-particle wave function is oscillatory inside the nucleus ($r < R$) and at large distances ($r > b$). Within the barrier region ($R \leq r \leq b$), it is a decreasing function of r. The barrier can be considered to be a set of narrow rectangular barriers of width dr, which the α particle has to penetrate in order to escape from the nucleus. For $r > R$, $V(r)$ varies as $1/r$, therefore, $R/b = Q/B$.

initial state 'looks like' the daughter nucleus in its final state plus an α particle. It is difficult to calculate P and, in our simple treatment, we shall assume it has a value of unity. In principle, it can be determined from experimental measurement; however, uncertainties in estimating the barrier penetration factor mean that an experimental determination of P may be uncertain by well over an order of magnitude.

Given that the α particle exists, we can imagine it moving back and forth inside the nucleus with a speed v given by its kinetic energy and presenting itself at the barrier with a frequency:

$$f \approx v/2R. \tag{3.24}$$

Quantum mechanically, the α particle is represented by a wave function ψ. Inside the nucleus, ψ is oscillatory with a wave number $k = \sqrt{2m(Q+U)/\hbar^2}$ where m is the mass of the α particle. Within the barrier region ($R < r < b$), the wave function is not zero and it takes the form of a falling exponential, if we assume (as is justified by a more rigorous calculation) that any reflected wave from the outer edge of the barrier at $r = b$ is negligible. Outside the nucleus, the wave function is again oscillatory with a wave number at large distances equal to $\sqrt{2mQ/\hbar^2}$.

For purposes of illustration, we can estimate the penetration probability by treating the barrier as a series of narrow rectangular barriers in one dimension. Each barrier element has height:

$$V(r) = 2Ze^2/4\pi\varepsilon_0 r \tag{3.25}$$

and width dr, as shown in Figure 3.10, where Z is the atomic number of the daughter nucleus. The wave function at each barrier has the form $e^{-\mathcal{K}r}$ where $\mathcal{K} = \sqrt{2m[V(r) - Q]/\hbar^2}$ and its value after the barrier is reduced to $e^{-\mathcal{K}(r+dr)}$. Therefore, the transmission probability through a single element of the barrier is $\exp(-2\mathcal{K}dr)$.

The probability of penetrating the complete barrier is the product of such infinitesimal probabilities calculated for the set of mini-barriers between $r = R$ and b. The product of infinitesimal exponentials is the exponential of the sum of their exponents, i.e. it is the exponential of the integral of the exponents and, therefore, we obtain for the total transmission probability:

$$T = \exp\left(-2\int_R^b \mathcal{K}dr\right) = \exp(-2G) \tag{3.26}$$

where G is called the Gamow factor. It can be evaluated, with $V(r)$ given by Equation (3.25), to give

$$G = \sqrt{\left(\frac{2m}{\hbar^2 Q}\right)}\frac{2Ze^2}{4\pi\varepsilon_0}\left[\cos^{-1}\sqrt{x} - \sqrt{x(1-x)}\right] \tag{3.27}$$

where $x = R/b = Q/B$ with B being the barrier height (see Figure 3.10). Combining the results of Equations (3.24) and (3.26) with the preformation probability, we obtain an expression for the α-decay rate of

$$\lambda = Pf\exp(-2G). \tag{3.28}$$

We will use our simple theory to estimate the dependence of the α-decay half-life on the decay Q value, shown in Figure 3.9. We take ^{238}U as a representative parent nucleus and calculate the penetrability factor e^{-2G} for $Q = 4.268$ MeV (which is the actual α-decay Q value of ^{238}U) and $Q = 5.268$ MeV, i.e. two values that differ by 1 MeV.

Rewriting Equation (3.27) slightly gives

$$G = 2\alpha Z \sqrt{\frac{2mc^2}{Q}}\left[\cos^{-1}\sqrt{Q/B} - \sqrt{(Q/B)(1 - Q/B)}\right] \tag{3.29}$$

where the quantity $\alpha = e^2/4\pi\varepsilon_0\hbar c$ is known as the fine-structure constant and is equal to 1/137. We also have $Z = 90$ and the α-particle rest-mass energy $mc^2 = 3727$ MeV. The Coulomb barrier height B is 27.87 MeV, where we have used Equation (3.25) with $r = R = 1.2(A_1^{1/3} + A_2^{1/3}) = 9.3$ fm for the effective separation distance between the daughter nucleus ($A_1 = 234$) and the α particle ($A_2 = 4$) at the peak of the barrier. This distance takes into account the finite sizes of both particles. Substituting values into Equation (3.29) gives $G = 44.391$ and 36.047 for $Q = 4.268$ and 5.268 MeV, respectively. Therefore, since $t_{1/2} \propto 1/\lambda$, we have $t_{1/2}(5.268)/t_{1/2}(4.268) = \exp(-16.69) = 5.7 \times 10^{-8}$. An increase of 1 MeV in Q changes $t_{1/2}$ by over seven orders of magnitude, which is similar to the dependence shown in Figure 3.9.

We can determine a half-life by estimating values for P and f in Equation (3.28). Studies of α-particle scattering on heavy nuclei indicate that the effective α-particle potential has a depth $U \approx 120$ MeV. Thus, in our simple treatment, the α particle has a kinetic energy of about $U + Q$ inside the nucleus. Using this, we obtain an estimate of 7.7×10^{22} fm s^{-1} for the speed of the α particle in the nucleus and a value for the frequency f from Equation (3.24) of about 4×10^{21} s^{-1}.

Combining this with the value for G, for $Q = 4.268$ MeV, and assuming a preformation probability of unity, we obtain a value for the half-life of

$$t_{1/2} = \frac{\ln 2}{\lambda} = \frac{\ln 2}{f e^{-2G}} = \frac{0.69}{4 \times 10^{21} \times e^{-88.78}} = 6.2 \times 10^{16}\,\text{s}$$

or about 2 billion years, which is to be compared with the measured half-life of 4.47 billion years.

This level of agreement is certainly as good as could be expected, given the crude nature of the calculation. Indeed, there are many approximations in the derivation, some of which we have already noted. Taking the preformation probability to be unity is surely a gross overestimate, and a reduction in this to a more realistic value would increase the half-life estimate substantially. The shape of the barrier is unrealistic and a different prescription could have been used for the choice of barrier radius R. This is particularly critical, since it determines the value of B that enters into the exponent G.

Despite these limitations, this simple analysis adequately illustrates the main physics underlying the α-decay process. As we have shown, it accounts very well for the strong dependence of $t_{1/2}$ on Q. The Z dependence shown in Figure 3.9 is similarly explained (see Problem 3.11) since B increases with Z, which means that the barrier will be harder to penetrate, resulting in an increase in the half-life.

3.4.2   Alpha-particle energies and selection rules

The energies of α particles emitted in radioactive decay often form a group of several discrete peaks, referred to as *fine structure*. They correspond to decay branches to a number of different states in the daughter nucleus, which are often closely spaced in energy. An example is shown in Figure 3.11 for the even–even nucleus ^{232}U. The intensities of the various branches are given as percentages by each level.

Note that, when the spins of the initial and final states are different, the α particle carries away some orbital angular momentum. The removal of angular momentum hinders emission, as we have seen it does in γ and β decays; however, it does not usually have a major impact on the α-decay probability. The variations in intensity, shown in Figure 3.11, are determined mainly by two factors: first, by the strong energy dependence of barrier penetrability, which decreases with increasing energy of the final state, and second, by differences in the structure of the initial and final states, which can profoundly influence the preformation probability. Compared with these, the additional effect due to the α particle carrying away a few units of angular momentum turns out to be small. This is in contrast to γ and β decays where, as we have seen, an increase in the transfer of even a single unit of angular momentum reduces the decay probability by a large factor.

There is a parity selection rule for α decay that depends on angular momentum. The angular part of the α particle's wave function is a spherical harmonic Y_L^M, where

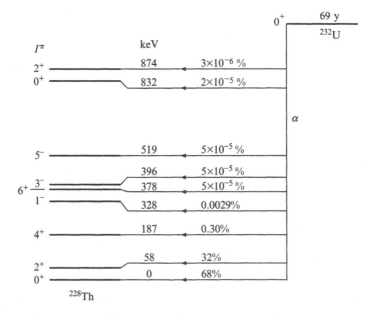

Figure 3.11   Alpha decay of ^{232}U. Relative intensities are indicated as percentages on the right-hand side of each energy level of the daughter nucleus ^{228}Th.

L and M are the orbital angular-momentum quantum numbers. As we have seen, this function has a parity of $(-1)^L$ and, therefore, since parity is conserved in α decay, the parities of the initial and final nuclear states will be different if L is odd and the same if L is even. Certain decays are forbidden by this rule. For example, a 0^+ parent state cannot decay to an odd parity state with even spin or to an even parity state with odd spin.

3.4.3 Transuranic nuclei

A recent use of α decay has been to identify and determine masses of the heaviest known transuranic nuclei. Nuclei near the upper mass limit of stability are the result of a fine balance between attractive and repulsive forces acting on the nucleons within them. Properties and even the very existence of these nuclei are sensitive to details of the nuclear force and constitute an important body of information with which current theories of nuclear structure can be tested.

Over the years, many new transuranic nuclei have been artificially created and studied. Their masses have generally been determined from measured energies of α particles emitted in decay chains which reach nuclei whose masses are known. Two measured decay chains are shown in Figure 3.12 for the (unnamed) nucleus $^{277}_{112}\text{X}$. These nuclei were synthesized in February 1996 during an experiment in which ^{208}Pb was bombarded with a beam of ^{70}Zn ions in the fusion reaction:

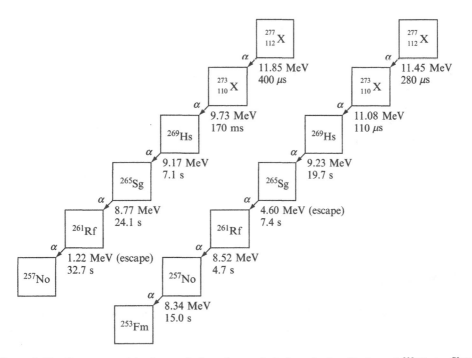

Figure 3.12 Two α-particle decay chains observed during the irradiation of ^{208}Pb by ^{70}Zn projectiles. The assignment of the measured signals indicates that the head of each chain is the nuclide $^{277}_{112}\text{X}_{165}$. Decay energies marked '(escape)' mean that the α particle deposited only part of its energy into the detector. From Hofmann *et al.* (1996).

$$^{208}_{82}\text{Pb} + ^{70}_{30}\text{Zn} \rightarrow ^{277}_{112}\text{X} + \text{n}.$$

In the experimental arrangement, the fusion products were separated from other, mainly beam particles and deposited into a sophisticated, position-sensitive detector. The energies of successive α decays were measured as well as the times when the decays occurred. The new element was identified as the head of a long decay chain of known daughter nuclei. There is enough information in such a set of measurements to enable an unambiguous assignment to be made even of a single detected nucleus. The element 112 has a 0.34 ms average half-life. The two events shown in Figure 3.12 were observed during 24 days of continuous bombardment.

The production cross section has been estimated to be about 1 picobarn (10^{-12} b). This is very small, but it was expected that spontaneous fission would become sufficiently important above $Z = 106$ that production cross sections of heavier elements would be even smaller than are observed. It is thought that the reason for its suppression is the stabilizing effect of shell closures, which, as described in Section 2.5, can occur in deformed nuclei and produce a second minimum in the potential energy versus deformation curve. The liquid-drop model without shell stabilization predicts no extra minimum, in which case it would be much easier for the deformation of the newly created nucleus to increase again, leading rapidly to spontaneous fission. Detailed calculations, including shell stabilization, have predicted the existence of an 'island of stability', near the magic numbers $N = 184$ and $Z = 114$, of so-called *superheavy* nuclei, some of which may be sufficiently close to being stable that they have half-lives of years or even centuries. Scientists have been trying to validate this prediction for a considerable time and, evidently, with the creation of element 112, a great deal of progress has been made.

More recently, it has been reported that experiments carried out in Russia and in the USA may have produced nuclei with $Z = 114$ and 118. The decay chains observed in these experiments do not connect to known nuclei and so there is some uncertainty in the assignments. However, if the results are confirmed, the goal to reach the region of superheavy nuclei will have been attained, opening up a new phase of research into the nature of nuclei at the extreme limits of stability. Unfortunately, cross sections remain exceedingly small, making the experiments very time consuming. Even if it becomes accepted, as now seems likely, that the island of stability exists, progress in its exploration could be frustratingly slow.

PROBLEMS 3

3.1 Calculate the recoil energy of a nucleus, mass m, after it has emitted a γ ray of energy E_γ. Hence, calculate the energy of the excited state of ^{16}O, which emits a 6128.63-keV γ ray when it de-excites to the ground state.

3.2 List all the possible multipolarities for the following γ-ray transitions, indicating, in each case, which is likely to be the most intense: (i) $3^- \rightarrow 2^+$; (ii) $\frac{5^+}{2} \rightarrow \frac{9^+}{2}$; (iii) $\frac{1^+}{2} \rightarrow \frac{1^-}{2}$; (iv) $\frac{3^+}{2} \rightarrow \frac{7^+}{2}$.

3.3 A nuclear excited state decays by an E2 transition to the $\frac{3+}{2}$ ground state. List the possible spin–parity (I^π) assignments of the excited state. If there is no evidence of decay by an M1 transition, what is the I^π of the excited state most likely to be?

3.4 The low-lying spectrum of ^{207}Pb is shown in Figure 2.11. Sketch the diagram and show the main γ-ray transitions for the first three excited states, indicating the most likely types and multipolarities in each case. Estimate the half-lives of the states and suggest which, if any, are likely to be isomers.

3.5 The ^{22}Na atom has a mass–energy of 20487.686 MeV and it decays, with a half-life of 2.6 years, from its 3^+ ground state by β^+ emission and electron capture (EC) to the 2^1 first excited state of ^{22}Ne (atomic mass–energy = 20484.844 MeV). (a) Write down expressions for the two types of decay; (b) calculate the β^+ decay Q value in MeV; (c) calculate the EC decay Q value in MeV.

3.6 Show that the Q value for the β decay of the neutron is 0.782 MeV. Give the type of transition and, hence, obtain an estimate of the half-life. (Hint: you will require the approximate value of $\log_{10} f$ for this transition, which can be found from Figure 3.8.)

3.7 A neutron at rest decays into a proton with a decay energy of 0.782 MeV. What will be the maximum kinetic energy of the residual proton?

3.8 From the information given for the β^- decay of ^{137}Cs in Figure 3.3, calculate the $\log_{10} ft_{1/2}$ values for the β transitions to the $\frac{11-}{2}$ and $\frac{3+}{2}$ states of ^{137}Ba. What type of transition would be expected in each case?

3.9 The α-decay Q value of ^{238}U is 4.268 MeV. Calculate the height B of the Coulomb barrier between the α particle and the daughter nucleus, assuming that the nuclear potential has a sharp edge at a radius of $1.4A^{1/3}$ fm (see Figure 3.10). Calculate the distance b beyond which the α-particle kinetic energy is positive.

3.10 Use the simple theory given in Section 3.4 to estimate the α-decay half-life of ^{209}Bi ($Q = 3.13$ MeV), given the known half-life of 2.1 min for the α decay of ^{211}Bi ($Q = 6.749$ MeV). Assume that the frequencies f, defined in Equation (3.24), and the α-particle preformation probabilities P have the same values for both isotopes. Use $R = 1.2(A_1^{1/3} + 4^{1/3})$ fm for the radius corresponding to the Coulomb barrier. Does your result justify the fact that ^{209}Bi is considered to be stable with a half-life in excess of 10^{19} years?

3.11 Estimate the Z dependence of the α decay of nuclei near ^{238}U by calculating the effect on the half-life of changing Z by 2 at an α-particle energy of 5 MeV. Compare your result with Figure 3.9. Assume that the values of f, defined in Equation (3.24), and the preformation probability P are the same for your two calculations and use $R = 1.2(A_1^{1/3} + 4^{1/3})$ fm for the radius corresponding to the Coulomb barrier.

4

Nuclear Reactions

4.1 INTRODUCTION

At the end of Chapter 1, we learned how the use of nuclear collisions enables scientists to obtain information about nuclei, create new ones and exploit the nuclear force for practical purposes. Basic concepts were presented with a few examples, mainly of the types of reaction that have found widespread application in other fields. In this chapter, we develop the ideas a little further and indicate how nuclear reactions are used to study different aspects of nuclear structure and the dynamic response of nuclear matter when it is subjected to the stress of a nuclear collision.

Many different reactions may occur during a collision between two nuclear particles. In elastic and inelastic scattering (see Section 1.6.5), the incoming and outgoing particles are the same:

$$a + A \rightarrow a + A \quad \text{(elastic)}$$
$$a + A \rightarrow a + A^* \quad \text{(inelastic)}.$$

In the case of inelastic scattering, the A^* indicates that the nucleus has been left in an excited state as a result of the collision and so the kinetic energy in the exit channel will be less than that in the entrance channel. In general, the particles emerging from a collision will not be the same as the initial ones and we write such a reaction as

$$a + A \rightarrow b + B + \cdots$$

indicating that there may be more than two particles in the exit channel. Examples include simple pickup and stripping transfer reactions (see Section 1.6.5) involving mass exchange between the initial particles. Many other types of reaction may occur in which one or both colliding nuclei are changed. For example, an energetic projectile may knock a piece out of the target or be broken up during the collision. A common reaction is one in which the projectile fuses with the target to form a compound nucleus, which later may emit one or more particles.

It is often possible to distinguish between different reactions because they lead to different final channels. However, this is not always the case. For example, a particle with a given energy, emerging from a nuclear collision at a particular angle, may be the result of a direct reaction or a compound-nucleus reaction. Data analysis is not straightforward in such situations because, in general, the different competing mechanisms will interfere with each other. A cross section is the square of a reaction amplitude, and all contributing amplitudes must be calculated and added together before the resulting cross section is determined. This can be done, but often there are ambiguities in interpretation and it is a great simplification if a given set of measurements can be considered to be due to a single reaction mechanism. We shall not consider the effects of interference between different reaction mechanisms in any of the examples presented here.

In the next section, we describe observable differences between compound-nucleus and direct reactions that can enable them to be distinguished from each other. This is followed by derivations of simple expressions, using classical and quantum-mechanical concepts, to illustrate some general features of reaction cross sections. In Section 4.3, we discuss briefly how measuring the elastic scattering of two of the simplest projectiles, electrons and protons, can give information about the sizes and shapes of nuclei. In Section 4.4, we show how direct reactions can be used to measure properties and obtain information about the character of nuclear states and in Section 4.5, the basic concepts of compound-nucleus reactions are outlined, including the excitation and decay of low-energy, discrete resonances in the compound system. Heavy-ion collisions are qualitatively different to those involving light ions and are discussed separately in Section 4.6.

Much of the discussion in this chapter will focus on types of nuclear reactions that are predominantly either direct or non-direct. In reality, however, there are many reaction processes that are intermediate in the sense that, while the combined system remains together long enough for a good deal of equilibration to take place, it is not so long that there is true independence between the initial and final channels. We shall refer to this again in Section 4.6.

The presentation is mainly qualitative. Sophisticated theories of nuclear reactions have been developed using full, quantum-mechanical or semi-classical approaches, but even a cursory review of them lies beyond the level and needs of this text.

4.2 GENERAL FEATURES OF NUCLEAR REACTIONS

4.2.1 Energy spectra

Different reaction mechanisms can often be distinguished from each other because they give rise to outgoing particles with different energies. We illustrate this in Figure 4.1, which is a typical energy spectrum of particles emerging from nuclear collisions occurring when protons are incident on a target of medium-mass nuclei at a bombarding energy a few times that of the Coulomb barrier.

At the upper end of the spectrum are a number of discrete peaks due to elastic scattering, inelastic scattering and transfer reactions. These different direct reactions can be distinguished from each other by identifying the nature of the outgoing particles in the detector that records their energies. At lower energies, the peaks correspond to more closely spaced levels in the final nuclei and are not fully resolved

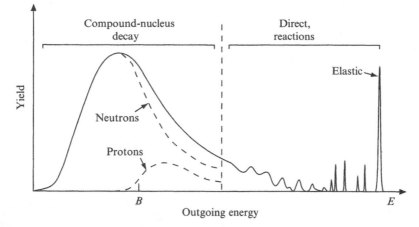

Figure 4.1 Energy spectra of charged particles and neutrons emitted from reactions induced by protons on medium-mass nuclei at a bombarding energy a few times that of the Coulomb barrier. Neutron and proton spectra are shown as separate contributions to the total in the low-energy region where compound-nucleus reactions dominate. The Coulomb barrier B for protons is labelled on the abscissa.

because of imprecision in the measurement of particle energy and spread in energy of the incident beam. At still lower energies, we see a broad continuum, which is due mainly to decays of compound nuclei formed by the fusion of protons with target nuclei.

In a compound-nucleus reaction, as we have noted in Section 1.6.5, the available energy becomes shared among many nucleons. An equilibrium situation is reached and the compound nucleus loses its energy over a period of time by emitting particles, mainly neutrons and protons, in a process analogous to the evaporation of molecules from a heated liquid. However, the yield of these evaporated particles is confined to low energies because the statistical probability of a particle acquiring a large fraction of the available energy is small. This suggests that, in order to study a direct reaction, the bombarding energy should be high enough so that the energies of particles from direct reactions are well above those expected of the same particles evaporated from a compound nucleus in a non-direct reaction.

In Figure 4.1, the neutron and proton evaporation contributions to the low-energy continuum are shown as dashed curves.[1] Below its maximum, the neutron spectrum reflects the energy dependence of the available phase space, which falls to zero as the neutron energy falls to zero. For protons (and other charged particles), the low-energy cut-off is above zero, because their emission is hindered by a Coulomb barrier. This means that a compound nucleus with only enough energy to emit one or two particles will usually emit a neutron rather than a charged particle. However, this may not be the case when the compound nucleus is highly neutron deficient, and much more energy is required to remove a neutron than a proton from the nucleus.

---

[1] Other charged particles may be evaporated as well but, generally, with lower yields.

4.2.2   Angular distributions

Representative differential cross sections, in the centre-of-mass system, of reaction products from direct and non-direct reactions are shown in Figure 4.2. The angular dependence of a direct reaction is forward peaked, since the interaction is restricted mainly to peripheral collisions from which the outgoing particle will carry most of the incident energy and momentum of the projectile. It is also common for the angular distribution to exhibit oscillations resulting from the wave nature of the particles, which gives rise to interference phenomena similar to those seen in the scattering of light by small objects. We show several examples of this in Sections 4.3 and 4.4.

   The characteristic angular distribution of particles evaporated from a compound nucleus is generally more isotropic and symmetric about 90°. If we imagine a classical collision between two spin-zero particles, the orbital angular momentum of their relative motion will be perpendicular to the velocity of the projectile. Therefore, the compound nucleus will be formed with its angular momentum L lying in the plane at right angles to the incident direction (zero degrees), since this plane contains all the possible directions of orbital angular momentum in the entrance channel. As the compound nucleus de-excites, it must lose this angular momentum, as well as excitation energy. Angular momentum is perpendicular to velocity therefore, evaporated particles, which reduce L most effectively, are emitted preferentially at right angles to L. This gives rise to increased particle emission in the two directions that are always at right angles to L, namely, at 0° and 180° relative to the incident beam. The anisotropy becomes more pronounced the higher the value of L. The angular distribution is isotropic if the compound nucleus has no angular momentum because, in that case, there is no preferred direction of emission.

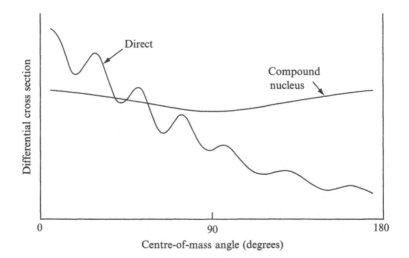

Figure 4.2   Representative differential cross sections for outgoing particles in a direct reaction and from the decay of a compound nucleus in a non-direct reaction.

4.2.3   Cross sections

Each of the different final states of a nuclear collision has its own probability or cross section. In general, it is difficult to calculate a particular cross section from first principles, and recourse is made to nuclear models some of which we refer to in later sections. Here, a few simple expressions are derived, which enable rough estimates to be made in certain circumstances. In some cases, a classical or semi-classical calculation can give a reasonable approximation. However, in general, quantum effects profoundly alter the classical picture, especially at low energies where the wavelength of the beam particle can be much greater than a typical nuclear size.

Classical estimates

In this discussion, we separate the total cross section into two parts: an *elastic scattering* cross section and a *non-elastic* or *reaction* cross section. Thus, $\sigma_t = \sigma_{sc} + \sigma_r$. We further consider that a nuclear reaction leading to σ_r can only take place if the colliding nuclei come within range of the nuclear force, which is short ranged. Accordingly, we develop the idea, introduced in Section 1.6.5, of an interaction radius R such that if the nuclei approach each other within this distance, i.e. if the distance between their centres $r \leq R$, a nuclear reaction can take place and if $r > R$, no reactions occur. The interaction radius depends on the interacting nuclei and can be represented as a sum $R_1 + R_2$ of the effective radii of the two nuclei as illustrated in Figure 4.3. The interaction radius and the effective radii are related to the geometric sizes of the nuclei, but do not have definite values because of the range of the nuclear force and because the distribution of nuclear matter in a nucleus has a diffuse surface. We shall see in Section 4.6 how an estimate of R can be made from experimental measurement.

For uncharged particles, the geometric cross section for the nuclei to collide is $\sigma = \pi R^2$, given by the cross-sectional area determined by the interaction radius (see Figure 4.3). However, in general, the colliding nuclei will be charged and the collision

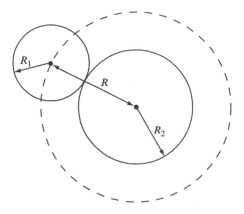

Figure 4.3   Diagram showing two nuclei with effective interaction radii R_1 and R_2. The dotted circle denotes the region representing the nuclear collision cross section. If the nuclei approach each other within the distance $R = R_1 + R_2$, they undergo a nuclear collision.

cross section will be modified because the nuclei experience an energy barrier due to the repulsive Coulomb force acting between them. In Figure 4.4, a charged particle is shown approaching the nucleus along a line which is separated from the line heading directly for the centre of the nucleus by a distance b. This distance is known as the impact parameter and if it is not zero, as is the case here, the Coulomb force will deflect the trajectory. Note also that the kinetic energy decreases as the distance between the particles decreases. In what follows, we shall assume, for simplicity, that the target is massive and remains at rest.

The collision cross section is given by πb^2, where b is the impact parameter for a distance of closest approach equal to R. Assuming the nuclei have sharp edges, the Coulomb energy at the distance of closest approach is given by the familiar formula

$$B = Z_1 Z_2 e^2 / (4\pi\varepsilon_0) R \tag{4.1}$$

where Z_1 and Z_2 are the atomic numbers of the two nuclei. The value of the impact parameter b is obtained by applying conservation laws.

At the distance of closest approach, the initial kinetic energy E appears partly as reduced kinetic energy E' and partly as Coulomb potential energy B. Hence,

$$E = E' + B.$$

Applying conservation of angular momentum gives

$$L = pb = p'R$$

where p and p' are the projectile momenta initially and at the distance of closest approach, respectively. We know that $E'/E = (p'/p)^2 = (b/R)^2$, from which it follows that

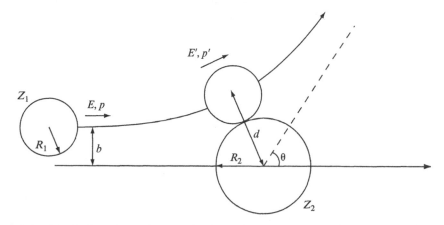

Figure 4.4   Classical grazing trajectory of a charged particle, effective radius R_1 and charge $Z_1 e$, incident with impact parameter b at energy E and momentum p and deflected through an angle θ by a target nucleus, effective radius R_2 and charge $Z_2 e$. The distance of closest approach d for this trajectory is the interaction radius $R (= R_1 + R_2)$ at which point the projectile has energy E' and momentum p'. The mass of the target is assumed to be large compared with the projectile mass.

$$\sigma = \pi R^2 \frac{E'}{E} = \pi R^2 \left(1 - \frac{B}{E}\right) \text{ for } E \geq B. \tag{4.2}$$

For the general situation, which takes target recoil into account, E is replaced by the centre-of-mass energy $E_{cm} = EM/(M + m)$ (see Problem 4.10), where E is the energy of the projectile (mass m) in the laboratory system and M is the mass of the target, initially at rest. Note that we obtain the value of πR^2 for uncharged particles when $B = 0$. Also, Equation (4.2) approaches this limit at high energy when the effect of the Coulomb force is small.

The cross section $\sigma = \pi b^2$ can be written in terms of the wave number $k \, (= p/\hbar)$ of the projectile and the angular momentum quantum number ℓ as

$$\sigma = \pi \left(\frac{L}{p}\right)^2 = \frac{\pi\ell(\ell+1)\hbar^2}{(\hbar k)^2} \approx \frac{\pi\ell^2}{k^2} \tag{4.3}$$

where we have assumed that $\ell \gg 1$ in the classical limit we are considering here.

It must be remembered that these equations represent an upper limit to the total reaction cross section, since it is assumed that there is a 100% chance of a reaction occurring if $r \leq R$. In general, there may be many different outcomes as a result of interactions that may take place during the collision and each has its own cross section proportional to its probability. In practice, these cross sections generally bear little resemblance to the geometrical nuclear size.

According to Equation (4.2), the cross section falls to zero when $E = B$ and it remains there at lower energies because the nuclei are always out of contact with each other, even in a head-on collision. In reality, however, the cross section remains finite when $E < B$ because, even though classically the particles remain energetically out of range, there is a probability of the Coulomb barrier being penetrated and a reaction taking place. It is an example of the effects of the wave nature of matter, which we explore a little further below.

Quantum mechanical considerations

Classical formulae are generally only useful when the de Broglie wavelength λ is small compared with the nuclear size (or interaction radius R). When λ is comparable with or greater than R, quantum effects become important; a proper trajectory cannot be defined and the simple relationships between the geometric cross section, impact parameter and angular momentum (see above) break down.

We can show how the situation is modified with the aid of a semi-classical picture, due to Blatt and Weisskopf, illustrated in Figure 4.5. Here, the interaction region, as seen by an approaching projectile, is divided into a set of ring-shaped zones concentric with the head-on collision point. The zones are labelled by the orbital angular momentum quantum number ℓ. The innermost ($\ell = 0$) zone corresponds to particles with impact parameters less than the reduced wavelength $\lambdabar \, (= \lambda/2\pi)$ of the beam particles, the next includes impact parameters between $\lambdabar$ and $2\lambdabar$ and so on. Impact parameters for the ℓth zone range from $\ell\lambdabar$ to $(\ell+1)\lambdabar$ and correspond, classically, to

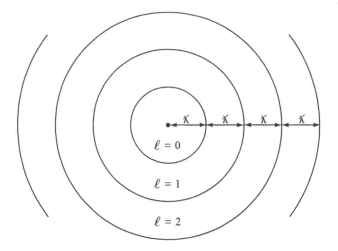

Figure 4.5 Angular momentum zones for an incident beam directed perpendicular to the plane of the figure. Particles with a given angular momentum quantum number ℓ, relative to the centre, are mainly confined to the ring-shaped zone corresponding to that value of ℓ.

particles with momentum p having angular momenta between $p\ell\lambdabar$ and $p(\ell + 1)\lambdabar$, i.e. between $\ell\hbar$ and $(\ell + 1)\hbar$. In quantum mechanics, only integral values of ℓ are allowed and, although the concept of impact parameter is imprecise, we interpret this picture to say that particles with angular momentum ℓ correspond to the ℓth zone. In this way, a plane wave, representing a parallel beam, can be split into a set of partial waves, or ℓ waves, each associated with a particular zone and partial-wave cross section σ_ℓ. The cross-sectional area of the ℓth zone is

$$\sigma_\ell = (2\ell + 1)\pi\lambdabar^2. \tag{4.4}$$

It represents an upper limit to the total reaction cross section since no more particles can be removed than are contained within the zone. Hence, the reaction cross section for the ℓth partial wave is

$$\sigma_{r,\ell} \leq (2\ell + 1)\pi\lambdabar^2. \tag{4.5}$$

It does not apply to the elastic scattering cross section.

Expanding the cross section into partial waves is most useful when we are dealing with low-energy collisions and the wavelength is comparable with the nuclear size. As the value of ℓ increases above zero, the overlap of the ℓ zone with the nucleus decreases and soon reaches the point where the effect of the nuclear interaction on the partial wave becomes negligibly small. Thus, we expect that only a few partial waves with $\ell \leq R/\lambdabar$ need to be taken into account in any analysis of the nuclear reactions that take place. In an extreme case, such as slow-neutron induced reactions, we have $\lambdabar \gg R$ and only the $\ell = 0$ wave need be considered.

Some general consequences of quantum-mechanical behaviour on the scattering and reaction cross sections can be derived by considering the form of the wave function in the exterior region only ($r > R$). To proceed further requires assumptions to be made about what happens at the nuclear surface and beyond into the interior region ($r < R$). As illustration, we shall consider the simple example of low-energy uncharged particles (neutrons) interacting with a central (spin-independent) force field, which can give rise to elastic scattering and absorption (removal of particles from the elastic channel).

General s-wave ($\ell = 0$) elastic scattering and absorption

Fluxes: The $\ell = 0$ part of a plane wave

$$\phi(r) = e^{ikz} \tag{4.6}$$

is the sum of spherical incoming and outgoing s-waves:

$$\phi_0(r) = \frac{\sin kr}{kr} = \frac{1}{2ikr}(-e^{-ikr} + e^{ikr}). \tag{4.7}$$

The effect of a scattering centre only changes the outgoing part ($\propto e^{ikr}$). This still propagates as a free wave when $r > R$, but its amplitude and phase may be altered by the interaction. Thus, the modified s-wave can be written:

$$\psi_0(r) = \frac{1}{2ikr}(-e^{-ikr} + \eta e^{ikr}) = \phi_0(r) + \frac{(\eta - 1)e^{ikr}}{2ikr} \tag{4.8}$$

where we have written $\psi_0(r)$ as the sum of the incident wave and a scattered wave:

$$\psi_{sc}(r) = \frac{(\eta - 1)}{2ikr} e^{ikr}. \tag{4.9}$$

The integrated outgoing s-wave flux, R_{out}, is equal to the rate at which s-wave particles pass outwardly through a sphere (of radius $r \geq R$), i.e.

$$R_{out} = \left| \frac{\eta e^{ikr}}{2ikr} \right|^2 4\pi r^2 v = \frac{|\eta|^2 \pi v}{k^2}. \tag{4.10}$$

Similarly, we can write the integrated incoming s-wave flux as

$$R_{in} = \frac{\pi v}{k^2} \tag{4.11}$$

and the integrated flux of scattered s-wave particles as

$$R_{sc} = \frac{|\eta - 1|^2 \pi v}{k^2}. \tag{4.12}$$

If we have only elastic scattering and no reactions (i.e. no loss of elastic flux), $R_{out} = R_{in}$ and $|\eta| = 1$. The incoming and outgoing amplitudes have the same magnitude and differ only in phase. If other reactions occur (which include inelastic scattering), then the outgoing integrated flux is depleted: $R_{out} < R_{in}$ and $|\eta| < 1$.

Cross sections: By definition, the s-wave elastic scattering cross section $\sigma_{sc,0}$ is the integrated flux of scattered particles per unit flux of the incident plane wave F_{inc}, which, from Equation (4.6), is equal to v. Therefore,

$$\sigma_{sc,0} = \frac{R_{sc}}{F_{inc}} = \frac{\pi|\eta - 1|^2}{k^2} = \pi\lambda^2|\eta - 1|^2 \tag{4.13}$$

where $k = 1/\lambda$.

The s-wave reaction cross section is defined as the depletion of the integrated flux divided by F_{inc}, i.e.

$$\sigma_{r,0} = \frac{R_{in} - R_{out}}{F_{inc}} = \pi\lambda^2(1 - |\eta|^2). \tag{4.14}$$

This is maximum when $|\eta|^2 = 0$, which is consistent with Equation (4.5) for $\ell = 0$.

As we have noted above, for $|\eta| = 1$ we have elastic scattering only and $\sigma_{r,0} = 0$. However, note that if $\sigma_{r,0}$ is greater than zero, i.e. $|\eta| < 1$, there is also elastic scattering. This is analogous to the situation in optics where the loss of light by a purely absorbing disc gives rise to diffraction with light appearing at finite angles away from the direction of the incident beam.

Matching the wave function at the nuclear surface: So far, we have considered the behaviour of the waves in the outside region ($r \geq R$). Solving the Schrödinger equation for a force field gives the wave function everywhere. However, in the inside region ($r < R$), the particles no longer propagate as free waves. At the surface ($r = R$), the interior and exterior wave functions, $\psi_i(r)$ and $\psi_o(r)$, must join smoothly, which means that the wave function and its derivative must be continuous at $r = R$. It is simpler algebraically to consider $u(r) = r\psi(r)$ and apply the condition that u and du/dr are continuous at the boundary, i.e.

$$f \equiv R\left[\frac{du_o(r)/dr}{u_o(r)}\right]_{r=R} = R\left[\frac{du_i(r)/dr}{u_i(r)}\right]_{r=R}. \tag{4.15}$$

We can evaluate f from $u_o(r)$, which we know from Equation (4.8). Also, we know something about η from observed cross sections. Therefore, Equation (4.15) tells us something about the wave function inside the nucleus and, in this way, we see that outside measurements provide information about the interior region.

Using the expression for $u_o(r)$, we obtain

$$f = R\left[\frac{du_o(r)/dr}{u_o(r)}\right]_{r=R} = ikR\left(\frac{\eta e^{ikR} + e^{-ikR}}{\eta e^{ikR} - e^{-ikR}}\right) \tag{4.16}$$

and solving this for η in terms of f gives

$$\eta = \left(\frac{f + ikR}{f - ikR}\right)\exp(-2ikR) \tag{4.17}$$

which, by substitution into Equations (4.13) and (4.14), gives expressions for the scattering and reaction cross sections in terms of f.

Equations (4.16) and (4.17) are general. A particular nuclear model allows us to calculate f and, hence, the cross sections. These can be compared with experiment in order to test the model or, if the model proves to be successful, it can be used to predict cross sections. As we shall see in Section 4.3.2, an example of such a model is the optical model for calculating the elastic scattering and total reaction cross sections.

Another simple model is the compound-nucleus model, which assumes that particles reaching the inside region fuse with the target nucleus and do not reappear. We can derive a useful general result by using this model. Again, we shall consider the case of low-energy (s-wave) scattering of neutrons.

Scattering via a compound nucleus: In this treatment we make use of the approximation that the nuclear potential energy experienced by an incoming particle can be represented by a mean field due to the attractive force of the nucleons in the target nucleus, as described in Section 1.4.1. For a neutron, this potential is the same as the shell-model potential (see Section 2.3.2) and has a depth V of about 50 MeV. We further assume that the nucleus has a well-defined surface and represent the edge of the nucleus by a potential step of height V at the interaction radius R. Part of the incident wave will be reflected at the discontinuity and part will be transmitted. The transmitted wave represents particles entering the nucleus, and the compound-nucleus hypothesis is that these particles rapidly exchange energy by colliding with nucleons in the target and are lost from the entrance (elastic) channel. Thus, inside the nucleus, the wave function is an ingoing wave only and we have

$$u_i(r) \propto e^{-iKr} \quad (r \leq R) \tag{4.18}$$

where K is the wave number of the neutron inside the nucleus. A low-energy neutron will gain about 50 MeV of kinetic energy after it crosses the surface, leading to a value of K of about $1.6\,\text{fm}^{-1}$.

Using the above expression for $u_i(r)$, we obtain from Equation (4.15):

$$f = -iKR. \tag{4.19}$$

Substituting this into Equation (4.17) gives

$$\eta = \left(\frac{K-k}{K+k}\right) e^{-2ikR}$$

(4.20)

whence, from Equation (4.14), we have

$$\sigma_{r,0} = \pi \lambda^2 \frac{4Kk}{(K+k)^2}.$$

(4.21)

Now $k \propto \sqrt{E}$ and $K \propto \sqrt{E+V}$, where V is the nuclear potential energy (approximately 50 MeV inside the nucleus) and E is the incoming neutron energy. Therefore, if $E \ll V$, as it will be for very slow neutrons, $k \ll K$ and K will vary very little with E. Then, since $\lambda = 1/k$, we obtain directly

$$\sigma_{r,0} \approx 4\pi/kK \propto 1/v.$$

(4.22)

This is the '$1/v$' law quoted in Section 1.6.5.

This result was obtained assuming that the nucleus is a passive absorber of waves that penetrate into it. As we shall see in Section 4.5, it does not apply when the neutron energy corresponds to exciting a resonance in the compound system. In that case, if the transmitted wave does reappear, the cross section is likely to be greatly changed.

4.3   ELASTIC SCATTERING AND NUCLEAR SIZE

Elastic scattering is the simplest of all nuclear reactions. Projectiles are scattered by a localized force field and either re-emerge with no loss of kinetic energy or are lost to the elastic channel by various direct or non-direct processes. The force field causing the scattering depends on the spatial distribution $\rho(r)$ of nucleons in a nucleus and also on the interaction force between the projectile and the nucleons comprising this distribution. Thus, by analysing the way particles are scattered, information can be obtained about the size and shape of the force field, and, hence, about $\rho(r)$. We saw an example of this in Section 1.6.5, where studies of the scattering of α particles indicated the distance at which they began to experience the nuclear force due to the target nucleus. However, once inside a nucleus, an α particle interacts strongly and there is a high probability for it to lose its identity and not to reappear in the entrance channel. The way they are scattered, therefore, depends essentially only on conditions in the nuclear surface. Determining the complete form of $\rho(r)$ requires a weakly absorbed particle which can 'sense' more than the edge of the nucleus. One of the best of these is the electron.

4.3.1   Electron scattering

The electron is an ideal probe because it is not absorbed by the nucleus and it interacts via the electromagnetic force, which is well known. The electromagnetic force is weak compared with the nuclear force and so the electron wave easily penetrates into the nuclear interior. Furthermore, electron beams with very high intensity can be produced and used, which makes it possible for a measurement to be made even when the cross section is very small.

As an example, consider the elastic scattering of 450-MeV electrons by ^{58}Ni. The differential cross section, shown in Figure 4.6(a), is measured in detail and covers many decades down to a very small value of about 10^{-13} b. It exhibits oscillations, which are due to diffraction of the electron wave and are related directly to the size of the target nucleus. These oscillations are clear, but are less pronounced than those seen in the diffraction of light by a black disc because the nucleus does not have a sharp edge or appear black to the electron wave.

The electron interacts mainly with the protons in the nucleus and, therefore, it is the distribution of nuclear charge $\rho_{ch}(r)$ that can be obtained from an analysis of results such as these. The curve in Figure 4.6(a) is a theoretical prediction assuming the particular form for $\rho_{ch}(r)$ shown in Figure 4.6(b). Varying $\rho_{ch}(r)$ varies the quality of the fit, and the uncertainty in $\rho_{ch}(r)$, obtained from the analysis of these data, is indicated by the varying thickness of the line in Figure 4.6(b). This shows that $\rho_{ch}(r)$ can be determined accurately, even at small radii, from high-quality, electron-scattering data.

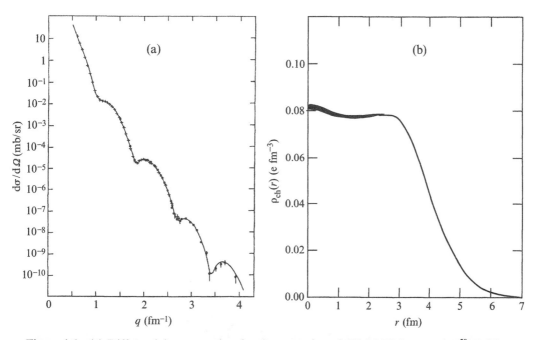

Figure 4.6   (a) Differential cross section for the scattering of 450-MeV electrons by ^{58}Ni. The data are plotted as a function of q, which is the momentum transferred to the target divided by $\hbar$. This quantity depends on the scattering angle θ, and it is easy to show (see Problem 4.4) that $q = 2k \sin(\theta/2)$, where k is the wave number of the electron. The solid line is a theoretical prediction obtained by using the charge density distribution for the ^{58}Ni nucleus $\rho_{ch}(r)$ shown in (b) and given in units of elementary charge per fm^3. The width of the line in (b) indicates the uncertainty in $\rho_{ch}(r)$ at different radii. From Sick (1975).

Unfortunately, electrons cannot be used to obtain the distribution of neutrons. Nuclear projectiles, such as neutrons or protons, will 'sense' the full matter distribution, but they are more likely to be absorbed as they pass through a nucleus, and the results must be interpreted using an effective force which is not known exactly. The model frequently used to analyse the scattering of nuclear particles is the *optical model*, so called because it is similar to the classical description of the scattering of light waves by an absorbing, refracting ball.

4.3.2   Optical model for nuclear scattering

In this model, the Schrödinger equation is solved using a complex potential

$$U(\mathbf{r}) = -V(\mathbf{r}) - iW(\mathbf{r}) \tag{4.23}$$

to represent the interaction between the projectile and target, where $\mathbf{r}$ is the distance between their centres. In general, the real part V is a sum of several terms: a repulsive Coulomb potential, an attractive, central nuclear potential V_n and a spin–orbit potential.

For nucleons, V_n is essentially the same as the central part of the shell-model potential (see Section 2.3) and is often represented by the Woods–Saxon (or Fermi) distribution, given in Equation (2.4), with adjustable parameters V_0, R and a. The imaginary term W is parametrized in a similar way, although forms other than the Woods–Saxon one are also commonly used.

The imaginary potential causes the projectile wave to be attenuated inside the nucleus. We can see this and estimate its effect by noting that a plane wave, propagating along the z-axis in a uniform potential $U = -(V + iW)$, has the form $\exp(ikz)$ with a wave number that can be written as

$$k = k_r + ik_i = \sqrt{2m(E - U)/\hbar^2} = \sqrt{2m(E + V + iW)/\hbar^2}. \tag{4.24}$$

If W is much less than $E + V$, we can expand the square root and take only the leading term in W to obtain

$$k_r = \sqrt{\frac{2m(E + V)}{\hbar^2}} \quad \text{and} \quad k_i = \frac{k_r W}{2(E + V)}. \tag{4.25}$$

Thus, $\exp(ikz) = \exp(ik_r z)\exp(-k_i z)$ is a wave with an exponentially decreasing amplitude. The mean free path d, over which the wave intensity is reduced by a factor of e^{-1} is $1/2k_i$. Substituting typical values for 30-MeV protons of $V = 50$ MeV and $W = 10$ MeV, gives a mean free path $d \approx 4$ fm, which means that these protons will penetrate into the nuclear interior and 'sense' more than the surface region.

An optical-model analysis of elastic scattering is carried out by performing repeated calculations and varying the parameters of the terms in the potential until the best fit to experimental data is obtained. A comparison of optical-model

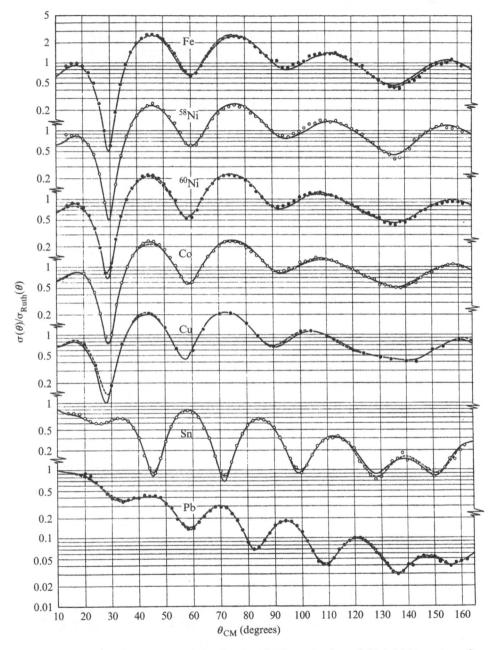

Figure 4.7 Differential cross sections for the elastic scattering of 30.3-MeV protons from different target nuclei plotted as a ratio of the Rutherford cross section. The data are shown as dots, and the solid lines are theoretical fits obtained by using the optical model. From G. R. Satchler (1967). Reproduced by permission of Elzevier Science.

predictions with experimental differential cross sections for the elastic scattering of 30.3-MeV protons by a number of different target nuclei is presented in Figure 4.7. It is an example of the high quality of fitting that can be achieved with this model.

The nucleon distribution in the target nucleus $\rho(r)$ can be derived from the potential $V_n(r)$, obtained from an analysis of nucleon scattering, by assuming a form for the effective nucleon–nucleon potential v_{nn} acting between the projectile nucleon and the nucleons in the target. Suitable forms for v_{nn} have been derived and nucleon distributions determined, mainly from analyses of accurate proton-scattering data taken at a number of bombarding energies. The results indicate that, for nearly all nuclei, the distribution of neutrons is similar to that of protons. For nuclei, with approximately equal numbers of protons and neutrons, this is perhaps not too surprising, but the difference is small, even in heavy nuclei, which may contain up to 50% more neutrons than protons. Part of the reason for this is that these extra neutrons are not bound more weakly than the most weakly bound protons. In fact, Coulomb repulsion reduces the depth of the attractive potential for protons relative to neutrons such that the binding energies of the most weakly bound protons and neutrons are almost identical. There is also a strong attractive nuclear force between protons and neutrons that resists any tendency for them to become separated from each other. The net effect is that the density of protons is less than the density of neutrons inside a heavy nucleus, but the root-mean-square radii of their respective distributions are almost the same.

By introducing the imaginary term W into the interaction potential, non-elastic processes are taken into account, and an optical-model calculation gives, in addition to a prediction for the elastic-scattering differential cross section, a value for the total reaction cross section σ_r, which can also be compared with experiment. However, W contains no information about what processes contribute to σ_r. In the next two sections, we consider particular aspects of the main reaction processes, which lead to loss of flux from the elastic channel, and what we can learn by studying them.

4.4   DIRECT REACTIONS

Direct reactions most commonly occur during the brief period of contact between the colliding particles in a peripheral collision. The simplest are those that proceed via a single-step process, such as the excitation of a nucleon in inelastic scattering from one shell-model state in a nucleus to another or the transfer of a single nucleon in a stripping or pickup reaction. In these reactions, if the final state is strongly populated, it is related in a simple way to the initial state by the reaction mechanism. It is for this reason that different direct reactions are used to excite and identify nuclear states that have different structures and properties. We cite examples in the following subsections.

4.4.1   Angular momentum transfer in direct reactions

In general, the transition between the initial and final states in a direct reaction requires the transfer of angular momentum between the projectile and target. This angular momentum comes from momentum transfer at the point of contact and it affects the angular distribution of the outgoing particle.

We can see how this might come about by using a simple, classical picture of a direct interaction, illustrated schematically in Figure 4.8. This shows an incident particle with momentum p_i interacting in the surface of a target nucleus, at a radius R, and an outgoing particle leaving with momentum p_o at an angle θ relative to p_i. The momentum transfer to the target is given by $p_t = p_i - p_o$ and we use the cosine rule to write

$$p_t^2 = p_i^2 + p_o^2 - 2p_ip_o\cos\theta = (p_i - p_o)^2 + 4p_ip_o\sin^2\frac{\theta}{2}. \tag{4.26}$$

The magnitude of the transferred angular momentum is given by $L = \sqrt{\ell(\ell+1)}\hbar \leq p_tR$. The direction θ of the outgoing particle depends on p_t and, therefore, θ will also depend on L.

As a simple, illustrative example, we consider inelastic scattering where the projectile energy is large compared with the energy transfer. In this case, we can simplify Equation (4.26) by noting that $p_i \approx p_o = p$, say. It then follows that

$$\sin\frac{\theta}{2} = \frac{p_t}{2p} \geq \frac{\sqrt{\ell(\ell+1)}\hbar c}{2pcR}. \tag{4.27}$$

Figure 4.9 shows two angular distributions for the inelastic scattering of 20-MeV tritons populating the 5^- and 3^- excited states in ^{90}Zr at 2.32 and 2.75 MeV, respectively. Since the ground-state spin of ^{90}Zr is zero, the ℓ transfer is given by the spin of the excited state. For 20-MeV tritons, $pc \approx 335$ MeV and, taking a point in the surface of ^{90}Zr at $R = 1.25 \times A^{1/3} \approx 5.6$ fm, Equation (4.27) predicts $\theta \geq 21°$ and $33°$ for $\ell = 3$ and $\ell = 5$, respectively. Given the crude nature of this picture of the reaction, these values are in remarkably good agreement with the observed maxima for these distributions.

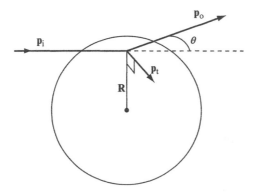

Figure 4.8   Schematic representation of a direct reaction occurring in the surface of a target nucleus at a radius R, where p_i, p_o and p_t indicate the incident, out going and transferred momenta, respectively.

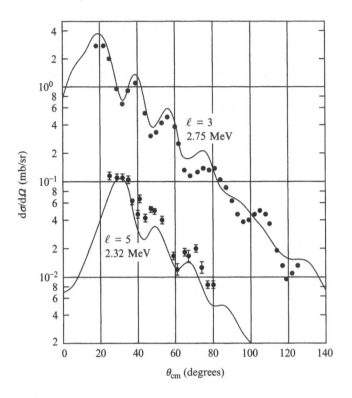

Figure 4.9   Differential cross sections for the inelastic scattering of 20-MeV tritons by ^{90}Zr exciting the 5^- and 3^- states at 2.32 and 2.75 MeV, respectively. The solid curves are theoretical calculations assuming $\ell = 5$ and $\ell = 3$ angular-momentum transfers. From Park and Satchler (1971).

The solid curves in Figure 4.9 are theoretical predictions obtained by using the so-called distorted-wave Born approximation (DWBA). These are quantum-mechanical calculations, which include effects of the interference of outgoing waves from different points on the nucleus and also take into account the distortion of incoming and outgoing waves by the optical-model potential. As can be seen, each of these curves satisfactorily reproduces the measured angular distribution and unambiguously indicates the ℓ value, which is a crucial piece of information for determining spins and parities of states populated by the reaction.

4.4.2   Selectivity in direct reactions

Inelastic scattering

In inelastic scattering, the target nucleus undergoes a transition from its ground state to an excited state. In principle, any state could be populated but, in many nuclear collisions, there is a strong preference for the reaction to excite collective states.

As we described in Section 2.5, a collective state involves the coherent motion of many nucleons. The microscopic wave function representing such a state, contains therefore, a large number of terms. These are all the different, possible shell-model configurations that have the spin and parity of the state. For a single-phonon state, each configuration is related to a configuration represented in the ground-state wave function by the promotion of a single nucleon from one shell-model state to another and corresponds to a path or reaction amplitude by which inelastic scattering can occur. Since there are many configurations, there are many reaction amplitudes and they must all be added together before being squared to give a quantity that is proportional to the cross section. When a collective state is excited, the individual reaction amplitudes tend to add constructively because they correspond to exciting nucleons that are all moving in phase with each other. Thus, inelastic scattering will excite a collective state much more strongly than a non-collective state or a state that is described in terms of the excitation of one or only a few nucleons.

A striking example of this selectivity, is seen in the inelastic scattering of 13-MeV deuterons by ^{209}Bi. An energy spectrum of scattered deuterons, plotted as a function of excitation energy in ^{209}Bi, is shown in Figure 4.10(a), and a prominent feature of this spectrum is the population of a group of states with excitation energies between about 2.4 and 2.8 MeV. In discussing the low-lying energy levels of ^{209}Bi in Section 2.4, we noted a number of them that are well described by the shell model as a single proton attached to a ^{208}Pb core in its (0^+) ground state (see Figure 2.11). For example, the ground state of ^{209}Bi is represented by the valence proton in the $h_{9/2}$ shell-model state. However, there is a group of seven states with excitation energies near 2.6 MeV, which correspond to the excitation of a one-phonon, collective octupole (3^-) state in the ^{208}Pb core coupled with the $h_{9/2}$ valence proton. It is these states that are strongly populated in this reaction because inelastic scattering preferentially excites the collective state of the core. By contrast, the $7/2^-$ single-proton state at 0.90 MeV, which would be excited by promoting the valence proton from the $h_{9/2}$ to the $f_{7/2}$ shell-model state, is not seen above the background in this reaction. The state labelled $13/2^+$ in Figure 4.10(a), near an excitation energy of 1.6 MeV, is indicated in Figure 2.11 as an $i_{13/2}$ single-proton state. However, the wave function of this particular state also contains a term arising from the 3^- core-excited collective state coupled with the $h_{9/2}$ proton to give a total spin of $13/2^+$ and it is the presence of this term that explains why the state is observed in this reaction.

Transfer reactions

Stripping or pickup of a single nucleon are the simplest examples of transfer reactions. They selectively excite states that are related to the initial state of the target by the addition or subtraction of a single nucleon.

As our illustrative example, we again consider the population of states in ^{209}Bi, but this time by the proton-stripping reaction ^{208}Pb(^{3}He,d)^{209}Bi. An energy spectrum of outgoing deuterons is plotted in Figure 4.10(b). In complete contrast with the inelastic-scattering spectrum in Figure 4.10(a), the states most strongly excited in (^{3}He,d) are the single-proton states, which are the ground state and excited states at 0.90, 1.61, 2.83, 3.12 and 3.63 MeV. The corresponding shell-model states are indicated by each

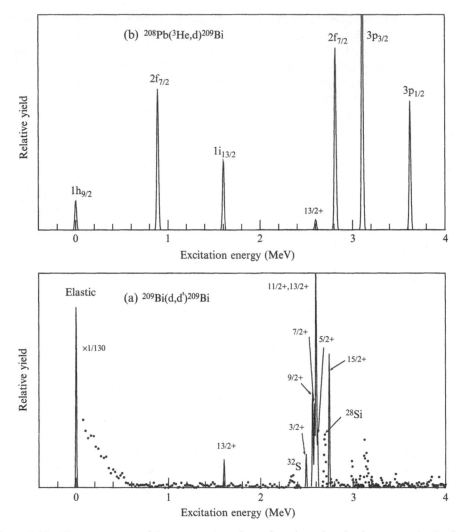

Figure 4.10   Energy spectra of deuterons plotted as a function of excitation energy in the final nucleus ^{209}Bi, for (a) the inelastic scattering of 13-MeV deuterons measured at a laboratory scattering angle of 150° and (b) the ^{208}Pb(^{3}He,d)^{209}Bi reaction measured at 110° and using 20-MeV incident ^{3}He ions. References (^{3}He,d): Ellegaard and Vedelsby (1968); (d,d'): Ungrin *et al.* (1971).

of these peaks in the figure. The group of states near 2.6 MeV, excited so strongly in inelastic scattering, are here seen only weakly, if at all. Populating these states would require excitation of the 3^- core state as well as the transfer of a proton and, normally, such a two-step mechanism is much less likely than a single-step one.

Reactions involving the transfer of more than one nucleon are more difficult to analyse because they can involve several steps and because there often are several different mechanisms leading to the same final channel. For example, the transfer of a neutron and a proton may occur as two, sequential single-nucleon transfers, or by the simultaneous transfer of the pair as a deuteron cluster. Also, breakup may occur either directly, by the projectile knocking a piece out of the target, or indirectly, by inelastic excitation to a state that has sufficient energy to dissociate into two or more fragments. Theoretical treatments of multi-step reactions have achieved some success, but calculations are complex, especially if there is a large number of different possible pathways connecting the initial and final channels.

4.5 COMPOUND NUCLEUS REACTIONS

Earlier in the chapter, we gave a qualitative description of reactions involving the formation of a compound nucleus. Here, we consider different exit channels, which can be populated when the compound nucleus loses its excitation energy.

A compound nucleus reaction is an extreme example of a multi-step reaction in which the detailed complexity of the successive steps is lost. Particles in the initial channel coalesce to form a compound system that stays together long enough for statistical equilibrium to be reached. Decay occurs some time later when, by chance, a nucleon or group of nucleons acquires enough energy to escape. In general, there may be many possible ways the compound nucleus can decay and the final channel is not unique. Each decay branch will have its own probability, which is often specified as an energy or decay width. Any excited state decays with a probability $\lambda = 1/\tau$, where τ is its mean lifetime. The finite lifetime leads to an imprecision in the energy of the state, according to the uncertainty principle. The spread in energy is called the energy width Γ of the state and it is equated to the decay probability multiplied by $\hbar$. The total decay probability is the sum of the probabilities of all the decay branches, i.e. $\lambda = \sum_i \lambda_i$ and, hence, the total decay width is the sum of all the partial widths:

$$\Gamma = \sum_i \Gamma_i = \lambda\hbar = \hbar/\tau. \tag{4.28}$$

Using this terminology, we can write the total cross section for the compound-nucleus reaction $A + a \rightarrow C \rightarrow B + b$ as

$$\sigma_{\alpha\beta} = \sigma_c \Gamma_\beta / \Gamma \tag{4.29}$$

where σ_c is the cross section for forming the excited, compound nucleus C and Γ_β/Γ is the fractional decay width into the final (β) channel B + b. Note that Γ includes Γ_α, which is the partial width for decay back into the entrance (α) channel A + a.

The relative probabilities of the different decay branches depend on the amount of available excitation energy. Normally, a neutron will be evaporated because there is no Coulomb barrier to hinder its escape, but charged-particle decay can compete if enough energy is available for the barrier to be overcome.

At higher excitation energy, several particles may be evaporated in succession. This is because, as we noted in Section 4.2.1, evaporation favours the emission of low-energy particles and so, as the excitation is increased above the threshold for emitting an extra particle, the probability for this to occur increases at the expense of the evaporation of fewer, high-energy particles. Figure 4.11, shows (α,n), (α,2n) and (α,3n) cross sections

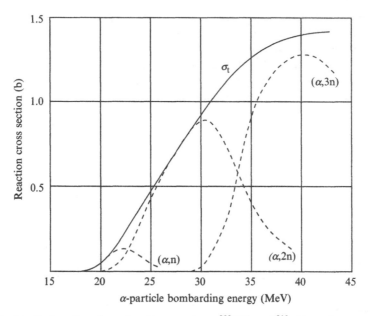

Figure 4.11   Excitation functions for the reactions ^{209}Bi$(\alpha, xn)^{213-x}$At, where $x = 1, 2, 3 \cdots$ (shown as dashed curves). The sum of the (α, xn) cross sections is shown by the solid line. Data from Barnett and Lilley (1974); and Kelly and Segrè (1949).

for reactions on a ^{209}Bi target plotted as a function of bombarding energy E_α. The total reaction cross section, shown by the solid line, rises from a small value as E_α crosses the 1n emission threshold but, as soon as the 2n threshold is exceeded, $\sigma(\alpha,n)$ falls and $\sigma(\alpha,2n)$ rises. At still higher energies, maxima occur in $\sigma(\alpha,2n)$, $\sigma(\alpha,3n)$ and beyond, as the compound nucleus is able to evaporate more neutrons.

Note that the total cross section varies smoothly with energy. This is because, in this example, the compound nucleus is created at high excitation, well into the continuum of its energy spectrum, where there are many possible states that can be excited, all overlapping in energy because of their finite decay widths. The probability of forming a compound nucleus depends on there being an energy level available at the required energy. In the continuum region, the large density of states means that this is always the case and so the formation probability will not be strongly dependent on bombarding energy. The situation is very different if the compound nucleus is formed in a region of excitation where its levels are well separated. In that case, as we noted in Section 1.6.5, the cross section will exhibit resonance behaviour, which we now consider in a little more detail.

4.5.1   Resonance in a compound nuclear reaction

When the bombarding energy E in the entrance channel is close to the energy E_r for exciting a state in the compound nucleus, the cross section for the reaction $A + a \rightarrow C \rightarrow B + b$ takes the form of a Lorentz distribution [Equation (1.32)] and is given by the Breit–Wigner equation:

$$\sigma_{\alpha\beta} = g_\alpha(J)\frac{\pi}{k_\alpha^2}\frac{\Gamma_\alpha\Gamma_\beta}{(E - E_r)^2 + (\Gamma/2)^2} \tag{4.30}$$

where α and β indicate the entrance (A + a) and exit (B + b) channels, and k_α is the wave number in the entrance channel.

The quantity $g_\alpha(J)$ is a statistical spin factor that takes into account the effects of angular momentum and spin. This factor takes the following form:

$$g_\alpha(J) = \frac{2J + 1}{(2i_a + 1)(2i_A + 1)} \tag{4.31}$$

where i_A, i_a and J are the spin quantum numbers of the nuclei A, a and the compound nucleus, respectively. In general, J is given by the vector sum of the orbital angular momentum and spins of a and A. If i_A and i_a are both zero, J is equal to ℓ the quantum number of the orbital angular momentum brought in by the fusion of a and A, and $g_\alpha(J) = (2\ell + 1)$.

The energy width of the resonance Γ is the total decay width, which is related to the lifetime of the compound nucleus according to Equation (4.28). The total width is equal to $\sum\Gamma_\beta$, where the sum is over all final (β) channels. The individual widths Γ_α, $\Gamma_\beta \cdots$ depend on the structure of the particular nuclear states. Both Γ_α and Γ_β need to be large for $\sigma_{\alpha\beta}$ to be large. We can determine upper limits for elastic scattering and absorption by evaluating their cross sections at the peak of the resonance.

Elastic scattering (for which $\alpha \equiv \beta$) is maximum when there is no absorption, i.e. when $\Gamma_\alpha = \Gamma$. This gives an upper limit for the integrated elastic scattering cross section of

$$\sigma_{\alpha\alpha}(\text{max}) = \sigma_\ell^{\text{el}} = (2\ell + 1)\frac{4\pi}{k_\alpha^2} \tag{4.32}$$

for the ℓth partial wave and where we have taken both projectile and target spins to be zero.

The total absorption cross section is proportional to $\Gamma_\alpha\sum_{\beta\neq\alpha}\Gamma_\beta = \Gamma_\alpha(\Gamma - \Gamma_\alpha)$, which has its maximum when $\Gamma_\alpha = \Gamma/2$. Thus, for spinless particles, the upper limit to the ℓth partial-wave absorption cross section (when $E = E_r$) is

$$\sigma_\ell^{\text{abs}}(\text{max}) = (2\ell + 1)\frac{\pi}{k_\alpha^2} \tag{4.33}$$

which is equal to the upper limit, given in Equation (4.5), derived by using a simple, semi-classical picture of wave absorption.

As an example of the use of this formalism, we consider resonance excitation induced by $\ell = 0$ (s-wave) neutrons. In this case, we normally find that charged-particle emission is strongly inhibited by the Coulomb barrier. Also, neutron emission is weak because it is unlikely to find all the available energy concentrated on a single neutron. Thus, gamma decay becomes important and Γ_γ is often greater than

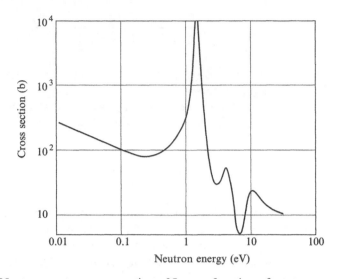

Figure 4.12   Neutron-capture cross section of In as a function of neutron energy, showing a strong resonance at an incident neutron energy of about 1.46 eV. From Goldsmith *et al.* (1947).

Γ_n. We now show how these decay widths can be obtained directly from measurements of the (n,γ) capture cross section on resonance.

When ^{115}In is bombarded with 1.457-eV neutrons [c-m (centre-of-mass) energy = 1.444 MeV], a $(J = 5)$ resonance, of total width $\Gamma = 0.075$ eV, is excited in ^{116}In (see Figure 4.12). On resonance, the ^{115}In$(n,\gamma)^{116}$In cross section is 38 100 b. Thus, from Equation (4.30), and setting $E = E_r$, we have

$$\sigma_{n\gamma} = g_n(J)\frac{\pi}{k_n^2}\frac{\Gamma_n\Gamma_\gamma}{(\Gamma/2)^2} = 38100 \text{ b}. \tag{4.34}$$

Substituting 1/2 and 9/2 for the spins of the neutron and the ^{115}In target into Equation (4.31), gives a spin factor $g_n(J = 5) = 11/20$. The wave number of a 1.444-eV neutron is 2.64×10^{-4} fm^{-1}, which gives $\pi/k_n^2 = 4.51 \times 10^5$ b and a value of 0.0384 for $\Gamma_n\Gamma_\gamma/\Gamma^2$ after substitution into Equation (4.34). Since we know that $\Gamma = \Gamma_\gamma + \Gamma_n = 0.075$ eV and $\Gamma_\gamma > \Gamma_n$, we obtain $\Gamma_\gamma = 0.072$ eV and $\Gamma_n = 0.003$ eV for the two branching widths.

Note that the total width $\Gamma = 0.075$ eV implies a lifetime of about 10^{-14} s. This is very long compared with the transit time of a neutron across the nucleus and consistent with the interpretation of the peak as a compound-nucleus resonance.

4.5.2   Low-energy, neutron-induced fission

Certain nuclei, known as fissile nuclei, will undergo fission when bombarded with low-energy (thermal) neutrons, a process that is of particular importance because of its exploitation in commercial nuclear reactors. The compound nucleus is formed by the fusion of a slow neutron with a heavy target, such as uranium. Decay is either by

Table 4.1   Values of neutron-binding energies, activation energies and thermal-neutron fission cross sections for several heavy isotopes.

Nucleus	Binding energy (MeV)	Activation energy (MeV)	Cross section (barns)
^{232}Th	4.8	6.7	$< 10^{-6}$
^{233}U	6.8	5.85	530
^{235}U	6.5	5.9	579
^{238}U	4.8	5.8	2.7×10^{-6}
^{239}Pu	6.5	6.3	742

neutron evaporation, γ emission or fission. As we have seen in the above example, Γ_n is much smaller than Γ_γ for slow neutron-induced reactions, so we need only consider competition between the γ and fission-decay branches.

The absorption of a slow neutron leaves the resulting compound nucleus at an excitation energy E^* very close to the neutron-binding energy. Decay by fission becomes likely if this energy exceeds the activation energy, which is the energy needed to overcome the fission barrier B_f (see Figure 2.5). If E^* falls below B_f, fission is strongly suppressed because the probability of penetrating even a small barrier, of the order of 1 MeV, is low when the mass of the outgoing particle is high, as it is in fission.

Neutron-binding energies and activation energies for several heavy nuclei, for which fission is energetically allowed, are listed in Table 4.1, together with the fission cross sections for thermal neutrons. Note that the values in the second and third columns are for the compound nuclei. For example, the third row in the table with ^{235}U refers to the reaction $n + {}^{235}U \rightarrow {}^{236}U^* \rightarrow$ fission. The binding energy and activation energy are for ^{236}U and the cross section is for fission of ^{235}U when bombarded with thermal neutrons.

The activation energies for all these nuclei are similar, yet three of them are fissile and two are not. Thorium-232 and ^{238}U are the exceptions because, for each of these nuclei, the neutron-binding energy is significantly lower than it is for the fissile nuclei and is below the activation energy. The binding energy of a neutron depends on whether the neutron number N in the target is odd or even. In the case of an odd N target (^{233}U, ^{235}U and ^{239}Pu), there is energy gain from neutron pairing in the compound nucleus, which has even N. This does not occur when the neutron is added to an already even-N nucleus, such as ^{232}Th or ^{238}U. It is an example of the consequence of the pairing term in the semi-empirical mass formula (see Section 2.2)

Of the three fissile nuclei, only ^{235}U exists in nature and it comprises only 0.7% of natural uranium. However, as we shall see in Chapter 10, ^{232}Th and ^{238}U can be converted into fissile material by neutron capture in a breeder reactor and are huge potential sources of nuclear fuel for the future.

4.6   HEAVY-ION REACTIONS

There are a number of aspects of heavy-ion reactions that distinguish them from those induced by nucleons or light ions ($A < 4$). The wavelength of the projectile is typically much smaller than nuclear dimensions, and the collision can often be well described classically or semi-classically. Large amounts of energy, angular momentum and nuclear material may be transferred and, if the collision leads to fusion, the resulting product lies far from the mass of either of the original particles. Heavy-ion reactions have been studied for many years and form a major part of current nuclear research. In this short section, we are only able to illustrate some general features of the main types of heavy-ion reaction processes with a few examples.

4.6.1   Elastic scattering and direct reactions

At low and moderate bombarding energies, the Coulomb force plays an important role in determining the course of a heavy-ion collision. This is illustrated schematically in Figure 4.13, which shows its effect on several trajectories near a target nucleus. Those with large impact parameters (1 and 2) miss the nucleus altogether and the particles are deflected away by the Coulomb field. For a grazing collision (path 3), the projectile just reaches the target and direct nuclear reactions begin to take place. At smaller impact parameters (path 4), the projectile collides with the target well inside the interaction radius and more complex reactions, sometimes leading to fusion, predominate.

The observed behaviour of heavy-ion elastic scattering, shown in Figure 4.14, is broadly in accord with this classical interpretation. Small angles correspond to trajectories with large impact parameters. These do not reach the nuclear surface, and the cross section $\sigma(\theta)$ follows the Rutherford formula $\sigma_{\text{Ruth}}(\theta)$ up to a certain angle. Beyond this point, $\sigma(\theta)$ falls below $\sigma_{\text{Ruth}}(\theta)$ because trajectories with smaller impact parameters, which would be scattered by a point charge to larger angles,

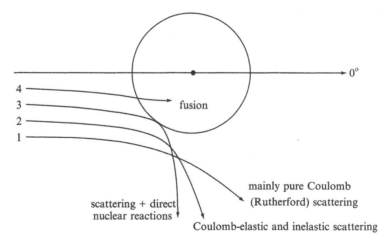

Figure 4.13   Effect of Coulomb repulsion on different trajectories in a heavy-ion collision. Trajectories 1 and 2 miss the nucleus and are scattered by the Coulomb field. Coulomb-inelastic excitation can also occur. On trajectory 3, the projectile grazes the nucleus during which simple, direct nuclear reactions can take place (as well as elastic and Coulomb-inelastic scattering). The projectile on trajectory 4 is shown entering the target nucleus and fusing with it.

interact with the target nucleus, and the onset of nuclear reactions begins to remove flux from the elastic channel. Larger angles correspond to more intimate contact, which leads to greater flux loss and a rapidly falling cross section. If the heavy ion is not too massive, i.e. if its wavelength is not small compared with its size, oscillations may be seen (as shown in Figure 4.14) superimposed on an otherwise classical distribution. They are due to diffraction of the projectile wave by the surface of the target nucleus acting as a single, absorbing edge.

When a heavy ion passes close to the target (path 2), the impulse due to the Coulomb force acting between the particles can be strong enough to lead to a significant amount of inelastic excitation, known as Coulomb excitation (or Coulex). Since the force causing the transition to an excited state is known, the cross section for Coulomb excitation can be calculated, given the wave functions of the initial and final states. By keeping the incident energy below the Coulomb barrier, heavy-ion Coulomb excitation can be measured without significant interference from nuclear inelastic scattering and is an important reaction for determining the properties of nuclear excited states.

Simple direct nuclear reactions occur when the impact parameter is such that the nuclei brush past each other with little extensive contact (path 3). These reactions do not perturb the path a great deal, and the angular distribution of the outgoing reaction product exhibits a bell-shaped form, which peaks at an angle θ_g corresponding to this grazing trajectory. For $\theta < \theta_g$, the nuclear surfaces do not approach as closely and the reaction probability is reduced. For $\theta > \theta_g$, there is greater overlap of the nuclei during the collision, more complex reactions leading to fusion predominate and the cross section for the simple direct reaction again falls. Angular distributions of single-neutron transfer, induced by ^{12}C ions incident on ^{208}Pb target nuclei, are shown in Figure 4.15 and clearly exhibit the bell-shaped form.

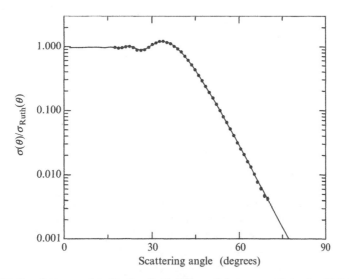

Figure 4.14 Ratio of observed to Rutherford differential cross sections for 70-MeV ^{16}O ions elastically scattered by a ^{60}Ni target plotted against angle in the centre-of-mass frame. The points are experimental data and the solid line is a theoretical prediction. From Keeley *et al.* (1995).

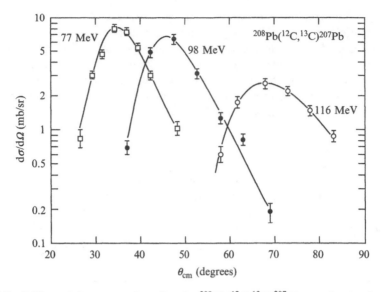

Figure 4.15   Differential cross sections for the ^{208}Pb(^{12}C,^{13}C)^{207}Pb reaction, at bombarding energies of 77, 98 and 116 MeV, populating the ground states of ^{207}Pb and ^{13}C. They exhibit bell-shaped curves characteristic of simple, direct reactions initiated by heavy ions at energies close to the Coulomb barrier. From Larson *et al.* (1972).

4.6.2   Fusion

When a heavy ion undergoes fusion, a large amount of material, energy and momentum is added to the target nucleus. This can lead to the formation of new nuclei and nuclear matter under extreme conditions of temperature and stress. Different combinations of colliding particles and energies produce different results.

For many years, heavy-ion fusion has been used to create nuclei much heavier than any that occur naturally on earth. In Section 3.4, we reported on the discovery of the element $^{277}_{112}$X. The tendency for these very heavy nuclei to undergo spontaneous fission increases with mass and, if the compound nucleus is to decay by nucleon evaporation and not by fission, it is important to produce it with the minimum amount of excitation energy. This is best achieved using projectile and target nuclei that are tightly bound. For example, the combinations ^{48}Ca + ^{208}Pb and ^{24}Mg + ^{232}Th both produce the same compound nucleus $^{256}_{102}$No, but the first one, using the doubly magic nuclei ^{48}Ca and ^{208}Pb, has a Q value which is 87 MeV less than the second one and thus is strongly preferred.

In general, the fusion of two heavy ions produces a nucleus, which is not only much heavier than either the original target or projectile, but is also much more proton rich. For example, $^{40}_{20}$Ca + $^{90}_{40}$Zr produces the compound nucleus $^{130}_{60}$Nd which has 12 neutrons less than the most proton-rich stable neodymium isotope. Moreover, the compound nucleus will be created with a certain amount of excitation energy and may even evaporate more neutrons, moving it even further away from the stability line. The discovery and measurement of properties of previously unknown, proton-rich nuclei has formed a major part of the programme of heavy-ion research for a number of years.

The high momentum of an incoming heavy ion means that the fused system can be created with very high angular momentum. For example, in the reaction:

$$^{40}_{20}\text{Ca} + ^{90}_{40}\text{Zr} \rightarrow ^{130}_{60}\text{Nd}$$

at a centre-of-mass bombarding energy about 50% above the Coulomb barrier of 100 MeV, a compound nucleus may be formed with angular momentum up to about 60 $\hbar$. Even after the evaporation of several particles, the residual nucleus will still be spinning very rapidly, but may have little or no internal (thermal) excitation energy. It will be in a highly excited, collective rotational state of the type discussed in Section 2.5.3. The study of high-spin rotational bands in nuclei is another very active field of nuclear spectroscopy research. Significant structural changes can occur in a nucleus when it is subjected to the centrifugal stress caused by its high spin, and these changes can be discerned from a detailed analysis of the sequences of energy levels in the rotational bands. An important example, discussed in Section 2.5.4, is the study of collective rotations of nuclei that are created in the superdeformed shape in a heavy-ion reaction.

At energies not too far above the Coulomb barrier, there are many cases for which Equation (4.2) is a good representation of the cross section for heavy-ion fusion. An example is shown in Figure 4.16, which plots the fusion cross section for $^{16}\text{O} + ^{27}\text{Al}$ as a function of $1/E_{\text{cm}}$. At low energies, the experimental data follow the solid line given by Equation (4.2). The intercept on the ordinate gives a value of πR^2 of about

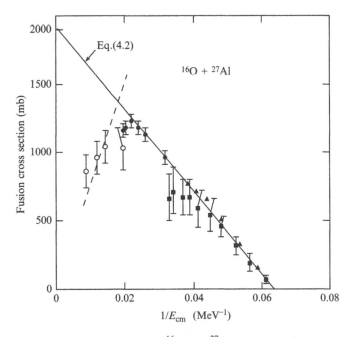

Figure 4.16   Cross section for the fusion of ^{16}O and ^{27}Al plotted against the reciprocal of the centre-of-mass energy. From Kozub *et al.* (1975) (open circles); Back *et al.* (1977) (black circles); Dauk *et al.* (1975) (black squares); Eisen *et al.* (1977) (black triangles).

2 b, from which we obtain an interaction radius R $(= R_1 + R_2)$ of about 8 fm. The nuclear matter distributions of these nuclei have been determined reasonably accurately and this example suggests that fusion begins to occur when their overlap is about 10% or greater.

4.6.3   Deep inelastic reactions and limits to fusion

In the example discussed above and shown in Figure 4.16, the points deviate below the line given by Equation (4.2) at higher energies (smaller values of $1/E_{cm}$). This is because other factors become important, such as the angular momentum brought in by the collision exceeding the amount that the fused system can sustain before it begins to fly apart. The impact parameter corresponding to this critical value of orbital angular momentum L_c is $b = L_c/p$, where p is the momentum. Therefore, the fusion cross section will be determined by the relation:

$$\sigma_f \sim \pi b^2 = \pi L_c^2/p^2 \approx \pi \ell_c^2/k^2 \propto 1/E_{cm} \qquad (4.35)$$

and not by Equation (4.2). This trend is indicated by the dashed line in Figure 4.16. In Equation (4.35), k is the wave number of the heavy ion and we have equated the angular momentum quantum number ℓ_c to $L_c/\hbar$, which is reasonable when L_c is large.

When two heavy nuclei collide and the orbital angular momentum exceeds L_c, a complex sequence of reaction processes can take place. Figure 4.17 illustrates the progress of such a collision. As the nuclei approach each other, a significant part of their relative kinetic energy is converted into Coulomb potential energy. This means that they may be moving slowly enough so that, during the period of contact as they pass each other, there is enough time for a large amount of the remaining relative kinetic energy to be converted into internal excitation. This energy dissipation occurs via nucleon–nucleon collisions and mass exchange through the region of contact, where a neck of nuclear matter may form as the two nuclei stick together and begin to rotate as a single unit. The angular momentum is too great for fusion to occur and, a short time later, the fragments separate with a certain amount of kinetic energy

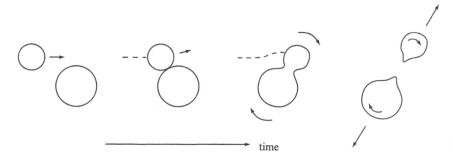

time

Figure 4.17   Sequences in the progress of a collision between two nuclei when the angular momentum is greater than can be sustained by the combined system.

corresponding to their relative angular momentum. They then acquire a considerable amount of extra, radial kinetic energy from their mutual Coulomb potential energy as they repel each other. During contact, nucleon transfer occurs in both directions, and the masses and charges of the final products may not be very different from those of the initial reactants. However, contact time does not last long enough for a complete rotation of the combination to take place and the angular distribution is peaked in the forward direction, characteristic of a direct reaction. Such highly dissipative reactions, often referred to as deep-inelastic reactions, are intermediate between simple direct and true compound-nuclear reactions. They become increasingly important in collisions with heavier projectiles which, for a given energy, bring more angular momentum into the combined system.

If we represent the total cross section as a function of angular momentum quantum number ℓ, the different reaction types fall roughly into distinct zones, as illustrated in Figure 4.18. Here, $d\sigma/d\ell$, obtained from the classical expression given in Equation (4.3), is plotted against ℓ. Figure 4.18(a) shows the situation for which the grazing value ℓ_g is less than the maximum (critical) value ℓ_c that the compound nucleus can carry. All ℓ values up to ℓ_g contribute to what, in this case, is the fusion cross section. For $\ell \approx \ell_g$, we have simple, direct nuclear reactions. Beyond ℓ_g, Rutherford scattering and Coulomb excitation are the only significant processes that take place, since the nuclei pass each other outside the range of the nuclear force. Figure 4.18(b) shows the situation at a considerably higher energy. The grazing angular momentum is now much higher and we find a deep-inelastic reaction zone for $\ell_c < \ell < \ell_g$. As the energy is increased further, the deep-inelastic zone extends to higher ℓ values and can constitute the major part of the total reaction cross section.

There is an additional limit to fusion that occurs when the energy brought into the collision is so large that it becomes comparable with the total nuclear binding energy. In such a violent collision, there is so much energy available that the combined system splits up by fission or fragmentation long before any form of equilibration can

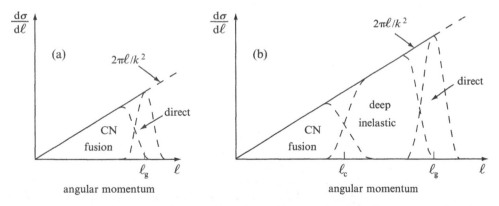

Figure 4.18 Schematic decomposition of the total heavy-ion reaction cross section into contributions from different partial waves when (a) the grazing angular momentum (quantum number ℓ_g) is below the critical angular momentum (quantum number ℓ_c) that can be carried by the compound nucleus, and (b) when ℓ_g exceeds ℓ_c. In both (a) and (b) the straight line is obtained from Equation (4.3) and the dashed areas indicate regions in which different types of heavy-ion nuclear reaction mechanisms predominate.

take place and the fusion cross section decreases with energy even faster than that shown by the dashed line in Figure 4.16.

Even though a true compound nucleus will not be formed, there is a great deal of interest in studying heavy-ion collisions by going to even higher energies. In a central collision between two massive nuclei, both moving at speeds close to the speed of light, the energy density and matter density reached as the nuclei merge together are both very great and it is thought that conditions may be created on a small scale similar to those that may have occurred at a very early stage of the Big Bang beginning of the Universe. In this tiny, but extremely hot, dense region of space, nucleons may overlap to such an extent that they lose their identity, forming a 'soup' or plasma of their constituent quarks for a fleeting moment. It is hoped that studies of the resulting explosion of particles emerging from these high-energy, heavy-ion collisions will give some insight into the nature of the Universe shortly after its creation.

PROBLEMS 4

4.1 The total reaction cross section for a certain nuclear collision is 1 b at a centre-of-mass bombarding energy which is twice that of the Coulomb barrier. Calculate the inter-action radius assuming that the collision can be treated classically.

4.2 Estimate the number of partial waves which will be important in nuclear collisions occurring between 9-MeV neutrons and ^{125}Sn nuclei. Assume that the interaction radius is given by $R = 1.2 \times A^{1/3}$ fm, where A is the mass number of the target.

4.3 Show that, in the limit, when the wavelength of an uncharged projectile is small compared with the nuclear size (i.e. $\lambda \ll R$), Equation (4.4) leads to the classical value of the collision cross section πR^2 for the maximum value of the total reaction cross section.

4.4 Derive the expression $q = 2k\sin(\theta/2)$, given in the caption to Figure 4.6. Hence, obtain (approximate) values for the scattering angles corresponding to the minima in the differential cross section plotted in Figure 4.6. Assume that the mass of the scattering nucleus is large compared with that of the projectile.

4.5 Derive a simple expression for the separation in angle $\Delta\theta$ of successive maxima (or minima) in a diffraction pattern given by the interference of particles with wave number k scattered from opposite sides of a nucleus, radius R. Assume that the particles are scattered through small angles.

   Using this and the results of problem 4.4, obtain an estimate of the size of a Ni nucleus from data given in Figure 4.6. (This estimate will be very approximate because the small-angle approximation is not a good one in this case.)

4.6 Taking the optical potential for 40-MeV protons to be $V = 50$ MeV, $W = 10$ MeV and for 40 MeV α particles, $V = 100$ MeV, $W = 30$ MeV, estimate the ratio of the mean free paths of protons and α particles in nuclear matter.

4.7 The energy released in a (d,p) reaction to a certain excited state of an even–even nucleus is +8 MeV. Assuming that the target nucleus remains at rest, estimate the bombarding energy for which the outgoing protons at forward angles will have approximately the same momenta as the incident deuterons.

If the angular distribution at this bombarding energy peaks at an angle of 40° and the target nucleus has a radius of about 4.1 fm, calculate the probable ℓ transfer and, hence, the spin and parity assignments of the excited state.

4.8 What is the half-life of the resonance in ^{113}Cd, shown in Figure 1.16, if it has a total width of 0.133 eV? Compare this with the collision time for a 0.17-eV neutron, which is approximately the time it takes for the neutron to pass the nucleus. Take the nuclear radius to be $R = 1.2A^{1/3}$ fm.

4.9 An s-wave ($\ell = 0$) resonance in the total cross section for neutrons incident on ^{238}U is observed at a centre-of-mass energy $E_n = 115$ eV. The total width of the resonance is 94×10^{-3} eV and the peak cross section is 19 200 b. Calculate the partial widths Γ_n and Γ_γ for neutron and γ emission (assuming that these are the only final channels populated) and the peak cross section for the (n, γ) capture reaction. Assume that $\Gamma_\gamma > \Gamma_n$.

4.10 A resonance in the ^{6}Li(n, α)t reaction occurs at a neutron bombarding energy of 244.5 keV. Calculate the excitation energy of the resonance state in the compound nucleus ^{7}Li. You will need to use (or derive) the result that the ratio of kinetic energies in the centre-of-mass and laboratory systems: $E_{cm}/E_{lab} = M/(M+m)$, where m is the projectile mass and M is the mass of the target (initially at rest).

4.11 Measured cross sections for the fusion of α particles by ^{209}Bi are as follows:

Energy (MeV) :	21	22	24.5	30	35	40
Cross section (mb) :	100	200	500	900	1210	1370

Deduce an effective interaction radius and Coulomb barrier from these data.

4.12 Using classical mechanics, show that the angular momentum L of two colliding particles is given by $L^2 = 2E_{cm}\sigma\, mM/[\pi(M+m)]$; σ is the geometric cross section, $= \pi b^2$, where b is the impact parameter. Note, you will need to work in the centre-of-mass system.

4.13 From the data given in Figure 4.16, and using the result of Problem 4.12, estimate the maximum angular momentum that the compound nucleus could have at a centre-of-mass bombarding energy of 33 MeV.

4.14 By extending the analysis of scattering via a compound nucleus given at the end of Section 4.2.3, show that the elastic scattering of neutrons via a compound nucleus at very low energies and far from any resonances is approximately independent of neutron energy. Assume that $\lambda\,(= 1/k) \gg R$, the nuclear radius.

PART II
INSTRUMENTATION AND APPLICATIONS

5

Interaction of Radiation with Matter

5.1 INTRODUCTION

Nuclear radiation normally consists of energetic particles or photons. Its interaction with matter lies at the the heart of all experimental work and applications of nuclear physics – detectors, material modification, analysis, radiation therapy, etc. It can damage materials, especially living tissue, and must be considered to be highly dangerous. However, because it is potentially so damaging, it is also easy to detect and can be controlled. The effects depend greatly on the intensity, energy and type of the radiation as well as on the nature of the absorbing material.

   In this chapter, we describe the main features of the interaction with matter of all types of nuclear radiation: charged particles, photons and neutrons. The interaction of any type of radiation always involves charged particles at some stage or other in the process. In the case of uncharged radiations (γ rays or neutrons) there is first a transfer of all or part of the energy to charged particles before there is any measurable effect on the absorbing medium. We begin, therefore, by describing the way charged particles interact in matter. Heavy charged particles (including protons) will be considered first. Electrons behave significantly differently because of their low mass and generally much higher speeds.

5.2 HEAVY CHARGED PARTICLES

A fast charged particle loses energy via the action of its electric field on the atoms it encounters as it passes through matter. In most cases, nearly all the energy is lost to electrons. Both nuclei and electrons feel the Coulomb force due to the transient field, but the impulse, exerted as the ion passes by, transfers momentum, and the much smaller mass of the electron means that it will carry away much more energy than the heavy nucleus. Also, unless the energy is very high, nuclear collisions are rare for most ions and have little effect on the overall energy loss process.

In a head-on collision with an ion of mass M and energy E, an electron (mass $m \ll M$) initially at rest emerges with a speed approximately twice that of the incident ion and its energy will be (see Problem 5.1):

$$\Delta E = E(4m/M). \tag{5.1}$$

For a 4-MeV α particle, this is about 2.2 keV. In general, the average energy loss will be much less than this. However, even this maximum energy transfer is small compared with E, and many collisions must occur before the ion comes to rest. The heavy particle is deflected very little by these interactions with electrons and moves in an almost straight line during the slowing-down process. Fluctuations in its energy loss tend to average out and it will have a definite *range* depending on its energy, mass, charge and the nature of the stopping medium.

The particle ionizes and excites many atoms in its passage through the stopping material. Also, the more energetic, recoiling electrons (known as delta rays) cause further ionization, generating more secondary electrons as they in turn lose their energy. The result is a trail of ionization and excitation of atoms and molecules along the path of the moving particle. This excitation and ionization in the medium is used to detect charged particles as described in the next chapter.

5.2.1   Bethe–Bloch formula

The rate at which a particle loses energy per unit path length is known as the stopping power of the medium. A quantum-mechanical derivation including relativistic effects, first carried out in 1930, is known as the Bethe–Bloch formula:

$$-\frac{dE}{dx} = \left(\frac{ze^2}{4\pi\varepsilon_0}\right)^2 \frac{4\pi Z\rho N_A}{Amv^2}\left[\ln\left(\frac{2mv^2}{I}\right) - \ln(1 - \beta^2) - \beta^2\right] \tag{5.2}$$

where $v = \beta c$ is the ion velocity and ze its electronic charge; m is the mass of an electron, N_A is Avogadro's number, and A, Z and ρ are the atomic mass number, atomic number and density of the stopping material, respectively. Many light particles (e.g. hydrogen or helium ions), moving at energies we shall generally be considering ($\gtrsim 1$ MeV u^{-1}), will be fully stripped of their electrons as they pass through matter and we can take ze to be equal to that of the bare nuclear charge. The terms in Equation (5.2) containing β^2 are added to take into account relativistic effects. However, for particles moving at speeds much less than c, β^2 is small and these terms may be ignored. The quantity I is the mean energy required to ionize an atom in the medium. It is difficult to calculate and, in practice, it is taken to be an empirical parameter approximately equal to 11 Z eV. For air, for example, $I = 86$ eV. Since the energy E of the particle decreases with distance travelled, dE/dx is negative and the rate of energy loss ($-dE/dx$) is positive.

We do not attempt to derive Equation (5.2) here, but consider briefly the main physics determining its form. This follows from the conceptual model of the energy-loss process outlined above.

The electrical impulse (force × collision time) transfers momentum p to an electron and so the energy transferred $(p^2/2m)$ will depend on the square of the electric force $[\propto (ze^2/4\pi\varepsilon_0)^2]$, the square of the transit time $(\propto 1/v^2)$ and $1/m$. The rate will also be directly proportional to the density of electrons in the material $(Z\rho N_A/A)$.

The amount of energy transferred depends on the distance of closest approach during the collision, or impact parameter b, assuming that the electron does not move significantly and the ion is not deflected from a straight-line path as it passes by. The first logarithmic term in Equation (5.2) comes from an integral over b in the derivation of the formula and contains the ratio of maximum and minimum energy transfers. The ion cannot approach closer than that distance which gives the maximum energy transfer $2mv^2$. This corresponds to the minimum value of b in the integral. Then, beyond a certain distance, the binding energy of the electron in the atom becomes important and the maximum value of b is that for which the energy transfer equals the average atomic electron ionization potential I. For impact parameters greater than this, the transient electric impulse gets progressively weaker and the probability of transferring enough energy to excite or ionize the atom falls.

Graphs and tabulations of range and stopping power versus energy for different ions in different media have been published.[1] Computer programs, which give the same information, also are available. Note that, in most cases, values are given in units of energy loss per mass per unit area, which is obtained by dividing Equation (5.2) by the density ρ. This is because stopping power is determined by collisions with atomic electrons and, therefore, varies with density ρ. Thus, $dE/\rho dx$ is independent of the density of the stopping medium. For example, it is the same for water, ice or steam. The range of a particle is also usually quoted in units of mass per unit area as the product of range (distance) × density.

Even when tables of range and stopping power are available, it is often useful and instructive to be able to use simple, approximate expressions and scaling formulae for estimating effects of changing a variable such as bombarding energy, particle type or stopping medium. A selection of these are presented below.

5.2.2 Energy dependence

Figure 5.1 shows the variation with energy of the stopping power of protons in aluminium. Its general form is typical for any charged ion in any medium. Note that both scales are logarithmic; the abscissa, in particular, covers a wide energy range. Apart from very low and very high energies, the energy dependence is dominated by the $1/v^2$ term in Equation (5.2). A much weaker dependence on v comes from the logarithmic term. Empirically, it can be seen that, from about 100 keV to 1 GeV, the variation is reasonably well represented by a simple power law and the expression

$$dE/dx = \text{const.}/E^k, \tag{5.3}$$

with $k \approx 0.8$, is a useful approximation where the energy dependence of the energy loss is needed in a calculation.

---

[1] For example, Littmark and Ziegler (1980).

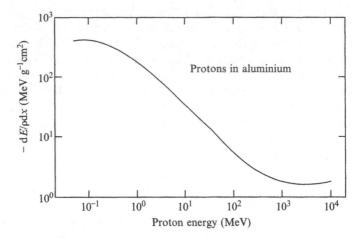

Figure 5.1 Variation of the rate of energy loss (or stopping power) divided by density as a function of energy for protons in aluminium.

For example, the range of the ion is given by:

$$R = \int dx = \int_E^0 \frac{dE}{(dE/dx)} \propto E^{1+k} \tag{5.4}$$

using the approximate expression of Equation (5.3) for the energy dependence of the stopping power.

The range R defined above is inversely dependent on ρ the density of the stopping medium, since dE/dx depends directly on ρ [see Equation (5.2)]. Therefore, range is commonly expressed as $R' = \rho R$, which has units of mass per unit area and does not depend on density.[2]

Beyond $E/A \approx 1\,\text{GeV}$ per nucleon, $-dE/dx$ passes through a point of minimum ionization and then begins to rise slowly due to the growing importance of the relativistic correction terms in the equation. At very low energies, the curve reaches a maximum and then decreases as E approaches zero. This is because the ion speed is now so low that the maximum energy transfer is approximately equal to I ($b_{max} \approx b_{min}$) and the logarithmic term tends to zero. Also, electron capture and loss by the ion begins to be significant and it no longer can be assumed that its charge z is that of the bare nucleus. This latter effect is more important for heavier ions, which require increasingly more energy to remove the inner electrons.

5.2.3 Bragg curve

According to Equation (5.3), the rate of energy loss $(-dE/dx)$ increases as the particle energy decreases and so the number of ions produced in the medium per unit distance

---

[2] Unfortunately, both R and R' are referred to as 'range' in the literature, which can lead to some confusion.

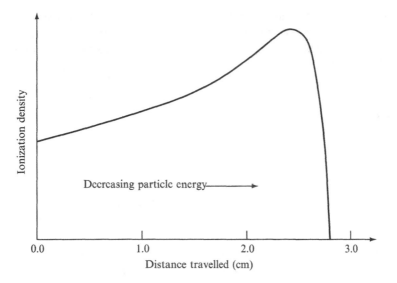

Figure 5.2 Variation of ionization density as a function of distance travelled by an α particle in air (the Bragg curve).

(ionization density) will increase along the path taken by the particle as it slows down. Near the end of the path, when most of its energy has been lost, $-dE/dx$ reaches a maximum, corresponding to the peak in energy loss at low energy (see Figure 5.1), after which it drops to zero as the particle comes to rest. An example of this behaviour is shown in Figure 5.2, which plots the changing density of ionization caused by an α particle travelling through air. Its form is known as the Bragg curve and it is characteristic of the way charged particles distribute their energy in matter. As the particle loses energy, the ionization density steadily increases with distance travelled until it reaches a pronounced peak (the Bragg peak) close to the end of the path. It then falls fairly abruptly to zero, enabling a range to be defined.

Advantage is taken of the enhanced ionization in the Bragg peak in treating certain, localized tumours by using heavy charged-particle radiation (see Chapter 9).

5.2.4 Projectile dependence

Equation (5.2) can be written:

$$dE/dx \propto z^2 f(v) \tag{5.5}$$

which explicitly shows the dependence on the charge of the projectile. A corresponding expression for the range is obtained straightforwardly from this equation. Since $dE = mv\,dv$, it follows that

$$R = \int_E^0 \frac{dE}{(dE/dx)} \propto \frac{m}{z^2} F(v). \tag{5.6}$$

Equations (5.5) and (5.6) are useful for estimating relative values since, for a given energy per nucleon (speed), dE/dx and R vary as z^2 and m/z^2, respectively. Thus, a 40-MeV α particle has four times the stopping power and the same range as a 10-MeV proton.

5.2.5   Stopping medium dependence

Dividing Equation (5.2) by the density of the stopping medium gives, for a fixed projectile charge,

$$\frac{1}{\rho}\frac{dE}{dx} \propto \frac{Z}{A}\ln\left(\frac{2mv^2}{I}\right) \tag{5.7}$$

in units of energy divided by mass per unit area; I depends on the Z of the medium, and both Z/A and the logarithmic term vary slowly with A.

A useful empirical relationship for estimating relative ranges of an ion in materials with different mass numbers (A) and densities (ρ) is the Bragg–Kleeman rule:

$$\frac{R_1}{R_2} \approx \frac{\rho_2\sqrt{A_1}}{\rho_1\sqrt{A_2}} \quad \text{or} \quad \frac{R'_1}{R'_2} \approx \frac{\sqrt{A_1}}{\sqrt{A_2}} \tag{5.8}$$

where R and R' denote the range in units of distance and mass per unit area, respectively.

Finally, it should be noted that the stopping process is statistical in nature. There is variation in the energy transfer per collision and in the number of ionized atoms for a given amount of energy loss or distance travelled. Hence, there is a spread in the observed range of monoenergetic particles. This is called straggling. For heavy charged particles, straggling is not a large effect; e.g for 5-MeV α particles, the standard deviation of the range distribution is about 1%.

5.3   ELECTRONS

In passing through matter, electrons lose energy to atomic electrons via the electric interaction in the same way as heavy charged particles. However, they are much lighter and so, for a given energy, their speeds are greater. As a result, $-dE/dx$ is much smaller, which means that electrons are more penetrating than heavy ions. For example, the range R (in units of distance) of a 1-MeV electron in aluminium is about 1800 μm compared with about 3 μm for an α particle with the same energy.

An electron colliding with other electrons loses a much greater fraction of its energy in a single collision than does a heavy ion. Its range, therefore, is much less well defined and its linear distance of penetration will be very different from the length of the path it actually follows through the medium. Indeed, an electron may exchange nearly all its energy in a single collision, and the electron which penetrates furthest into the stopping material may not be the incident one. In addition, the sudden changes in direction and speed cause *bremsstrahlung* radiation to be emitted. All accelerated charged particles radiate bremsstrahlung energy. The electron, being the lightest, radiates much more prolifically than any other charged particle and, at very high energies, most of the energy loss is by radiation. Thus, the stopping power for electrons is the sum of two terms: one due to collisions with electrons in the medium and the second due to radiated energy loss. Figure 5.3 shows the relative contributions for aluminium and lead. For typical β-decay energies of a few MeV (or less), there is very little variation in collision energy loss with either energy or type of material. Radiation loss, on the other hand, is much more strongly dependent on the Z of the stopping medium and rises rapidly with energy eventually becoming the dominant term. Bethe and Heitler estimated the critical energy at which the two terms are equal as $E_c \approx 1600\,mc^2/Z$, with $mc^2 = 0.511$ MeV. For lead ($Z = 82$), this gives $E_c \sim 10$ MeV, in agreement with the data in the figure.

A range equal to the total path length covered by an electron could be obtained by integrating an expression for dE/dx [as in Equation (5.4)]. However, the large statistical variations in energy loss complicates this procedure. Also, as already pointed out, electrons scatter easily and there is a large difference between the total path length they travel in matter and their average penetration depth.

A qualitative sketch of the transmitted intensity of a beam of electrons passing through matter is given in Figure 5.4. Although the electrons are not being stopped, the intensity begins to decrease slowly, even for thin absorbers, because of the significant probability of large-angle scattering which effectively removes electrons from the incident direction. Eventually, the intensity falls to zero when the penetration

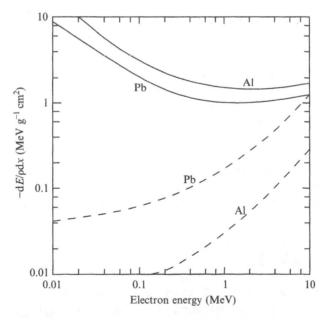

Figure 5.3 Plots of specific stopping power versus energy for electrons in aluminium and lead. Solid lines give the contribution due to collisions with electrons; dashed lines give the radiation loss due to bremsstrahlung.

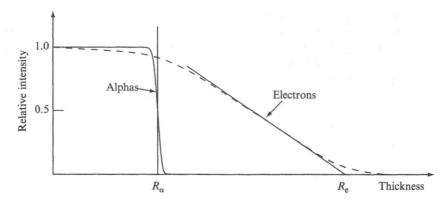

Figure 5.4   Relative transmitted intensity of collimated beams of α particles and electrons as a function of absorber thickness. For α particles, the mean range R_α is the point where the intensity has fallen to half its initial value; R_e is the extrapolated range for electrons (see text). The horizontal scale is not the same for the two types of radiation.

depth is comparable with the total path length (defined above) which is the distance required to stop the electrons.

Contrast this behaviour with the transmission of a beam of heavy charged particles, such as α particles, which is also shown in Figure 5.4. In this case, the maximum fractional energy loss per collision and the scattering from the incident direction are both much smaller than they are for electrons. Statistical fluctuations in energy loss are smaller, and all particles of a given energy travel about the same distance and in approximately the same direction. Therefore, the intensity of a beam of α particles will change very little with absorber thickness until near the end of the range when it drops quickly to zero. As shown in the figure, the range R_α is defined as the distance at which the intensity has fallen to half its initial value.

It is possible to define an extrapolated range R_e for electrons by extending the linear part of the transmission curve as shown in Figure 5.4. Figure 5.5 shows the range–energy relationship for electrons in aluminium obtained in this way from experimental data on the transmission of beams of monoenergetic electrons. These range–energy data can be used in conjunction with relative values of stopping power to estimate ranges in other materials. For example, the data in Figure 5.3 show that, at energies where the radiation loss is not important, $-dE/\rho dx$ for lead is approximately 0.7 times that of aluminium. Therefore, the extrapolated range R' of electrons in lead (in units of mass per unit area) will be $1/0.7$ or about 1.4 times that in aluminium.

5.4   GAMMA RAYS

Gamma-ray photons emitted in nuclear transitions are usually energetic and, in this section, we will consider the interaction of photons with energies greater than a few tens of keV and wavelengths shorter than most atomic dimensions.

There are three primary processes by which γ rays or X-rays interact with matter. These are the photoelectric effect, Compton scattering and pair production. Their

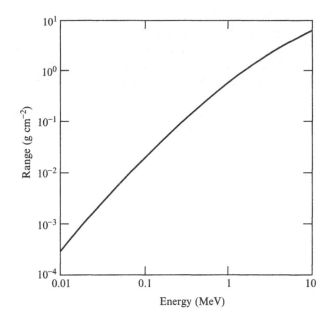

Figure 5.5 Range–energy relationship for electrons in aluminium derived from measure-ments. This curve is approximately correct for any material, since the range expressed in terms of mass per unit area varies only slightly with the atomic number Z of the stopping medium. From L. Katz and A. S. Penfold (1952).

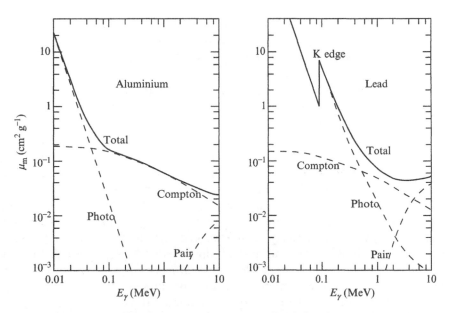

Figure 5.6 Photon mass-attenuation coefficients for aluminium and lead as a function of photon energy. Dashed lines show the separate contributions due to the photoelectric effect, Compton scattering and pair production. From Siegbahn (1966), pp. 832 and 838.

relative importance is illustrated in Figure 5.6 which plots, as a function of photon energy E_γ, the total mass attenuation coefficient for aluminium and lead. The mass attenuation coefficient μ_m, defined in Equation (5.17), is a measure of the probability of the γ ray interacting in the material. It depends on the atomic number Z of the atoms of the absorbing medium and varies strongly with γ-ray energy E_γ. It has contributions from all the three interaction processes that can take place in the absorbing material. Each has its own characteristic energy dependence, and the Z dependence, referred to above, can be seen by comparing the two sets of curves for aluminium and lead. Contributions from the photoelectric effect and, to a lesser extent, pair production are relatively more important in Pb ($Z = 82$) than in Al ($Z = 13$). The contribution from Compton scattering shows a very weak dependence on Z. Brief descriptions of each of these processes are given below.

5.4.1 Photoelectric effect

Photoelectric absorption occurs in the vicinity of an atom where the photon's energy is wholly converted into releasing an electron from its site in the material, which is usually one of the inner atomic shells. This so-called photoelectron emerges with kinetic energy given by

$$T = E_\gamma - B_\mathrm{e} \qquad (5.9)$$

where B_e is the binding energy of the electron.

The atom which has lost the electron may de-excite by releasing other, less tightly bound electrons. Electrons emitted by this process are called Auger electrons. Alternatively, an electron from a higher shell may fill the vacancy in the inner shell with the emission of a characteristic X-ray photon. This is known as X-ray fluorescence. This X-ray in turn may interact and be absorbed by the medium.

The photoelectric mass attenuation coefficient, shown in Figure 5.6, varies strongly with E_γ and Z. A sharp rise at low energy for lead is due to the contribution of the innermost, K-shell electrons which in lead have a binding energy B_K of about 90 keV. Below this energy, these electrons do not contribute to the photoelectric absorption because there is not enough energy to eject them from the atom. The discontinuity in the energy dependence of μ_m is known as the K edge; its energy is approximately proportional to Z^2. At lower photon energies, photoelectric absorption edges due to less bound shells (L, M $\cdots$) become evident.

Note that the relative contribution of the two K electrons is much greater than that of all the rest of the electrons in the lead atom put together. This is at first surprising since they occupy only a relatively small fraction of the atom's volume. However, the need to conserve both energy and momentum forbids the photoelectric effect from occurring on a free electron. The process must take place in the vicinity of an atom, which takes away the excess momentum but, because of its much greater mass, acquires very little of the original energy. The smaller the electron's binding energy relative to E_γ the more it appears like a free electron and the smaller will be the probability of the photoelectric effect. This is why the most tightly bound electrons contribute most to photoelectric absorption (for $E_\gamma > B_\mathrm{K}$, where B_K is the binding energy of the K-shell electron) and why there is such a strong dependence on energy and atomic number. An approximate expression, illustrating these dependencies, gives for the photoelectric cross section:

$$\sigma_{\text{pe}} \propto Z^5 / E_\gamma^{3.5} \tag{5.10}$$

for γ-ray energies of a few hundred keV.

5.4.2 Compton scattering

In this interaction, the photon, with incident energy E_γ, scatters from an electron normally regarded as being free. The result is a photon of lower energy E'_γ and an electron recoiling with an amount of kinetic energy T which depends on the scattering angle. The process is illustrated schematically in Figure 5.7(a). Using relativistic kinematics, energy conservation gives for the kinetic energy of the electron:

$$T = E_\gamma - E'_\gamma = E - mc^2 \tag{5.11}$$

where E is the total energy of the recoil electron including its rest mass energy mc^2. Momentum conservation requires that the momenta $p_\gamma = E_\gamma/c$, $p'_\gamma = E'_\gamma/c$ and p the momentum of the electron must be added vectorially to form a closed triangle [see Figure 5.7(b)]. Using the cosine rule, we can write

$$(pc)^2 = E_\gamma^2 + (E'_\gamma)^2 - 2E_\gamma E'_\gamma \cos\theta = E^2 - m^2 c^4 \tag{5.12}$$

where we have substituted for p_γ and p'_γ and used the relation $E^2 = p^2 c^2 + m^2 c^4$ for the electron. After eliminating E from Equations (5.11) and (5.12), we obtain an expression for the scattered photon energy:

$$E'_\gamma = \frac{E_\gamma}{1 + (E_\gamma/mc^2)(1 - \cos\theta)} \tag{5.13}$$

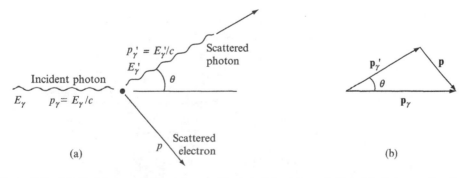

(a)

(b)

Figure 5.7 (a) Geometry of Compton scattering; (b) vector relationship between the three momentum vectors $\mathbf{p}_\gamma$, $\mathbf{p}'_\gamma$ and $\mathbf{p}$.

which varies from E_γ (for $\theta = 0°$) to a minimum value E'_γ (min) at $\theta = 180°$; E'_γ (min) approaches $mc^2/2$ (approximately 256 keV) when E_γ is large. The corresponding electron kinetic energy varies from near zero for a glancing collision to a maximum (always less than E_γ) when $\theta = 180°$.

The probability of Compton scattering is less strongly dependent than the photo-electric effect on E_γ and Z. It does depend directly on the density of electrons in the medium and, therefore, on the density of the material multiplied by Z/A. However, over a large range of the periodic table, Z/A is approximately constant and so the Compton contribution to the mass attenuation coefficient, which measures the probability as a function of mass per unit area and takes into account the density, is almost independent of Z.

The differential cross section for Compton scattering is given by a formula derived by Klein and Nishina. It predicts that the total Compton cross section, σ_c, should decrease with E_γ, as is shown in Figure 5.6 by the curves depicting the Compton contributions to μ_m, which are proportional to σ_c/ρ. The angular variation of the differential cross section is shown in Figure 5.8 for a range of values of E_γ. It is symmetric about $90°$ when E_γ is small and becomes increasingly more forward peaked at higher energies. This is an important consideration in designing effective shielding for high-energy γ rays.

5.4.3   Pair production

In this process, the entire photon energy is converted in the field of an atom into the creation of an electron–positron pair with total kinetic energy given by

$$T_- + T_+ = E_\gamma - 2mc^2. \tag{5.14}$$

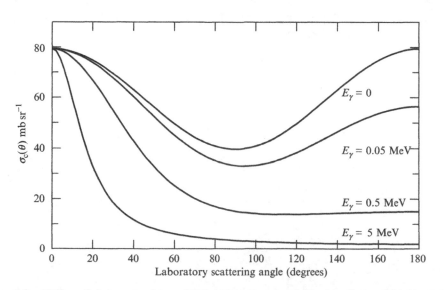

Figure 5.8   Differential cross section $\sigma_c(\theta)$ for Compton scattering by electrons in the energy range $E_\gamma = 0$ to 5 MeV, given by the Klein–Nishina formula. See Siegbahn (1966), p. 51.

Like the photoelectric effect, pair production requires the presence of a heavy body in order to conserve both energy and momentum. This means that there is some dependence of its probability on Z, as can be seen by comparing pair production for Al and Pb in Figure 5.6. There is an energy threshold of $2mc^2 = 1.022$ MeV and the pairproduction cross section does not become important until E_γ exceeds several MeV.

Note that, initially, only part of the γ-ray energy is converted into kinetic energy of charged particles, which gets transferred to the absorbing medium. The rest is in the form of the rest masses of the electron and positron. The positron is an anti-electron and, after it slows down and almost comes to rest, it will be attracted to an ordinary electron. Annihilation then takes place in which the electron and positron rest masses are converted into two γ rays, each with an energy of 0.511 MeV. These annihilation γ rays are emitted in opposite directions in order to conserve momentum and they may in turn interact in the absorbing medium by either photoelectric absorption or Compton scattering.

5.4.4 Attenuation

If a photon beam is well collimated, all three interaction processes described above cause it to be attenuated as it passes through matter. Photons which undergo photoelectric absorption or pair production disappear altogether and most of those that are Compton scattered are deflected away from reaching a detector placed in the forward direction. A schematic arrangement to measure this attenuation is illustrated in Figure 5.9(a). The situation is more complicated if the beam is uncollimated, as shown in Figure 5.9(b), because photons, which, initially, are not heading towards the detector, can be scattered into it when an absorber is put in place.

The dependence of photon intensity on absorber thickness is most easily calculated for a well-collimated beam. Consider such a beam, of intensity I photons per second, incident on a thin section (thickness dx) in a slab of material whose atomic density is N. If the total interaction cross section per atom of material is σ, the rate of removal of photons from the beam is given by

$$dI = -N\sigma I dx. \tag{5.15}$$

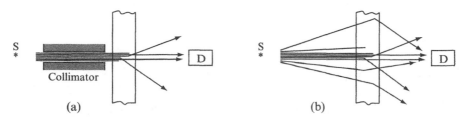

Figure 5.9 Arrangements for measuring the transmission of (a) a collimated or (b) an uncollimated beam of γ rays, from a source S, incident on a slab of material. In (a), both scattering and absorption reduce the number of photons reaching the detector D; in (b), more photons are detected because some are scattered into the detector.

The total cross section has contributions from all three interaction processes: i.e. $\sigma = \sigma_{pe} + Z\sigma_c + \sigma_{pp}$, where σ_{pe} and σ_{pp} are the atomic photoelectric and pair production cross sections, respectively, and σ_c is the Compton scattering cross section for a single electron. Integrating Equation (5.15) gives an exponential variation of γ-ray intensity with depth:

$$I = I_o \exp(-N\sigma x) = I_o \exp(-\mu x) \tag{5.16}$$

where I_o represents the incident intensity, x the depth in units of length and we have introduced the linear attenuation coefficient μ $(= N\sigma)$, which has dimension of $(\text{length})^{-1}$. The mass attenuation coefficient μ_m is defined as the linear attenuation coefficient divided by the density of the medium:

$$\mu_m = \mu/\rho. \tag{5.17}$$

A commonly quoted quantity for an absorber is the half thickness, equal to $\ln 2/\mu$, which reduces the intensity to half its initial value. Note that μ (or μ_m) is independent of x only for monoenergetic photons. This is because different energies are attenuated at different rates and, therefore, the total intensity of a beam of mixed energies E_γ will not vary according to a single exponential.

In general, it is much more difficult to provide an effective shield for γ rays than for charged particles. The half thickness for a 1-MeV γ ray in aluminium is about 4.2 cm. This is about 20 times thicker than the extrapolated range of a 1-MeV electron and over 10 000 times thicker than the range of a 1-MeV α particle (see Section 5.3). Furthermore, whereas it is possible to stop completely an intense beam of charged particles with an absorber slightly in excess of the range, the attenuation of γ rays is exponential and many half thicknesses are required to provide a safe shield for a strong γ-ray source.

It is also important to emphasize that Equation (5.16) gives the transmitted intensity for a collimated beam. As noted above, if the beam is uncollimated [Figure 5.9(b)], there will be a contribution from Compton-scattered γ rays which otherwise would not reach the detector. This gives rise to the concept of a buildup factor, which is the ratio of the actual intensity reaching the detector to that given by Equation (5.16). The buildup factor is close to unity for thin absorbers, but it can be large (>10) for a thick shield where a large attenuation is required.

5.5   NEUTRONS

Neutrons interact in matter via nuclear reactions, which depend strongly on energy and the particular nuclei with which the neutrons collide. Neutrons may be scattered – in which case energy is transferred to recoiling (charged) nuclei. Alternatively, they may be absorbed in a variety of different processes as outlined in Chapters 1 and 4.

At energies of a few MeV, with which we will mainly be concerned, fusion is the main contributor to the absorption cross section. In fusion, a neutron is first captured to form a compound nucleus excited to an energy approximately equal to the initial kinetic energy plus the neutron's binding energy which is, typically, 7 to 8 MeV. This energy is subsequently released in the form of reaction products, which may be γ rays, charged particles or neutrons. Reactions that result in the emission of energetic charged particles are important because they enable the presence of neutrons to be detected (see Section 6.6).

Scattering slows the neutron down as energy is lost in successive collisions. If the absorption cross section is relatively small, the neutron may be slowed down until it comes into thermal equilibrium with its surroundings. Such neutrons are called thermal neutrons and represent a unique type of radiation. A single, thermal neutron with only about 0.025 eV of energy can easily be detected because it releases many MeV of energy as other radiation when it is captured. Indeed, they are particularly easy to detect because, as pointed out in Chapters 1 and 4, neutron absorption cross sections for many materials are highest for low neutron energies. Some examples of slow-neutron detectors are described in Section 6.6.

5.5.1 Attenuation

When traversing a slab of material, neutrons, like γ rays, can be absorbed or scattered through large angles. Therefore, if we assume that the neutron beam is well collimated in a geometry similar to that in Figure 5.9(a), both the scattering and absorption cross sections (σ_s and σ_a) contribute to the loss in transmitted intensity I and, by following a derivation very similar to that given in the previous section for Equation (5.16), we obtain the relation:

$$I = I_o \exp(-N\sigma x) \tag{5.18}$$

where $\sigma = \sigma_s + \sigma_a$ and the other symbols have the same meaning as in Equation (5.16).
Equation (5.18) is often written as

$$I = I_o \exp(-\Sigma x) = I_o \exp(-x/\lambda) \tag{5.19}$$

where $\Sigma = N\sigma$ is called the macroscopic total cross section, and $\lambda = 1/\Sigma$ is the mean attenuation length. As was the case for γ rays, these equations refer only to monoenergetic radiation that is collimated.

It is straightforward to show that the attenuation length is equal to the mean free path, which is the average distance travelled by a neutron before it interacts in the medium. The probability of finding a neutron between x and $x + dx$ is proportional to $\exp(-x/\lambda)$ and, therefore, we obtain the mean free path (average distance travelled) as

$$\frac{\int_0^\infty x \exp(-x/\lambda)dx}{\int_0^\infty \exp(-x/\lambda)dx} = \lambda. \tag{5.20}$$

We can define a scattering mean free path $\lambda_s = 1/\Sigma_s$ as the average distance between successive scatterings, and an absorption mean free path $\lambda_a = 1/\Sigma_a$ as the average distance travelled before the neutron is absorbed. We also have $\Sigma = N(\sigma_s + \sigma_a) = \Sigma_s + \Sigma_a$. Therefore, the total mean free path is given by

$$\frac{1}{\lambda} = \frac{1}{\lambda_s} + \frac{1}{\lambda_a}.$$ (5.21)

A full analysis of the effect on the life of a neutron of all nuclear interaction processes within a material is complex even for moderate or low energies. As already noted, it depends strongly on the relative magnitudes of the scattering and absorption cross sections. If $\sigma_a \gg \sigma_s$, then, before any scattering occurs, neutrons are removed according to Equation (5.18) with $\sigma_t \approx \sigma_a$. If $\sigma_s \gg \sigma_a$, many scatterings occur with successive energy losses, slowing the neutron to low, possibly thermal, energies before it is finally captured. This slowing-down process is called *moderation*. As described in Chapter 10, it is one of the critical factors governing the design of a thermal fission reactor.

5.5.2   Neutron moderation

Consider a non-relativistic, elastic collision between a neutron (mass m), with energy E_0 and speed v_0, incident on a target nucleus (mass M) initially at rest. This event is illustrated in Figure 5.10 for the laboratory (lab) and centre-of-mass (c-m) frames.

In the lab frame, the scattered neutron has energy E_1 and speed v_1. The speed of the centre of mass V_{cm} is defined by

$$mv_0 = (m + M)V_{cm}$$ (5.22)

since M is initially at rest. In the c-m frame, the centre of mass is at rest and, therefore, the speed of the target nucleus w^* is equal to V_{cm}. Also, since kinetic

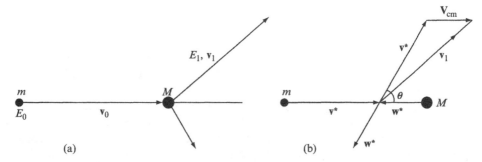

(a)   (b)

Figure 5.10   Elastic scattering kinematics for an incident particle of mass m, initial energy E_0 and speed v_0, colliding with a particle of mass M in (a) the laboratory and (b) the centre-of-mass frames of reference. The particle of mass M is initially at rest in the laboratory frame.

energy is conserved in an elastic collision, the initial and final speeds of m and M in the c-m frame (v^* and w^*) are unchanged by the collision.

Velocities in the lab system are related to the corresponding ones in the c-m frame by the addition of the c-m velocity. Thus, $v_0 = v^* + V_{cm}$ from which, after using Equation (5.22), we obtain for the speed of the neutron in the c-m frame:

$$v^* = v_0 M/(m+M). \tag{5.23}$$

The relation between the scattered neutron velocities in the two frames is

$$\mathbf{v_1} = \mathbf{v^*} + \mathbf{V}_{cm}$$

as is shown in Figure 5.10(b). Squaring this equation gives

$$v_1^2 = (v^*)^2 + (V_{cm})^2 + 2v^* V_{cm} \cos \theta \tag{5.24}$$

where θ, the angle between v^* and V_{cm}, is the scattering angle in the c-m frame. After substituting for v^* and V_{cm} [using Equations (5.22) and (5.23)] and noting that $E_1/E_0 = (v_1/v_0)^2$, Equation (5.24) gives for the lab energy of the scattered neutron:

$$E_1 = E_0 \left(\frac{M^2 + m^2 + 2Mm\cos\theta}{(M+m)^2} \right). \tag{5.25}$$

If $\theta = 180°$, this gives

$$E_1(\text{min}) = E_0 \left(\frac{M-m}{M+m} \right)^2 = \alpha E_0 \tag{5.26}$$

which defines α. For neutron scattering from a nucleus of mass number A, we substitute $m = 1$ and $M = A$ in the above equations.

Next, to determine the effect of the scattering on the average neutron energy, we calculate the probability distribution $P(E_1)dE_1$ of the scattered neutron having an energy between E_1 and $E_1 + dE_1$. From Equation (5.25) there is a one-to-one correspondence between E_1 and the c-m scattering angle θ. Hence, the probability distribution $P(E_1)dE_1$ in energy also defines a probability distribution $p(\theta)d\theta$ in the angle and

$$-p(\theta)d\theta = P(E_1)dE_1. \tag{5.27}$$

The negative sign on the left-hand side of Equation (5.27) allows for the fact that the energy decreases as the angle increases.

If we assume that the scattering angular distribution is isotropic in the c-m system, which is approximately valid for neutrons in a fission reactor, then

$$p(\theta)d\theta = \frac{2\pi \sin\theta d\theta}{4\pi} = \frac{1}{2}\sin\theta d\theta.$$

From Equation (5.25), after substituting for M and m, we find

$$\frac{dE_1}{d\theta} = -\frac{2AE_0 \sin \theta}{(A+1)^2}.$$

Hence,

$$P(E_1)dE_1 = \frac{(A+1)^2}{4AE_o} dE_1 = \frac{dE_1}{(1-\alpha)E_0}. \tag{5.28}$$

The distribution in energy will be uniform between $E_1(\min) = \alpha E_0$ and E_0, as shown in Figure 5.11. Note that the area under the rectangle in this figure is unity as it should be. The quantity $(1-\alpha)E_0$ is the range of all post-collision energies, the average neutron energy after the collision is $\overline{E}_1 = \frac{1}{2}(1+\alpha)E_0$, and the average energy loss is $\frac{1}{2}(1-\alpha)E_0$. It is important to note that the average fractional energy loss per collision is independent of energy. After a second collision, each of the neutrons in the distribution loses on average the same fraction of its energy again, and so on. After n collisions, the average energy becomes

$$\overline{E}_n = E_0 \times \left(\frac{\overline{E}_1}{E_0}\right)^n. \tag{5.29}$$

However, Equation (5.29) is not the most useful measure of the average neutron energy since the distribution after n collisions has a long high-energy tail and most of the neutrons have energies well below $\overline{E}_n$.

A quantity which better characterizes the distribution is the *logarithmic energy decrement* ξ, defined as the average value of $\ln(E_0/E_1)$ after the first collision:

$$\xi = \overline{\ln(E_0/E_1)} = \int_{\alpha E_0}^{E_0} \ln(E_0/E_1) P(E_1) dE_1. \tag{5.30}$$

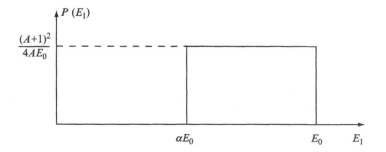

Figure 5.11   Probability distribution for neutrons (with initial energy E_0) having a final energy E_1 after being elastically scattered by a stationary target with mass number A. It is assumed that the scattering is isotropic in the centre-of-mass frame; $\alpha = [(A-1)/(A+1)]^2$.

Substituting for $P(E_1)$ from Equation (5.28) and putting $x = E_1/E_0$ gives

$$\xi = \frac{1}{(1-\alpha)} \int_1^\alpha \ln x \, dx = 1 + \frac{\alpha}{(1-\alpha)} \ln \alpha. \qquad (5.31)$$

As we see from this equation, the logarithmic energy decrement:

$$\xi = \ln E_0 - \overline{\ln E_1} \qquad (5.32)$$

is independent of initial energy E_0. Thus, with each collision, the average value of $\ln E$ decreases by ξ and, after n collisions, the average value of $\ln E_n$ is given by

$$\overline{\ln E_n} = \ln E_0 - n\xi. \qquad (5.33)$$

In terms of the mass number A, ξ can be written as

$$\xi = 1 + \frac{(A-1)^2}{2A} \ln\left(\frac{A-1}{A+1}\right). \qquad (5.34)$$

Expanding the logarithm in Equation (5.34) in powers of $1/A$, one obtains a simpler, approximate expression:

$$\xi = \frac{2}{A} - \frac{4}{3A^2} \qquad (5.35)$$

which is accurate to better than 1% for $A \geq 6$. Table 5.1 lists values of ξ for a number of different materials.

Equation (5.33) can now be used to estimate the number n of collisions required to slow the neutrons down from their initial energy E_0 to their median energy after n collisions, E'_n, which is defined by

$$\ln E'_n = \overline{\ln E_n}. \qquad (5.36)$$

Table 5.1 Scattering properties of several nuclei.

Nucleus	α	ξ	n (to thermalize)
^{1}H	0	1.00	18
^{2}H	0.111	0.725	25
^{4}He	0.360	0.425	43
^{12}C	0.716	0.158	115
^{238}U	0.983	0.0084	2200

Combining Equations (5.33) and (5.36) one obtains

$$n = \frac{1}{\xi}\ln\frac{E_0}{E'_n}. \tag{5.37}$$

Values of n to slow neutrons from an average energy typical of fission neutrons ($\sim 2\,\text{MeV}$) to thermal energy neutrons ($\approx 0.025\,\text{eV}$) are listed in the right-hand column of Table 5.1 for a number of different scattering nuclei.

This simple way of estimating n is not reliable at near-thermal energies since the scattering nuclei cannot be regarded as stationary and can cause the neutron energy to increase as well as decrease. Also, for a molecular gas and where energies are comparable with atomic binding energies, the collisions cannot be considered to be strictly elastic.

PROBLEMS 5

5.1 Calculate the maximum energy a 5-MeV α particle can transfer to an electron initially at rest. Assume the α particle is infinitely massive compared with the electron.

5.2 A fast charged particle (charge ze) passes by an initially stationary electron (mass m). Calculate the energy transferred to the electron given that the impact parameter is b and the speed of the particle is v. Assume that the electron does not move significantly as it receives the impulse due to the passing charged particle and that the charged particle is not deflected from its path.

5.3 If the charged particle in Problem 5.2 passes by an initially stationary carbon nucleus ($A = 12$, $Z = 6$), calculate the ratio of the energy transferred to the nucleus compared with that transferred to an electron if the impact parameters are the same. You may make the same assumptions as in Problem 5.2.

5.4 The specific rate of energy loss ($-\mathrm{d}E/\rho\mathrm{d}x$) of a 5-MeV proton in silicon is 59 keV mg^{-1} cm^2 and its range R' is 50 mg cm^{-2}. Calculate values of ($-\mathrm{d}E/\rho\mathrm{d}x$) and range R' for deuterons, tritons, ^{3}He and α particles, all of which have the same speed as the proton.

5.5 (a) If, in problem 5.4, we know that $\mathrm{d}E/\mathrm{d}x$ is proportional to $-1/E^n$, calculate n.
(b) What fraction of its range would the proton have travelled when its energy has been reduced to half its initial value?

5.6 A 10-MeV α particle has a specific rate of energy loss of 233 keV mg^{-1} cm^2 in silicon. Estimate, using the Bethe–Bloch formula, the specific rates of energy loss for a proton, deuteron and ^{3}He nucleus, each with the same initial energy as the α particle. In this question, it is sufficiently accurate to use mass ratios $m_\alpha/m_x = 4$, 2 and 4/3, where x = p, d and ^{3}He, respectively.

5.7 The energy of the photon, after being Compton scattered (from an electron at rest) through 60°, is half its initial energy. Calculate its initial energy and, hence, deduce in what part of the electromagnetic spectrum the photon originated.

5.8 What is the maximum energy a γ ray can have after being Compton scattered through 180°?

5.9 Calculate the velocity of recoil electrons when light of wavelength 1 μm is Compton scattered through 90°. Hint: Are the electrons relativistic?

5.10 What thickness of concrete (density 2200 kg m^{-3}) is needed to attenuate a collimated beam of 1-MeV γ rays by a factor of 10^6? The mass attenuation coefficient of concrete $\mu_m = 0.064$ cm^2 g^{-1}.

5.11 A pulse of 10^{18} 100-keV X-ray photons per m^2 is directed at right angles on to a 5 mm thick slab of iron. Calculate the temperature increase of the slab. Density of iron = 7870 kg m^{-3}; mass attenuation coefficient for 100 keV photons on iron = 0.04 m^2 kg^{-1}; specific heat capacity of iron = 106 J kg^{-1} K^{-1}.

5.12 When a slab of material is inserted between a collimated ^{60}Co source and a detector, it is found that the fluxes of 1.17 and 1.33 MeV γ rays are reduced, respectively, to 62 and 65% of their values with no absorber. Calculate the ratio of the attenuation coefficients of the material for the two energies. What would be the reduction in the fluxes if two slabs were used?

5.13 Calculate the thickness of cadmium ($A = 112.4$, density = 8650 kg m^{-3}) that would attenuate the intensity of a collimated beam of thermal neutrons by a factor of 1000. The average absorption cross section for thermal neutrons = 3000 b. In this problem, the scattering cross section is small and you may neglect it.

5.14 (a) If the mean attenuation length of uranium ($A = 238$) for a collimated beam of neutrons of a certain energy is 2 cm, calculate the total microscopic cross section σ_t for the removal of neutrons from the beam. Density of uranium = 18.9 g cm^{-3}.
(b) If the neutron-scattering cross section is six times the absorption cross section, what is the mean total distance travelled by the neutrons (absorption mean free path) in uranium?

5.15 Calculate the average number of collisions needed for neutrons of average initial energy 2 MeV to be thermalized (average energy 0.025 eV) when scattered by beryllium ($A = 9$), iron ($A = 56$) and lead ($A = 207$).
      In each case, how many collisions would be needed to reduce the average energy to be about half the initial energy?

6

Detectors and Instrumentation

6.1 INTRODUCTION

Experimental methods and instrumentation developed for nuclear physics are widely used in many other branches of science and technology. In this chapter, we consider the principles of systems for detecting radiation and accelerators for producing controlled beams of radiation.

Any detector of radiation generates its signal from the results of the interaction of radiation with matter, as described in the previous chapter. There are a number of ways this can be done: collecting and measuring the charge released by ionization of a gas or excitation of electrons in a semiconductor, observing fluorescent photons emitted when de-excitation takes place in a scintillating material, or making the ionization trail visible in some form of a track chamber. A detector may simply record the presence of nuclear radiation, but more often it also gives information about its energy and type. Many specialized varieties of detector exist and continue to be developed. Some can be used with different types of radiation, but often they are designed for specific tasks, such as detecting thermal neutrons or heavy charged particles. Detector technology is highly sophisticated and the reader is directed to the references in the bibliography for a comprehensive coverage of the subject. Here, we can describe only the basic principles of the most commonly used types with a few illustrative examples.

In the next section, we describe detectors which use a gas as the detecting medium. They are simple to construct and do not suffer from radiation damage. However, they are not well suited for measuring weakly ionizing particles, which deposit very little energy per unit path length and they are not efficient for high-energy γ rays or neutrons. More suitable for these purposes are detectors made of solid or liquid materials. The most widely used of these are the scintillation and semiconductor detectors. They are described in Sections 6.3 and 6.4 and their relative performance for detecting γ rays is discussed in Section 6.5. A neutron must interact with a nucleus before its presence can be detected and, in Section 6.6, we indicate briefly how gas and scintillation detectors have been adapted specifically to detect neutrons. Detectors are often required to indicate the nature of the radiation, as well as its presence, and some common ways whereby this is accomplished are outlined in Section 6.7.

Modern accelerators can deliver beams of charged particles of any desired type and with a wide range of possible energies and intensities. The availability of such beams has enabled researchers over the years to probe and learn a great deal about the nature and internal structure of the nucleus and the nuclear force. Increasingly, accelerators are being used for a large and growing number of applications and we conclude the chapter with a brief introduction to the main types of accelerator that are used for these purposes.

6.2   GAS DETECTORS

6.2.1   Ionization chamber

One of the simplest detectors conceptually is the ionization chamber, which measures the ionization produced when charged particles pass through a gas. The basic version, shown schematically in Figure 6.1(a), consists of two electrodes forming a parallel-plate capacitor (capacity C) between which a voltage is applied, normally through a large bias resistor R. The electric field, which is set high enough to prevent recombination, sweeps electrons to the anode and positive ions much more slowly to the cathode. The resulting current is proportional to the rate of production of ionization between the plates. The average energy required to ionize a gas atom or molecule depends on the gas, but, except for helium, it is about 30 eV per ion pair. In air, for example, it is 34 eV per ion pair. Thus, if particles entering an air-filled detector deposit an average of 1 GeV s^{-1} in the gas, the average current flowing through the chamber will be

$$\frac{10^9 (\text{eV s}^{-1}) \times 1.6 \times 10^{-19} (\text{C per ion})}{34 (\text{eV per ion})} \approx 5 \times 10^{-12} \text{A}$$

which is measurable with a sensitive electrometer.

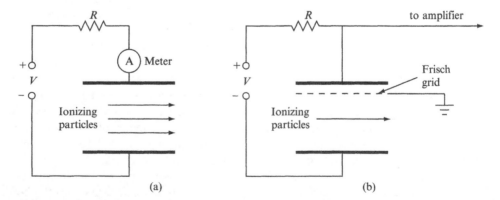

(a)   (b)

Figure 6.1   Gas ionization chamber using (a) simple parallel-plate geometry, and (b) with the addition of a Frisch grid for pulse counting.

A single, charged particle entering the detector will produce a pulse of current and, if the time constant RC is large compared with the pulse duration, there will be a change in the voltage on the chamber as the electrons and ions are collected. The total voltage change is q/C, where q is the charge collected. For example, the capacity of a chamber with electrodes $10\,cm \times 10\,cm$, separated by $2\,cm$, is about 4.5 pF and, if a 5-MeV α particle is stopped in it, the voltage pulse will be about $5\,mV$ (see also Problem 6.2). This is rather small and must be amplified considerably before it can be analysed with conventional electronics.

As the electrons and ions move towards their respective electrodes under the influence of the applied field, they undergo many collisions with gas atoms or molecules on the way and quickly acquire drift velocities which depend on the voltage gradient. For a voltage gradient of $100\,V\,cm^{-1}$, ions move at about $1\,m\,s^{-1}$ in air at STP and would take about $0.02\,s$ to cross the $2\,cm$ gap. Electrons move about 1000 times faster and are collected very much more quickly. The slow collection time for the positive ions means that the simple ionization chamber is not suitable for detecting individual particles (pulse operation) where a count rate of several kilohertz may be required. However, a more serious complication arises because the time dependence of the voltage induced on the electrodes depends on the movement of the charges as they are collected. Electrons and ions move at very different speeds and, since the shape of the output pulse depends on the time taken for the electrons and ions to reach their respective electrodes, it also must depend on where the ionization was formed between the plates.

A chamber designed for pulse operation is the gridded ion chamber, shown in Figure 6.1(b), which has a fine mesh, called a Frisch grid, placed close to the anode. This serves to screen the anode from the effects of the movement of charge within the main body of the chamber until the electrons reach the gap between the grid and anode. The full pulse, measured between the grid and the anode, is then developed quite quickly and the shape of the voltage pulse is determined only by the electron transit time between the grid and anode, which, typically, is a few microseconds. The amplitude of the voltage pulse is proportional to the charge collected and, therefore, to the energy deposited in the gas.

The choice of detector gas is important and depends on a number of factors that influence the time and efficiency for collecting electrons. For example, while air is adequate for a detector used in continuous (current) mode, it is not suitable for pulse operation. This is because oxygen has a high electron affinity, which means that it has a high probability for capturing electrons and thus degrading the fast part of the electronic pulse. Common gases used for pulse operation include inert gases (helium, argon) and certain organic gases, such as isobutane. It is not possible or appropriate here to give a full discussion of the many technical aspects of gas detectors and we refer the interested reader to more advanced texts on detectors and instrumentation.

6.2.2 Proportional counter

A common way of amplifying the signal from a gas-filled detector is by increasing the electric field until the electrons can gain sufficient energy between collisions with gas

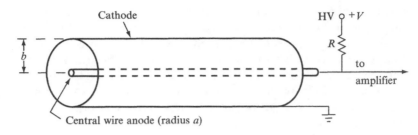

Figure 6.2   Schematic representation of a gas-filled ionization chamber with cylindrical geometry for use as a proportional counter.

atoms to cause further ionization. The original electrons and the newly released ones can then go on to produce further ionization, and a rapid amplification of the original cascade occurs in what is called a Townsend avalanche. The multiplication factor can be large (10^3–10^5) and, if the field in the chamber is not too high, the size of the output pulse remains proportional to the amount of the original ionization. Such a chamber is called a proportional counter.

To make it easier to achieve the required field strength, the geometry is usually cylindrical, as shown in Figure 6.2, with a thin anode wire (radius a) and a concentric outer cylinder (inner radius b) forming the cathode. The electric field in such an arrangement varies radially according to the formula:

$$\mathcal{E}(r) = \frac{V}{r \ln b/a} \qquad (6.1)$$

Typical values for b and a are 10 and 0.01 mm and, with an applied voltage V of 1 kV, the field is over 10^7 V m^{-1} near the wire.

Since nearly all the ionization is produced close to the central wire where the electric field is strongest, the output pulse is independent of where the initial ionization occurs. Electron collection times (1–10 ns) are much more rapid than in the ionization chamber, which is important if timing information is required. Positive-ion collection times are still slow, however, and this limits the ultimate count rate at which the detector can be operated.

The variation of the logarithm of pulse height with applied voltage V for a gas detector is sketched in Figure 6.3. Two curves are shown in the figure corresponding to different amounts of energy deposited in the gas. Initially, the output signal increases with V as recombination of ions and electrons in the gas is reduced. This is followed by a plateau region where there is full charge collection and the output is independent of V. As the voltage is increased further, charge multiplication begins to occur and we enter the proportional region where the output pulse increases rapidly with voltage, but still is a measure of the energy deposited in the detector. At the upper end of this range, the proportionality of the output to deposited energy becomes progressively weaker until, finally, a point is reached where the output is very large, but becomes independent of the input energy. This is the region where the Geiger–Mueller counter is operated, which we now describe.

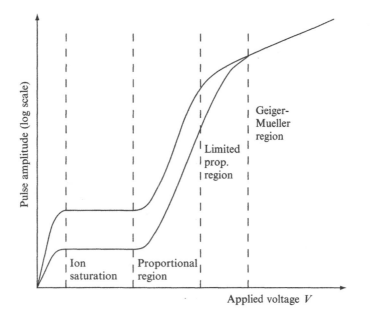

Figure 6.3 Variation of the output signal (on a log scale) from a gas-filled ionization chamber as a function of applied voltage. The two curves correspond to different amounts of energy deposited in the chamber gas.

6.2.3 Geiger–Mueller counter

A Geiger–Mueller counter uses the same electrode geometry as a proportional counter but, by increasing the applied voltage, the electric field near the wire becomes so high that electrons can gain sufficient energy between collisions to excite inner electrons of gas atoms. Ultra-violet photons from the de-excitation of these atoms are energetic enough to ionize other atoms throughout the volume of the chamber releasing more electrons. These electrons trigger further avalanches and the whole process grows until a discharge extends along the whole length of the central wire. Electrons are collected quickly, leaving positive ions behind and the discharge stops when the buildup of positive charge near the wire reduces the electric field below the critical level for Geiger action. The resulting voltage pulse is large. It can easily be counted without further amplification, but it contains no information about the amount of initial ionization.

After the discharge, the positive ions drift outwards towards the cathode to be neutralized. However, if they were to get there, there is a significant probability of electrons being released from the surface, which would trigger a further discharge in the tube. This is prevented by introducing a molecular gas, called a quenching gas, which has a high probability of charge exchange with the atomic ions of the primary gas. Thus, only molecular ions reach the cathode, and the excess energy released as they are neutralized causes molecular dissociation and not the release of free electrons. The quenching period can be hundreds of microseconds and this limits the counter for high-rate applications. Examples of quenching gases are ethanol or a halogen, and a typical gas mixture consists of 90% argon and 10% ethanol. Gradually, the quenching gas is used up, which means that the Geiger tube has a finite operating lifetime. However, it is a cheap, rugged detector and is widely used in situations where information on the energy and type of radiation is not required.

6.3   SCINTILLATION DETECTORS

The most familiar example of the use of the scintillation mechanism is the TV screen, which emits visible photons under electron bombardment in a cathode-ray tube. It is also one of the oldest techniques used for detecting radiation. In the early years of nuclear physics, Rutherford and his co-workers detected α particles using a microscope to view scintillations from a zinc sulphide screen. The modern scintillation detector followed from the production of materials that are transparent to their own fluorescent light and the development of the photomultiplier tube (PMT) as an extremely sensitive detector of this light. A schematic arrangement for γ-ray detection is shown in Figure 6.4.

There are several stages in a scintillation detector during which the energy carried by the radiation is transformed before being in the form of a signal suitable for data processing. First, after entering the block of the scintillator, the photon transfers all or part of its energy to an electron or an electron–positron pair by one of the processes described in Section 5.4. Multiple collisions with other electrons in the material then dissipate the energy into even smaller amounts with the net result that many electrons in the medium are raised to excited states. Another stage occurs when these excited states lose their energy. In a scintillator, some of the energy transferred to the medium is emitted in the form of a large number of visible or near-visible photons as fluorescent radiation. As many of these photons as possible are directed on to the photosensitive surface (photocathode) of a PMT, which releases photoelectrons.

In a PMT, a resistor chain establishes a voltage gradient between the photocathode and a series of electrodes (dynodes) further down the tube to the anode. Electrons from the photocathode are focused and accelerated by the resulting electric field and driven into the surface of the first dynode. Each arriving electron has enough energy to knock out several secondary electrons depending on the voltage applied to the tube. These secondary electrons are accelerated to the next dynode where further

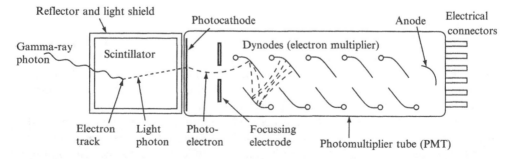

Figure 6.4   Schematic diagram of a scintillation detector.

multiplication takes place and the process is repeated through all the stages to the final anode. The gain per stage may be as high as five, giving an overall amplification for a 10-stage tube of 5^{10} or about 10^7. The result is a brief pulse of current at the output, which can be amplified and processed by subsequent electronics.

There are two basic types of scintillator material: organic and inorganic. Organic scintillators consist of molecules which interact with each other rather weakly. Energy from incoming radiation can raise electrons to higher electronic states and also excite molecular vibrations. The energy to excite an electron is generally much higher (eV) than it is to excite a molecular vibration ($\approx 0.1\,\text{eV}$). Figure 6.5(a) illustrates this with several molecular vibrational states ($S_{01}, S_{02} \cdots$ and $S_{11}, S_{12} \cdots$) built on each of two electronic states: S_0 and S_1. It shows a number of possible upward transitions to excited states, which could be populated by the absorption of energy from the radiation. The molecular excitation is lost quickly (in a few picoseconds) in transitions to the lowest vibrational state (S_{10}) of the electronic excited state S_1. A fluorescent photon is emitted when this state decays to the ground electronic state S_0 or to one of several possible molecular-vibrational states built on S_0. In only one case is the resulting photon energy sufficient to cause re-excitation. Thus, the scintillator will be transparent to a significant fraction of its own radiation, which is a very useful property for a detector.

A common inorganic scintillator is a crystal of an alkali halide, such as sodium iodide (NaI). Electrons in a crystalline solid occupy levels which are grouped into bands separated by forbidden regions called energy gaps or band gaps. The uppermost bands are called the valence and conduction bands. In an insulator, the valence band is full of electrons that are weakly bound to their individual atoms or molecules. The conduction band is generally empty. Incoming radiation causes many electrons to be excited into the conduction band leaving behind (positively charged) vacancies, called holes, in the valence band. If, during de-excitation, an electron drops back

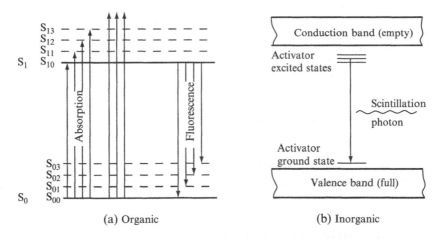

(a) Organic   (b) Inorganic

Figure 6.5 (a) Molecular energy-level structure in an organic scintillator showing electronic excitation states (S_0 and S_1) and more closely spaced molecular vibrational states (dashed lines) superimposed on each. (b) Electronic band structure in inorganic scintillator material illustrating the presence of activator states at energies within the band gap between the valence and conduction bands of the pure crystal.

Table 6.1   Properties of some common scintillators.

Name	Type	Density $(g\,cm^{-3})$	Maximum emission wavelength (nm)	Time constant (ns)	Light output[a]
Stilbene	Organic solid	1.16	384	3–8	0.20
NE 102A[b]	Plastic (organic solid)	1.03	425	2.5	0.28
NE 213[b]	Organic liquid	0.87	425	3.7	0.34
NaI(Tl)	Inorganic crystal	3.67	410	250	1.00
BGO	Inorganic crystal	7.13	480	300	0.10

[a] Relative to NaI(Tl).
[b] Nuclear Enterprises Ltd, Edinburgh, UK.

from the conduction band to the valence band, the emitted photon is energetic enough to re-excite an electron from the valence band, and the crystal, therefore, is not transparent to its own radiation. In order to make the crystal into a useful scintillator, a small amount of impurity (activator) is often added – such as thallium to make NaI(Tl). These impurity atoms provide states within the energy gap, as illustrated in Figure 6.5(b). Now, when energy is deposited in the crystal, a fraction of the excited electrons and holes migrate to the activator sites. Since the energy difference between the activator states is less than the band gap, the energy of photons emitted when these states lose their energy is below that needed to re-excite the main crystal and so the scintillator is transparent to them. The addition of the thallium shifts the wavelength of maximum emission from 333 nm in pure NaI to 410 nm in NaI(Tl). The longer wavelength is better matched to the wavelength response of most PMTs.

Properties of several, commonly used scintillator materials are listed in Table 6.1. Important properties are light output, the efficiency for stopping γ rays and the time constant, which is a measure of how quickly fluorescent light is emitted after the scintillator has been excited. Organic scintillators are effective for electrons and fast neutrons, but have low γ-ray detection efficiency because they consist mainly of low Z elements. However, they, generally, give a faster response than most commonly used inorganic scintillators and, therefore, are useful if it is important to measure the time of arrival of the radiation. Organic scintillators are available as crystals, liquid solutions and in plastic form. Inorganic scintillators generally contain high Z elements, which improves their efficiency for stopping γ rays (see Section 5.4). Bismuth germanate (BGO) is particularly effective since it contains bismuth ($Z = 83$) and has a high density; 1 cm of BGO has the stopping equivalent of about 2.5 cm of NaI.

6.4   SEMICONDUCTOR DETECTORS

Semiconductor detectors are widely used in research and industry where an accurate measurement of energy is needed. A detector consists of a single crystal of a semiconductor, usually either silicon or germanium, across which a voltage is applied between two attached electrodes. Relevant properties of these materials are listed in Table 6.2. The device operates like a solid ionization chamber. The transfer of radiation energy to the crystal creates electron–hole (e–h) pairs, as in an inorganic scintillator (see previous section). The electrons in the conduction band are mobile and, under the influence of an applied electric field, move through the crystal at a speed determined by their mobility (see Table 6.2). Vacancies or holes in the valence band also move, but in the opposite direction, in a manner analogous to that of a bubble in a liquid, which rises in the opposite direction to the gravitational force acting on the surrounding liquid. The movement of electrons and holes in a semiconductor constitutes a current and if they arrive at their respective electrodes without recombining or becoming trapped at impurity sites in the crystal lattice, the result is a pulse of current proportional to the energy deposited in the crystal.

The average energy to create an e–h pair is about 3.6 eV for silicon and 3 eV for germanium. Thus, many more e–h pairs are created per MeV than ion pairs in a gas, for which about 30 eV per ion pair is required. However, the signal size is still very small and considerable post-amplification is needed.

The total time to collect the positive charge in a semiconductor detector is, generally, much shorter than in a typical gas chamber and, therefore, a semiconductor detector can be operated at a higher rate. The drift velocity is given by $v_d = \mu V/d$, where μ is the mobility, V is the applied voltage, and d is the detector thickness. For a typical voltage gradient V/d of 5 kV cm^{-1} and using the data in Table 6.2, the time to collect holes in silicon at room temperature across 0.5 mm is about 20 ns.

The main technological requirement for a semiconductor detector is the extremely high purity (about 1 part in 10^{12}) required in the sensitive region in order to reduce trapping and recombination to a negligible level as the electrons and holes are swept through the material. In principle, a single crystal of many insulators could be used as a detector, but it is the special techniques developed for semiconductors that enable single crystals to be produced with the required purity from which a detector can be constructed.

Table 6.2   Physical properties of silicon and germanium.

	Silicon	Germanium
Atomic number Z	14	32
Mass number A	28.09	72.59
Density (g cm^{-3})	2.33	5.33
Energy gap (eV)	1.17	0.75
Electron mobility (cm^{2}V^{-1}s^{-1})		
300 K	1350	3900
77 K	21 000	36 000
Hole mobility (cm^{2}V^{-1}s^{-1})		
300 K	480	1900
77 K	11 000	42 000
Energy / e–h pair (eV)	3.62 (at 300 K)	2.96 (at 77 K)

Two types of semiconductor detector are in common use. One is called a p–n junction detector and the other an intrinsic detector. The former is usually made of silicon and the latter of germanium.

6.4.1   The p–n junction detector

This type of detector is a diode, which is formed at the boundary between two different types of semiconductor. During manufacture, impurity atoms are introduced into the semiconductor material (e.g. boron or phosphorus into silicon), which have a valency one less or one more (3 or 5) than the semiconductor atoms. Phosphorus atoms incorporated into a silicon lattice are 'donor' sites with extra (donor) electrons occupying states which lie close to the conduction band, as shown in Figure 6.6(a). These electrons are not strongly bound to these sites and are easily excited into the conduction band where they are mobile. Such material is called n-type because the mobile charge carriers have negative charge. Boron atoms, on the other hand, occupy 'acceptor' sites because they require an extra electron to form a full set of covalent bonds within the crystal. Electrons from the valence band can easily be promoted to these acceptor states, as indicated in Figure 6.6(b), leaving holes in the valence band, which are also mobile. A hole is effectively a positive-charge carrier and, hence, the material is called p-type.

When a junction is made between the two types of semiconductor, electrons from n-type material migrate across the boundary and combine with holes in the p-type material, creating a zone of intrinsically much purer (compensated) material depleted

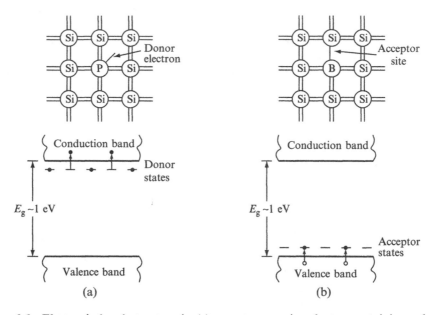

(a)   (b)

Figure 6.6   Electronic band structure in (a) an n-type semiconductor containing valence-5 atoms within the lattice, which provide donor states containing weakly bound electrons close to the conduction band, and (b) a p-type semiconductor in which valence-3 atoms provide acceptor states close to the valence band.

of either weakly bound electrons or trapping sites. The movement of the charges sets up a voltage (bias) across the junction which eventually stops the net migration. The situation is illustrated in Figure 6.7(a). The depleted zone is the region of the detector sensitive to radiation. Electrons and holes created by the radiation move under the influence of the electric field across the region and form a current pulse, which is the detected signal. The depleted zone can be made deeper by applying an external voltage V in the reverse bias direction [see Figure 6.7(b)]. The depth of the zone also depends on the purity of the material. Reverse bias increases the electric field in the depletion zone and, therefore, also increases the rate and efficiency with which electrons and holes are collected at the electrodes. However, it also gives rise to a small current, known as leakage current, which contributes to electronic noise and a consequent degradation in the signal-to-noise ratio of the output pulse. Much of this leakage can arise from thermal generation of e–h pairs in the semiconductor and is overcome by cooling the material. Thermal excitation is much more likely in germanium than in silicon because of its narrower band gap (0.75 eV) and, for this reason, germanium detectors are usually cooled to near liquid-nitrogen temperature (≈ 77 K). Silicon detectors are normally operated at room temperature unless electronic noise is seriously limiting their performance.

Silicon junction-diode detectors are commercially available in a wide range of areas and thicknesses up to several mm. They can also be made very thin (down to about 10 μm) and it is often possible to use several of them stacked together as a detector telescope, which can measure the stopping power of a charged particle passing through the stack, as well as its energy. As we shall explain in Section 6.7.1, this is important if it is desired to identify the nature of the detected particle. Silicon junction diodes are ideal for detecting charged particles and, when cooled, for low-energy photon detection. Their main advantages, relative to gas detectors, are

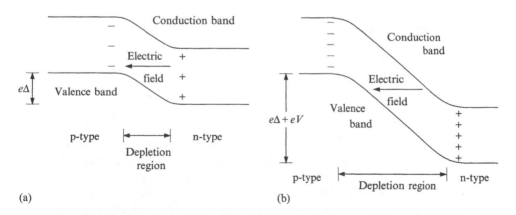

(a)   (b)

Figure 6.7 (a) Depletion region formed in a p–n junction between p- and n-type semiconductors when extra electrons from donor (n-type) material migrate to combine with holes in p-type material. The migration sets up a positive charge on the n-type side of the junction and a negative charge on the p-type side resulting in a reverse bias voltage and an electric field across the depletion region. The difference in the band energies across the junction is $e\Delta$. (b) Increased depth of the depletion zone when an external, reverse-bias voltage V is applied across the p–n junction.

their small size, faster response time and better energy resolution. Multiple detectors can be formed on a single slice of silicon with any desired geometry and they can be made position sensitive. However, all semiconductor detectors suffer from radiation damage since neutrons and other heavy particles can knock atoms from their sites creating imperfections in the crystal lattice where electrons and holes can be trapped. This means that there is a finite, integrated flux of particles of a given type which can be detected before the energy resolution becomes noticeably degraded.

6.4.2   The intrinsic detector

An intrinsic detector is made from a single crystal of hyperpure (intrinsic) material. Special contacts are made to it by which voltage can be applied and the output signal extracted. A p–i–n junction is created by making a p-type impurity contact on one side of the intrinsic (i) material and an n-type contact on the other. Lithium is diffused into the surface to form the n contact and boron is implanted for the p contact. Voltage is applied in the reverse bias direction, as on a p–n junction diode.

   The technology of the hyperpure detector is most highly developed for germanium because its higher Z makes it better than silicon for stopping γ rays, and large crystals are needed for high efficiency. Germanium detectors are available with crystal diameters and lengths up to about 100 mm. As we shall see in the next section, they give much better energy resolution than scintillation detectors, but they are more expensive and are less convenient to operate since they need to be cooled with liquid nitrogen. Also, they are more prone than scintillation detectors to radiation damage.

6.5   DETECTOR PERFORMANCE FOR GAMMA RAYS

6.5.1   Response to monoenergetic photons

Figure 6.8 shows the responses of a NaI(Tl) scintillation detector (upper graph) and a Ge detector (lower graph) to monoenergetic (662 keV) γ rays from a ^{137}Cs source. In producing the spectra, the detector output signal (pulse height) is converted electronically into an integer, which is indicated as channel number in the abscissae of the graphs. Although these pulse-height spectra are qualitatively different, for reasons we discuss below, the main features on each are the same. The peak at the upper end of each graph, called the full-energy peak, corresponds to events in which the γ ray transfers all its energy E_γ to electrons in the crystal, either in a single photoelectric conversion or by Compton scattering one or more times before undergoing photoelectric absorption. A plateau region or continuum at lower pulse heights is due to Compton scattering with the scattered γ ray escaping undetected. The upper limit to the plateau is known as the Compton edge and its energy is easily obtained from Equations (5.11) and (5.13) as the maximum energy of the recoil electrons:

$$T(180°) = E_\gamma - E'_\gamma(180°) = \frac{2E_\gamma^2}{(mc^2 + 2E_\gamma)} \qquad (6.2)$$

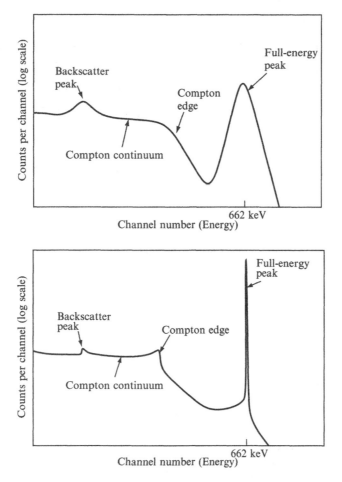

Figure 6.8 Schematic pulse-height distributions (energy spectra) recorded with a scintillation detector (upper graph) and a Ge detector (lower graph) of 662-keV γ rays from a ^{137}Cs source. The channel number is a measure of the amplitude of the signal out of the detector (pulse height), which, for the cases shown, is a measure of the amount of energy deposited in the detector.

The small peak, labelled 'backscatter peak', originates from γ rays that were Compton scattered by surrounding materials through large angles back into the detector where they then underwent photoelectric absorption.

If the energy of the γ ray exceeds 1.022 MeV, it is possible for pair production to occur (see Section 5.4.3) in the detector crystal. The electron and positron created by this process rapidly slow down and transfer their energy to the detector material after which the positron annihilates with an electron resulting in two 511-keV γ rays. If pair production is significant, two additional peaks, called 'double-escape' and 'single-escape' peaks, appear in the detector response, as is shown schematically in Figure 6.9. The former is due to events in which both annihilation γ rays escape from the crystal. It occurs at a pulse height corresponding to an energy 1.022 MeV below the full-energy peak. The 'single-escape' peak, 511 keV higher in energy, arises from events where one of the annihilation γ rays escapes and the other is captured in the

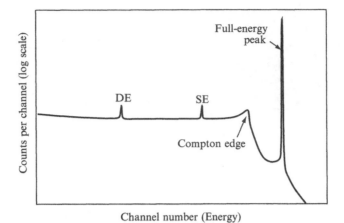

Channel number (Energy)

Figure 6.9   Schematic energy spectrum illustrating the response of a Ge detector to mono-energetic γ rays of about 2 MeV. The peaks labelled DE (double escape) and SE (single escape) are due to pair production occurring in the crystal and correspond to events in which one (SE) or two (DE) γ rays (produced when positrons annihilate in the crystal) escape undetected.

detector. If both γ rays are captured, all the energy of the original γ ray is transferred to the crystal, and the event contributes to the full-energy peak.

An ideal detector would give a single, sharp full-energy peak for each γ ray entering it. This is not possible because of the background of Compton-scattering events, which can mask the presence of other peaks. However, different detector arrangements give different detector performances, which, therefore, can be optimized to some extent. Two important factors are: energy resolution, which affects the prominence of the full-energy peak above the background, and peak-to-total ratio, which we define here as the fraction of the total number of γ rays converted in the detector that contribute to the full-energy peak.

6.5.2   Energy resolution

Energy resolution is measured in terms of the full width at half maximum (FWHM) of the full-energy peak and is expressed either in units of energy or as a percentage of the peak energy. The former is usually quoted for Ge detectors and the latter for scintillation detectors. The peak is often fitted to a Gaussian distribution of the form: $\exp[(E - E_\gamma)^2/2\sigma^2]$ where E_γ is the peak energy and σ is the standard deviation of the distribution. The FWHM for a Gaussian distribution is approximately equal to 2.35σ.

The spectra in Figure 6.8 show that the resolution of a semiconductor (Ge) detector is far superior to that of a NaI(Tl) scintillation detector. One reason for this is the difference in the statistical variation on the critical number of events which determines the signal size in each case. Taking the average energy to create an e–h pair in a Ge crystal to be 3 eV, we find that a 660-keV γ ray excites $n \approx 220\,000$ pairs; then, assuming that fluctuations in this number obey Poisson statistics, the standard deviation σ is the square root of n or about 470. Using this, we estimate the FWHM to be $2.35 \times 470 \times 3(\text{eV}) = 3.3\,\text{keV}$ or about 0.5% of the γ-ray energy.

A gas detector is sometimes used to detect low-energy photons, but its resolution will normally be poorer than that of a semiconductor detector because some tens of eV of energy are required to create an ion pair, depending on the gas. This means that the number of ion pairs will be smaller than the corresponding number of e–h pairs in a semiconductor and so the percentage energy resolution (which is proportional to $1/\sqrt{n}$) will be correspondingly greater.

In a scintillation detector, the transfer of γ-ray energy populates a large number of excited states in the crystal (see Section 6.3). However, only a fraction of these de-excite by emitting photons and there are further losses because not all these photons reach the PMT. Also, the probability that an electron will be emitted from the photocathode per incident photon (photocathode efficiency) is much less than unity. The result is that a few hundred eV of initial γ-ray energy is needed per photoelectron (see Problem 6.6) and the statistics on the total number of photoelectrons contributes an energy spread which is about an order of magnitude greater than that in the Ge detector.

There are several non-statistical factors, which significantly affect the energy resolution of these detectors. For example, the probability of a photon reaching the photocathode depends on where it originated in the scintillator, and the photocathode efficiency varies by several per cent across its surface. Both these effects worsen the resolution of the scintillation detector to a value which, typically, is 7–8%, or about 50 keV for a 660-keV γ ray. Also, Poisson statistics overestimates the spread in the number of e–h pairs in the Ge crystal, and the resolution of a Ge detector is generally considerably better than the rough estimate outlined above.

6.5.3   Peak-to-total ratio

The fraction of detected γ rays that contribute to the full-energy peak is very dependent on the size and shape of the detector and on atomic number Z and γ-ray energy E_γ because of the relative dependence on these quantities of the photoelectric, Compton-scattering and pair-production cross sections (see Section 5.4). The probability of full-energy conversion in the detector material may be close to unity for low E_γ ($\lesssim$ 150 keV), but it becomes progressively lower at higher energies as the importance of photoelectric absorption decreases relative to Compton scattering. Large crystals give a performance better than that of small ones because their greater volume-to-surface ratio increases the probability of converting any Compton-scattered γ rays within the crystal and, hence, the relative number of events contributing to the full-energy peak.

The peak-to-total ratio can be further improved by surrounding the main detector with an efficient outer detector. A Compton-scattered γ ray from the main crystal will have a high probability of converting in the outer shield, and the signal is used to veto the event, thus reducing the background. This is called an escape-suppressed detector. It is a particularly effective arrangement for improving the performance for detecting γ rays with energies above about 200 keV, which are likely to undergo Compton scattering. However, it is more expensive and requires more complex electronics than for a single, unsuppressed detector and, generally, is chosen only when the ultimate in detector performance is required.

6.6   NEUTRON DETECTORS

A neutron detector does not record the presence of a neutron directly but responds to secondary radiation (generally fast charged particles) which is emitted when the neutron undergoes a nuclear reaction in the detector medium. For slow and thermal neutrons, the (n,p), (n,α) or (n, fission) reactions on light nuclei are among those most commonly used in detectors. Many of these reactions exhibit a $1/v$ dependence at low energy, giving high cross sections for thermal neutrons. For fast neutrons of several MeV, scattering off a light target can give enough energy to a recoiling nucleus for detection. Figure 6.10 shows the energy dependence of some of these reactions.

6.6.1   Slow-neutron detection

The isotope ^{10}B is commonly used in the form of BF_3 gas inside a proportional counter. The BF_3 is both the target for the nuclear reaction and the counter fill gas. The reaction ^{10}B(n,α)^{7}Li has a Q value of 2.79 MeV, which is shared between the outgoing α particle and the ^{7}Li ion. These charged particles cause ionization in the detector gas, which gives rise to an output signal. Most of the time (94%), the residual ^{7}Li nucleus is left in an excited state at 0.48 MeV. This reduces the sum of the α and ^{7}Li kinetic energies to 2.31 MeV, which is also easily detectable.

In the ^{3}He proportional counter, the reaction used is ^{3}He(n,p)t, which has a positive Q value of 0.764 MeV. The energy released is smaller than for ^{10}B capture, but the signal is still easily detectable and the reaction cross section is almost 40% higher.

The $1/v$ dependence of these reaction cross sections has the advantage that the count rate is proportional to the neutron density at the detector and independent of

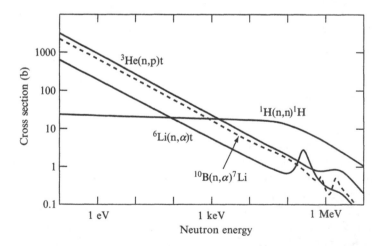

Figure 6.10   Energy dependence of cross sections for neutron reactions on ^{3}He, ^{6}Li and ^{10}B and for total scattering by H.

the functional form of the neutron velocity distribution. To see this, we consider an isotropic flux of neutrons with speeds between v and $v + dv$, interacting inside a detector containing N nuclei. If the density of neutrons in this range is $n(v)dv$, the flux at the detector will be $vn(v)dv$ and the detection rate will be

$$dR = N\sigma v n(v)dv \qquad (6.3)$$

where σ is the cross section for the reaction. The total count rate in the detector is given by the integral over v: $R = \int N\sigma v n(v)dv$ and, since $\sigma \propto 1/v$, this is proportional to $\int n(v)dv$, which is the neutron density.

6.6.2   Fast-neutron detection

At higher energies, the neutron's kinetic energy adds to the reaction Q value and increases the energies of the reaction products in either of the detectors described above. The signal from the ^{3}He detector is cleaner because it gives a single peak at full energy and also because the reaction Q value is relatively low so the effect of the neutron's energy on the signal is much greater than it is in the BF_3 detector and, therefore, is easier to measure. However, the $1/v$ cross-section dependence means that the detector efficiency becomes very low for faster neutrons.

It is more common to use a detector for fast neutrons consisting of a plastic or liquid organic scintillator. These materials have high hydrogen content and the signal comes from the energy of recoiling protons scattered by neutrons within the scintillator itself. The high density of hydrogen in the scintillator material and the larger interaction cross section means that these solid or liquid detectors are much more efficient for fast neutrons than any of the gas detectors based on neutron-induced reactions.

Following the discussion of neutron moderation in Section 5.5, we find that if monoenergetic neutrons are scattered by protons, the energy spectrum of the recoiling protons should be a continuum extending from zero, when the scattering angle is small, to full energy for a head-on collision when the neutron transfers essentially all its energy to the proton. Figure 6.11(a) illustrates the expected

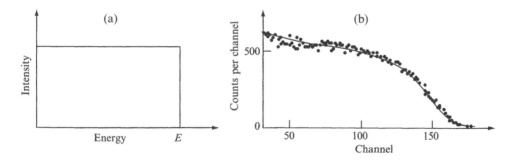

Figure 6.11 (a) Calculated energy spectrum of recoiling protons scattered by neutrons of energy E, assuming that the differential scattering cross section is isotropic in the centre-of-mass system. (b) Measured pulse-height spectrum for 2.6-MeV neutrons incident on an organic-scintillator (stilbene) detector. The solid line is a calculated response function. From Bormann *et al.* (1970).

form, assuming that the scattering cross section is isotropic in the centre-of-mass system. An example of the response of an actual detector is shown in Figure 6.11(b). It is well reproduced by a detailed calculation, given by the solid curve, which takes into account the effects of detector resolution and scintillator non-linearity. Although the detector response contains information about the neutron energy, it is difficult in practice to unfold the continuum spectrum to obtain this information if the incident neutron flux contains several different energy components.

6.7   PARTICLE IDENTIFICATION

The signal from a single, charged-particle detector generally does not determine the nature of the particle causing it. However, there are many ways the particle can be identified, which, generally, become easier the more energy it has. Indeed many complex systems using arrays of multiple detectors have been devised for particle physics research where the energy of a single violent collision can result in the simultaneous creation of a large number of fast particles all requiring identification and energy measurement. These lie outside the scope of this text. Here, we review three common methods for particle identification, which involve the use of several detectors and/or magnetic analysis and which are often incorporated into the multi-element systems referred to above.

6.7.1 $E - \Delta E$ counter telescope

A counter telescope consists of two or more detectors through which a charged particle passes in sequence, usually stopping in the last one. The fraction of the energy ΔE it loses in the passing detector(s) is a measure of the stopping power which, according to the Bethe–Bloch formula [Equation (5.2)], varies approximately as z^2/v^2 or mz^2/E, where v is the speed of the particle of mass m and charge ze. The energy E is obtained by summing the signals from all the detectors, and the product $E \times \Delta E$ is roughly proportional to mz^2. A graph of ΔE versus E gives a family of hyperbolae, each corresponding to a different value of mz^2. For light ions with sufficient energy, this often is enough to identify the ion uniquely. However, the method is limited by the finite energy resolution, particularly of the passing detector, which causes the hyperbolae to merge at low energies and at high m and z values.

6.7.2   Time of flight

A complementary technique, more suitable for slower particles, requires the extraction of timing signals from the passage of the particle through two detectors that are some distance apart. The time difference between the two signals is a measure of the particle's speed which, together with the total-energy signal, enables the mass of the ion to be determined.

The time-of-flight (TOF) technique does not require more than one detector if the detected radiation comes from a nuclear reaction induced by a pulsed beam on a thin target. Measuring the arrival time at the detector relative to the beam pulse also enables the speed to be determined. In this way, the TOF method is often used to measure neutron energies which, as noted in the last section, generally cannot be obtained from the detector signal itself. It is particularly useful at low energies where flight times are large and more accurately measurable, but where the $E - \Delta E$ method is less suitable.

Timing techniques are used in other ways to identify particles. For example, a radioactive nucleus, after stopping and signalling its arrival in a detector, may then decay after a short time, marking this event with a second signal in the same detector. The decay energy and (approximate) lifetime, obtained from these signals, are characteristic of the nuclide. This method has proved particularly useful for identifying new, highly unstable nuclides close to the boundaries of stability (see, e.g. Section 3.4.3). In this case, a sequence of signals, all carefully timed, identify a decay chain as the unknown nuclide transforms into other unstable, but known, radioactive nuclides closer to the region of stable nuclei.

6.7.3 Magnetic analysis

Where a very precise determination of the mass or energy of a charged particle is required, the best technique is to measure its deflection in a magnetic field via the action of the Lorentz force $\mathbf{F} = q\mathbf{v} \times \mathbf{B}$. Electric fields can be used but, for particles with energies greater than a few MeV, it is generally easier to cause a given deflection by using a magnetic field. In a magnetic field, the radius of curvature r is equal to mv/qB. Thus, the field disperses particles according to momentum mv divided by charge q and, hence, a measurement of r gives the energy if the particle type and its charge state are known.

A magnetic spectrometer accepts particles emerging with a given momentum from an object point and bends and focuses them to an image point after the instrument. Different momenta will form at different image points, the locus of which defines the focal plane.

Many different designs for field shapes exist, which act to focus and disperse particles along the focal plane. One of the simplest consists of a uniform magnetic field in the shape of a wedge as shown in Figure 6.12(a). Monoenergetic particles, emerging from the source point over a range of angles ($\pm\varepsilon$) defined by an entrance slit, follow different paths through the field, three of which are shown. The angle of the wedge can be arranged so that, to first order in the angle ε, the different paths re-emerge and converge horizontally to a common image point at the focal plane. It can be shown, for example, that, if the ray c enters and leaves a 90° wedge magnet at right angles to the field boundaries, the horizontal-focusing condition gives $u \times v = r^2$, where u and v are, respectively, the distances of the source and image points from the field boundary [see Figure 6.12(a)]. However, there is no vertical focusing in this case.

Modern spectrometers are designed to focus vertically as well as horizontally. This can be arranged to occur in the boundary region where the field lines are curved as

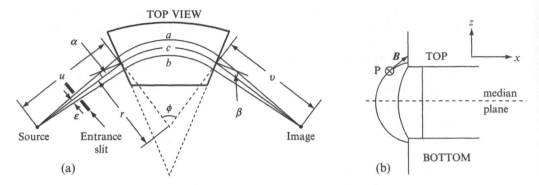

Figure 6.12   (a) Magnetic spectrometer consisting of a uniform-field, wedge-shaped dipole magnet. The ray c passes through the centre of the entrance slit, and the other rays (a and b) deviate from it by small angles ($\pm\varepsilon$) as shown. The central ray enters the magnet at an angle α to the normal to the magnet edge and its exit angle is β. (b) A side view showing field lines near the boundary region at the entrance to the magnet. A particle which has a velocity component into the page at **P** experiences a component of force in the vertical plane ($-z$ direction) due to the x component of the field **B**.

shown in Figure 6.12(b). On the median plane, the field has only a vertical component B_z (in the z direction). Above the median plane, at point P for example, the field has a horizontal component B_x normal to the field boundary as well as a vertical component. Particles entering the magnetic field of the spectrometer in the x direction are unaffected by B_x. However, if a particle enters at an angle α to the normal to the field boundary [see Figure 6.12(a)], it has a velocity component in the y direction, which is directed into the page in Figure 6.12(b). This particle will experience a force, due to B_x, directing it towards the median plane. Below the median plane, the direction of the force is reversed (since B_x is reversed) and particles again are directed back towards the median plane. Thus, vertical focusing occurs at the entrance boundary. There will be defocusing at the exit if the exit angle β is as shown in Figure 6.12(a), However, the net effect can be focusing and, by a suitable choice of angles (α, β and bending angle ϕ), it can be arranged that the vertical and horizontal image foci are coincident.

The distribution of particles arriving at different points on the focal plane is recorded in a detector, such as a photographic plate or position-sensitive proportional counter with its wire lying along the plane. The spectrometer can be calibrated enabling r and, hence, mv/q to be determined directly from the position measurement. A powerful detection system consists of a spectrometer with a focal-plane detector, consisting of a position-sensitive device, for recording a signal giving mv/q, and a gas ionization chamber divided into sections to give ΔE and E signals. Such a combination enables m, z and E to be determined for many charged particles over a wide range of energies. Timing techniques can also be used to provide further information for discriminating between different types of particle. A spectrometer is also excellent for detecting rare events in the presence of intense background if the background events are dispersed to a position on the focal plane where they can be eliminated.

6.8   ACCELERATORS

For its operation, an accelerator requires a source of charged particles and an electric field to provide the accelerating force. There are two main types. In an electrostatic (DC) machine the accelerating field is obtained by applying a high DC voltage across an accelerating region. A limit is imposed by the maximum voltage that can be attained and this type of machine is used for producing low and moderate energy continuous (DC) beams. The second type is the radio-frequency (AC) accelerator in which the accelerating field is applied in a series of relatively small repetitive steps. These machines produce beams which are pulsed in time. Examples are the linear accelerator (linac), the cyclotron and the synchrotron. They are not voltage limited and the synchrotron, in particular, can achieve very high energies. We shall not discuss high-energy accelerators based on the synchrotron principle here, since they are used almost exclusively for research into particle physics, which lies outside the scope of this text.

6.8.1   DC machines

In a single-stage DC accelerator, or electrostatic generator, a high voltage V is maintained on a terminal enclosure housing a source of ions or electrons which will form the beam. Particles are extracted from the source, focused and then accelerated by the electric field as they pass through an evacuated tube to a lower, usually ground, potential. The energy gain is equal to $q \times V$, where q is the charge on the particle. The DC voltage is obtained either by using a conventional transformer–rectifier unit for voltages below 1–2 MV or by using the charging system developed by Robert Van de Graaff in 1931, which can produce voltages an order of magnitude higher.

The Van de Graaff method is based on the familiar principle in electrostatics that any charge Q on a hollow conductor resides on its outer surface such that the electric field is zero in the interior. The electric potential of the conductor is equal to Q/C, where C is its capacity. If a charged object (A) is connected to the inside surface of the conductor (B), the charge will be transferred from A to B, irrespective of the potential of B. This process can be repeated and the potential of B raised steadily to the point where electrical breakdown begins to occur in the surrounding medium. In the Van de Graaff generator, shown schematically in Figure 6.13, charge is brought to the inside of a hollow conducting terminal on the surface of an endless moving belt of insulating material or by means of a necklace of conducting beads or bars insulated from each other and forming a flexible chain. Positive charge is put on to the chain by electrostatic induction or on to a belt by corona discharges from a mesh of sharp, metallic points or needles, which are connected to a positive high-voltage supply and are positioned either close to or in light contact with the belt. The electric field at the tip of a sharp point can be sufficiently intense to cause local ionization in the gas surrounding it. The positive ions are repelled from the points and on to the surface of the belt, which carries them to the inside of the terminal where the process is reversed. The maximum potential of the terminal is limited by electrical breakdown either in the surrounding gas or within the evacuated accelerator tube through which the

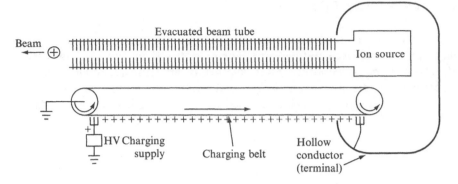

Figure 6.13   Schematic diagram of an accelerator for positively charged particles using the Van de Graaff charging principle. The charged particles are extracted from an ion source housed inside the high-voltage terminal and accelerated down an evacuated tube to ground potential.

beam particles from the ion source are accelerated. To maximize this limit, specially constructed accelerator tubes are used consisting of a series of closely spaced electrodes separated by insulating spacers. The electric field within the tube is minimized by maintaining a uniform potential gradient along its length. The probability of external electric discharge is reduced by enclosing the entire machine in a pressure vessel filled with an insulating gas with a high breakdown potential, such as sulphur hexafluoride (SF_6), at pressures up to 10 bar.

The early electrostatic generators were single-ended machines with a source of positive ions in the terminal and a single accelerator tube. They are still widely used to provide high current (mA) beams at modest energies for applications such as material modification and analysis (see Chapter 8), where a change in ion species is rarely needed. Another common use of low-energy charged particles is to produce neutrons as a secondary beam. The d(t,α)n reaction, for example, has a large positive Q value of 17.6 MeV with the reaction energy being shared between the outgoing neutron and α particle in the ratio of 4:1. In a commercially available machine, a beam of up to 100 μA of 100-keV deuterons incident on a tritium target produces over 10^8 neutrons per second at 14.1 MeV. The neutrons may be pulsed by pulsing the deuteron beam. The d(d, ^{3}He)n reaction has a lower Q value of 3.27 MeV, and the same machine can be used with a deuterium target as a source of 2.5-MeV neutrons.

A very useful development of the DC machine for higher energies is the tandem accelerator, which uses two stages of acceleration of ions from a source located outside the pressure vessel at ground potential where it is much more accessible than an ion source inside the terminal of a single-ended machine. The layout of a tandem accelerator is shown in Figure 6.14. The ion source produces singly charged, negative ions (e.g. H$^-$, He$^-$ or ^{16}O$^-$), which are accelerated towards and into the positively charged terminal. There they pass through a stripper, consisting of a thin carbon foil or a tube containing gas, which removes two or more electrons. The emerging ions, now with a positive charge ne, gain a second stage of acceleration and attain a final energy of $(1 + n)eV$. In many facilities, a double-focusing bending

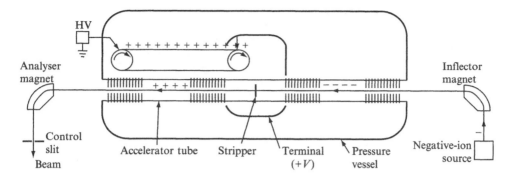

Figure 6.14 Schematic representation of a tandem Van de Graaff accelerator. Negative ions from an external ion source are injected into the accelerator to the positively charged terminal where a stripper removes several electrons. The resulting positively charged ions are accelerated from the terminal to ground potential where they are bent by a second magnet to pass between the jaws of a control slit.

magnet, similar to that illustrated in Figure 6.12, is used as a mass analyser at the low-energy end to select, focus and direct the desired ion species into the accelerator. This avoids injecting unwanted beam produced by the source into the accelerator tube where it could trigger local instability and voltage breakdown. At the high-energy end, a second double-focusing magnet is used to analyse the beam which, at the desired energy, passes through a slit between two insulating jaws. If the beam deviates off axis, there will be a difference in the current intercepted by the two jaws and this is used to form a correction signal to maintain the terminal voltage at a desired value.

Tandem accelerators provide stable DC beams with excellent energy resolution and stability to a few parts in 10^4. The beam energy and ion species are easily changed. Currents, typically, are of the order of $1\,\mu A$ but are very dependent on the ease with which a given ion species will form a negative ion. Many tandem accelerators exist worldwide. The largest currently in operation is at Oak Ridge, USA. It achieves voltages on its terminal in excess of 20 MV and at this voltage delivers beams of 40-MeV protons, 60-MeV He^{2+} ions and 180-MeV $^{16}O^{8+}$ ions. These are energetic enough for many nuclear structure investigations but, if higher energies are required, an AC type of accelerator must be used.

6.8.2 AC machines

Linear accelerator

Acceleration in a linear accelerator (or linac) is achieved by repeated application of a high-frequency (r.f.) voltage. In the Sloan–Lawrence scheme, shown in Figure 6.15, the beam passes through a series of hollow, conducting drift tubes, separated by small accelerating gaps. The tubes are alternately connected to a r.f. supply and particles gain energy if they cross the gaps when the field is in the accelerating direction. During the reverse part of the cycle, they move inside the drift tubes

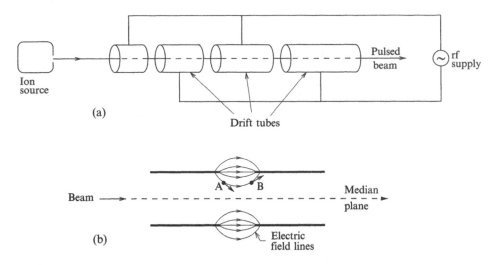

Figure 6.15   (a) Schematic representation of the Sloan–Lawrence, drift-tube linear accelerator. (b) Focusing effects on charged particles in a linear accelerator due to the electric field between two cylindrical drift tubes. The arrows at A and B show the direction of the electric force acting on positively charged particles passing through the region at those points.

where they are screened from the field. Thus, the beam is formed into bunches and is pulsed at twice the frequency of the r.f. supply (typically 30 MHz).

The length of the drift tubes must vary in such a way that the particles remain correctly in phase with the accelerating voltage This is accomplished if the ion drifts for half a r.f. cycle between one accelerating gap and the next and so the length of a given drift tube is proportional to the speed of the ion passing through it. After being accelerated across n gaps, a particle of charge q has kinetic energy $E_n = nqV$, where V is the voltage gain per gap. The non-relativistic speed of the particle (mass m) is given by $v_n = \sqrt{2nqV/m}$ and, therefore, the length of the nth drift tube L_n increases as $\sqrt{n}$. Once the particles become relativistic, v_n approaches the speed of light and L_n is approximately constant.

For successful acceleration, particles must remain focused both spatially and in phase with the r.f. voltage. Phase focusing occurs for particles which cross the gaps within a particular range of the r.f. cycle. The principle is illustrated in Figure 6.16. Assume that the linac is set so that particles which cross a given gap at a phase angle corresponding to the point a on the cycle arrive at the next gap at exactly the same phase point a'. Those which arrive at the first gap earlier (e.g. at point b) receive a smaller impulse and so arrive at the next gap relatively later at a phase point b', which is closer to a' than b is to a. Similarly, late particles arriving at the first gap at point c, receive a larger accelerating pulse and arrive at the next gap relatively earlier at c'. Particles within a range of phase, therefore, are trapped and accelerated; those outside this range are lost.

There is some radial defocusing of the beam because the acceleration occurs on an increasing part of the voltage cycle. This can be seen by reference to Figure 6.15(b), which shows the form of the electric field lines as particles cross the gap. They experience radial focusing as they enter the gap and defocusing at the exit. If the

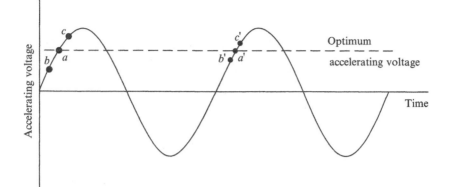

Figure 6.16 Phase stability in a linear accelerator. Particles crossing one gap at point *a* arrive at the next at the same point on the next cycle (point *a'*). Particles arriving earlier at point *b* receive less acceleration and arrive relatively later at point *b'* at the next gap. Particles arriving later at *c* receive more acceleration and arrive relatively earlier at the next gap at *c'*.

field were constant, this would lead to net focusing but, because the field is greater on exit than at entry, the net effect, in general, is to defocus the beam. Radial stability in linacs is maintained by incorporating magnetic-focusing elements within the drift tubes.

The linac is well suited for accelerating electrons which, above a few 100 keV, rapidly become relativistic with speeds close to that of light. In this case, it is more efficient to accelerate them on the crest of an electromagnetic travelling wave inside a waveguide accelerator. The electron bunches remain in phase because, once they become relativistic, their speed does not change very much as they gain energy. Radio frequency power must be fed into the waveguide at intervals to compensate for resistive losses as the wave travels down the guide and gives energy to the particles it is carrying.

Linacs are capable of producing intense beams up to high energies. For example, the linear accelerator at Los Alamos in the USA can deliver a 1-mA beam of 800-MeV protons, and at Stanford, California, a 2-mile-long linac produces 20 mA of 20-GeV electrons. A disadvantage for experiments measuring events in coincidence is the bunched nature of the beam current, which is very high during the pulse and increases the random, accidental coincidence rate compared with that due to a DC beam at the same average current. It is also much more difficult to vary the beam energy or the ion species being accelerated and, in these respects, the electrostatic generator or cyclotron (see below) is superior.

Cyclotron

In a cyclotron, charged particles are constrained to orbit repeatedly within a magnetic field as they are accelerated. In the schematic arrangement shown in Figure 6.17, particles move inside two semi-circular, conducting electrodes, called 'dees' because of their shape, which are connected to a r.f. supply. Acceleration will occur

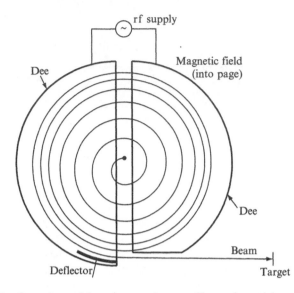

Figure 6.17   Path of accelerated ions in a cyclotron. Charged particles are injected into the machine near its centre, and beam pulses spiral outwards as they receive accelerating impulses each time they cross the gap between the 'dee'-shaped electrodes. At the maximum radius, the beam pulses are deflected outwards and emerge from the machine.

if particles cross the gap between the dees when the field is in the right direction and they will remain in phase if they reach the next gap half a r.f. cycle later.

   The cyclotron principle, conceived by E. O. Lawrence at the University of California, Berkeley, in 1929, is that, for non-relativistic particles and a fixed magnetic field B, the period of an ion orbiting in a magnetic field is independent of its speed (or energy). This important principle means that, when the particles are orbiting inside the cyclotron in phase with the accelerating voltage, they will remain in phase as they continue to be accelerated. This follows from the result, given in Section 6.7.3, that a particle, mass m, charge q and speed v, directed at right angles to a magnetic field B, will move in a circular path with radius of curvature:

$$r = mv/qB. \tag{6.4}$$

Its orbital frequency, called the cyclotron frequency, is given by

$$f = v/2\pi r = qB/2\pi m. \tag{6.5}$$

During operation, the cyclotron has to be tuned to this frequency but, once that has been done, particles remain in phase during acceleration and gradually spiral outwards to a radius R where an extraction electrode deflects them out of the machine. At this radius, the ions have their maximum speed, which, according to Equation (6.4), is

$$v = \frac{qBR}{m} \tag{6.6}$$

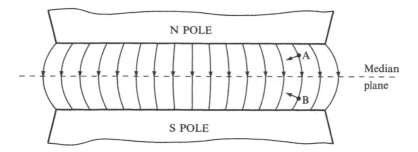

Figure 6.18 Magnetic field lines in a cyclotron in which the magnetic field decreases radially. The resultant forces on ions circulating above and below the median plane (e.g. as shown at A and B) give vertical focusing.

giving a final energy of

$$E = \frac{1}{2}mv^2 = \frac{q^2 B^2 R^2}{2m}.$$ (6.7)

For example, if $B = 1.5$ T, a proton extracted at $R = 0.3$ m would have an energy of about 10 MeV.

Although it is attractively simple conceptually, the uniform-field cyclotron has several drawbacks. One is the need for vertical focusing to avoid the loss of ions that have a component of velocity in the vertical direction. This can be achieved if the field is allowed to decrease radially as shown in Figure 6.18. The field lines then bow outwards slightly and particles orbiting above or below the median plane will experience a vertical focusing force. Unfortunately, this destroys the cyclotron principle because the orbit frequency will now depend on the radius of the orbit r, since B depends on r. This is made even worse by the relativistic increase in mass of the accelerating particles, which would require B to *increase* with r in order to maintain the orbiting frequency at a fixed value.

Both these problems can be overcome by using a magnet design that divides the magnetic field into sectors of alternating high- and low-field regions. Particles orbiting in such a field cross the field boundaries at non-normal angles, as shown in Figure 6.19. As a result, an orbiting particle experiences a vertical component to the Lorentz force, as described in Section 6.7.3, each time it crosses a boundary. These forces can be arranged to provide sufficient vertical focusing to prevent particles from being lost vertically, even though the average field increases radially to compensate for the relativistic increase in mass, noted above. Sector-focusing or AVF (azimuthally varying field) cyclotrons have been built capable of accelerating protons to nearly 1 GeV and can deliver very high-intensity beams in excess of 1 mA. Such a beam can be used to produce a copious, pulsed flux of fast neutrons for a wide range of uses. A possible application of this for nuclear waste disposal is described in Chapter 10.

The cyclotron can accelerate negative as well as positive ions. Negative-ion extraction is easily done with high efficiency simply by setting a thin foil at the extraction radius. As the ions pass through the foil, electrons are stripped off the ion and the sign of its charge changes. This reverses the curvature of the orbit and the magnetic

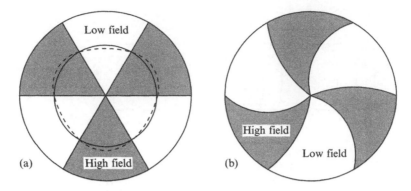

Figure 6.19   High- and low-field regions for azimuthally varying-field (AVF) cyclotron mag-
nets. In (a), an $N = 3$ straight-ridge pole profile is shown, and (b) shows an $N = 3$ spiral-ridge
pole profile. The dashed line in (a) gives the shape of a closed orbit of a charged particle
moving in the field, relative to a circular orbit (solid line).

field now helps to direct the beam out of the machine. A superconducting cyclotron
often uses this principle because the orbit spacing in them is small, making conventional extraction from a single orbit more difficult. A superconducting magnet
requires much less power than one operated at room temperature, generates a
much higher field ($\simeq$ 5T) and can be used to construct a very compact machine. A
superconducting cyclotron manufactured by Oxford Instruments accelerates protons
to about 10 MeV for radioisotope production. The extraction radius for $B = 5$ T is
less than 10 cm! The machine does require a rather bulky refrigeration system to
maintain the low temperature necessary for superconductivity, but it is still small,
the power supplies are modest and it is not too difficult to move it from one location
to another for providing short-lived radioisotopes close to the point of need.
A particular application of such a machine for nuclear medicine is described in
Chapter 9.

PROBLEMS 6

6.1 A 50-μCi source emitting 5-MeV α particles is suspended between the plates of a large
gas ionization chamber. If the emitted α particles lose all their energy in the chamber
gas, calculate the current measured at the output. Assume that the average energy to
create an ion pair in the gas is 34 eV and that all the charges produced in the chamber
are collected.

6.2 When a 4-MeV α particle is stopped in an ionization chamber, the voltage across the
detector output undergoes a step change of 2 mV. If the energy to produce an ion pair
in the gas is 34 eV, calculate the capacity of the chamber (in pF). Assume the time
constant of the system is long compared with the pulse duration.

6.3 The diameters of the wire and cylindrical outer electrode of a proportional counter
are 0.1 mm and 2 cm, respectively. If the mean free path λ of an electron in the counter

gas is $5 \mu m$ and ionization by collision begins to occur when an electron can gain 10 eV of energy between successive collisions with the atoms of the gas, estimate the applied voltage needed to achieve a multiplication factor $M = 1024$.

6.4 When a Geiger counter registers a count, a certain amount of time must elapse before it will register a subsequent count. This is known as the paralysis time or dead time τ. A Geiger counter, with $\tau = 200 \ \mu s$, is used to detect radiation from a radio-active source. (a) If the observed count rate $m = 1000 \ s^{-1}$, what is the true count rate n? (b) What would be the observed count rate if the source strength were increased by a factor of 10?

6.5 The 4.4-MeV γ rays from the decay of the first excited state of ^{12}C to the ground state are observed using a NaI(Tl) scintillation detector. (a) Explain why three peaks are observed instead of just one full-energy peak. What can you say regarding the relative heights of these peaks in relation to the size of the scintillation crystal? (b) The pulse-height spectrum of a radioactive source, known to emit only monoener-getic γ rays, shows three peaks at channel numbers 1650, 1252 and 854. What is the γ-ray energy?

6.6 You are given the following information about a scintillator: the average energy to excite an electron from the valence band to the conduction band in the crystal is 4.5 eV; 10% of the electron de-excitations result in the emission of fluorescent photons and 90% of these photons reach the photocathode of a PMT, which has an efficiency of 0.15 photoelectrons per incident photon. Calculate the radiation energy E (in eV) deposited in the scintillator per photoelectron.

The PMT has 10 dynodes, each of which has a secondary electron emission coefficient of 4. If the capacity C of the anode is 50 pF, calculate the output voltage pulse if a photon of energy $E = 200 \ keV$ is stopped in the scintillator crystal.

6.7 Estimate the detector efficiency ε for neutrons incident on a plastic scintillator detector (density $1.1 \ g \ cm^{-3}$) of thickness $t = 2 \ cm$. Assume that the detector is set to respond only to scattering from protons. The plastic consists of carbon and hydrogen only in the form $(CH)_n$. Take the neutron-scattering cross sections for carbon and hydrogen (σ_C and σ_H) to be the same and equal to 2b. Assume that the absorption cross sections are small compared with the scattering cross sections.

6.8 Two detectors, placed symmetrically on either side of a fission source, record two fission fragments f_1 and f_2 from a fission event. If the flight time for f_1 is 20% greater than that for f_2, calculate the ratio of the masses of f_1 and f_2. Which has the most energy?

6.9 In the set up of Problem 6.8, one of the detectors records a γ ray and 14 ns later the other records a neutron, both having been emitted simultaneously from the same fission event. If the separation of the detectors is 0.6 m, calculate the energy of the neutron in MeV.

6.10 For a 90° uniform-field, wedge magnetic spectrometer with normal entry and exit angles [see Figure 6.12(a) with $\phi = 90°$ and $\alpha = \beta = 0°$], show that, for equal object (u) and image (v) distances, $u = v = r$, where r is the radius of curvature of particles in the field. Hint: Draw a diagram. Consider the location of the centre of curvature of the central ray ($\varepsilon = 0°$) in the field and then consider a second ray with $\varepsilon \neq 0°$.

6.11 The path length of the central ray in the field of a 90° bending magnet is 1 m. (a) Calculate the magnetic field strength to bend a 10-MeV proton through 90°; (b) calculate the energy E' of inelastically scattered protons if they can be bent through 90° by reducing the field strength by 1%. A non-relativistic calculation is adequate.

6.12 What is the length L of the longest drift tube in a linear accelerator of the Sloan–Lawrence type, which operates at a frequency f of 30 MHz and will accelerate ^{16}O ions to 80 MeV?

6.13 Alpha particles are accelerated in a superconducting cyclotron operating with a magnetic field of 4 T. If the energy of extracted α particles is 12.5 MeV, calculate the value of the extraction radius in cm.

6.14 Calculate the cyclotron frequency in Problem 6.13.

7

Biological Effects of Radiation

7.1 INTRODUCTION

We are complex biological organisms constantly exposed to nuclear radiation. When our tissue is exposed, the primary physical interactions are similar to those in any other material, but there are important secondary effects involving chemical reactions. In addition, and most important is the fact that living tissue actively responds to any damage and has repair mechanisms which have ensured that life has been able to develop and, indeed, thrive in a low-level radiation environment over geological time.

The sequence of events leading to a radiation injury is complex and involves a number of stages during which transformations take place each with its own approximate time scale. Initial interactions occur very quickly (10^{-12} to 10^{-8} s) as radiation energy is transferred to the tissue. Some of this causes direct damage to biologically sensitive molecules and some leads to the generation of highly active chemicals which cause additional biomolecular damage in a time scale of 10^{-7} s to several hours. Further consequences occur over a period of several days to weeks and, depending on the extent of damage done earlier, may result in the death of cells and even the death of the animal itself. More subtle biological changes, leading to the onset of cancer and the transmission of genetic defects to future generations, may take many years and even centuries to manifest themselves.

The biological effects of radiation have been studied for many decades and are continually being reassessed. A good deal of information about the effects on humans has been obtained from survivors of two nuclear bomb explosions over Japan in 1945 and from major nuclear accidents. As a result, our knowledge of the results of massive human exposure is considerable. Relatively little is known about the more subtle effects of small doses. However, as we describe in the next two chapters, radiation is widely used for many purposes and it is important to assess the risk of its negative effects in order to establish acceptably safe levels of exposure for beneficial use, e.g. to patients undergoing radiation treatment, to occupational radiation workers and to the general public.

We begin this chapter with a section summarizing the initial physical and chemical processes that take place when different radiations interact with living tissue. This is followed by two sections which present first, the ways a radiation dose and its biological effectiveness are quantified and then examples of how critical parts of a living cell and organisms as a whole respond to different amounts of radiation. In Section 7.5, we describe the ways humans are exposed in different conditions to radiation from the natural environment and from man-made sources, and how doses from specific external and internal sources are estimated. Finally, in Section 7.6, we outline how the risks to humans of radiation-induced cancer and genetic damage are assessed.

7.2   INITIAL INTERACTIONS

7.2.1   Direct and indirect physical damage

The immediate result of the interaction of any radiation with matter is the deposition of energy causing ionization and excitation of atoms and molecules. As we have seen in Chapter 5, the extent to which this occurs depends very much on the nature of the radiation and its energy and we review below, for each radiation type, the relevant details pertinent to living tissue.

- *Heavy charged particles* (protons, α particles, heavy ions): These interact directly with the tissue, losing energy according to the Bethe–Bloch formula [Equation (5.2)], primarily by colliding with electrons. The density of ionization and excitation generally is high and varies with the square of the charge on the particle for a given speed. It reaches a maximum at the Bragg peak near the end of the range (see Figure 5.2). In tissue, 1-MeV α particles travel only a few tens of micrometres and thus are easily stopped by skin. However, they can cause a great deal of damage if an α-emitting isotope is ingested and more sensitive, internal organs are irradiated.

- *Electrons*: Electrons also lose energy by colliding with electrons but, being light, they are easily scattered and so, unlike heavy charged particles, do not travel in approximately straight lines through matter. Also, their energy loss per unit distance travelled is much smaller than that of α particles of the same energy, which means that they are more penetrating. A few millimetres of metal is needed to shield 1-MeV β particles. Electrons also lose energy by emitting bremsstrahlung radiation as photons (see Section 5.3), which may be absorbed elsewhere in the tissue. The net result is that the electron's energy is distributed within a much greater volume and the resulting ionization density is much smaller than that caused by a heavy, charged particle.

- *Neutrons*: Neutrons do not cause ionization directly. They are uncharged and interact only with other nuclei. Low-energy neutrons (0.025–100 eV) lose energy primarily through the (n, γ) capture reaction. Since living tissue has a high density of hydrogen atoms, mostly in the form of water, the main capture process is $n + p \rightarrow d + \gamma$, releasing a 2.2-MeV photon. At high energies ($\gtrsim$ keV), neutrons lose energy through elastic collisions. Again, because of the high hydrogen content in tissue, elastic scattering by protons predominates. Proton and neutron masses are very similar and, on average, a neutron loses a large fraction of its energy in collision with a proton (see Section 5.5). The proton recoils, producing a dense trail of ionization as it slows down. Neutrons also collide with other nuclei in tissue, such as carbon and oxygen, which then cause even greater ionization density over a very short range. A 2-MeV neutron travels about 6 cm before being thermalized and thus fast neutrons can deposit regions of high ionization density to a considerable depth in living tissue. At intermediate energies (0.1–20 keV), both scattering and capture are important. However, as the neutrons slow down, capture predominates because the capture cross section generally varies inversely with the speed of the neutron. In the energy range 0.2–0.025 eV, neutrons can also induce molecular excitation, but the biological damage this may cause is still an open question.

• *Photons*: As described in Section 5.4, photons transfer their energy to electrons via Compton scattering, the photoelectric effect and pair production. The attenuation coefficient for photons depends strongly on energy and the atomic number Z of the material. Tissue contains mainly low Z elements (C, H and O) and Compton scattering is the most important interaction process for photon energies above about 40 keV to several tens of MeV. The scattered photons may interact further by a second Compton scattering or by photoelectric absorption, depending on the energy. Overall energy deposition in general is not localized. It varies approximately exponentially with depth and can reach deep into tissue. Gamma rays of a few MeV are particularly penetrating and require many centimetres of lead to provide an effective shield.

7.2.2   Indirect chemical damage

The result of all the physical interaction processes between the tissue and the incident radiation is a trail of ionized and excited atoms and molecules. Some of the radiation energy is transferred by direct interaction on to biologically sensitive material at critical sites and, for highly ionizing particles, this direct action is an important cause of irreversible, biological damage. However, most of the primary interactions of radiation in tissue result in the ionization of simpler molecules and the creation of chemically active *free radicals*. A free radical is an electrically neutral atom or molecule that has an unpaired electron. There is a strong tendency for this electron to pair with a similar one in another radical or to eliminate the odd electron in an electron-transfer reaction. Thus, free radicals are chemically extremely reactive and can be electron acceptors (oxidizers) or donors (reducing agents). Free radicals are important since they can diffuse far enough to reach and induce chemical changes at critical sites in biological structures. It is this secondary, chemical damage which can dominate the total biological disruption arising from exposure to radiation.

Most direct ionization occurs on water since it comprises about 80% of tissue. For a proper understanding of radiobiological effects, therefore, the radiochemistry of water is of crucial importance.

Ionization of a water molecule produces a free electron and a positively charged molecule:

$$H_2O + \text{radiation} \rightarrow H_2O^+ + e^-$$

The released electron is most likely to be captured by another water molecule converting it into a negative ion:

$$e^- + H_2O \rightarrow H_2O^-$$

Both these ions are unstable and dissociate as follows:

$$H_2O^+ \rightarrow H^+ + OH^\bullet$$

$$H_2O^- \rightarrow H^\bullet + OH^-$$

creating free radicals, which we denote as $OH^\bullet$ and $H^\bullet$.

By abstracting hydrogen from organic molecules, which we represent as RH, these free radicals can generate organic free radicals ($R^\bullet$):

$$RH + OH^\bullet \rightarrow R^\bullet + H_2O$$

$$RH + H^\bullet \rightarrow R^\bullet + H_2$$

The organic free radicals may react with and disrupt other molecules that may be part of a biologically more complex system, such as a chromosome, possibly disabling it and leading to the death of a cell. Alternatively, they may modify the genetic information that is passed on to future generations (a genetic mutation). Evidently, biologically important molecules can be changed both directly, by the radiation itself, and indirectly, by the action of secondary free radicals. It is the indirect, chemical damage that accounts for the majority of the changes. However, as we shall see in Section 7.4, at low radiation levels most of this indirect chemical damage and that caused by relatively weakly ionizing radiation (photons and β particles) is reparable and does not lead to permanent damage.

Reactions with oxygen in tissue lead to other chemical reactions of biological significance because they produce damaging peroxide radicals ($HO_2^\bullet$) and hydrogen peroxide (H_2O_2). In addition, another set of processes involving organic free radicals is possible in an environment rich in oxygen:

$$R^\bullet + O_2 \rightarrow RO_2^\bullet \text{ (organic peroxy radical)}$$

$$RO_2^\bullet + RH \rightarrow RO_2H + R^\bullet$$

These reactions not only fix the molecular change, but result in another free radical which can continue the process in a chain reaction.

The increased effectiveness of radiation in the presence of oxygen is known as the *oxygen effect* and is discussed further in Section 7.4. It has the consequence that irradiated cells have a lower chance of survival in tissue rich in oxygen than in tissue less rich in oxygen. This is unfortunate for the treatment of many tumours, which generally have an inferior blood supply than normal tissue and thus are less well oxygenated.

7.3   DOSE, DOSE RATE AND DOSE DISTRIBUTION

The amount of a radiation dose depends on the intensity and energy of the radiation, the exposure time, the area exposed and the depth of energy deposition. Various quantities have been introduced in the past which attempt to specify the dose received and the biological effectiveness of a given dose. These are called the absorbed dose, the effective dose and the equivalent dose. Both old and new units will be given in this section, but, hereafter, modern units will be used exclusively.

7.3.1   Absorbed dose

One of the earliest known effects of radiation is its ability to ionize a gas. The unit, introduced for measuring this, is called the *roentgen* (R), which is now defined as the amount of exposure that will create 2.58×10^{-4} C of singly charged ions in 1 kg of air at STP. Since about 34 eV of energy is needed to produce one ion $(1.602 \times 10^{-19}$ C) in air, 1 R corresponds to an energy absorption per unit mass of 0.0088 J kg^{-1}.

Materials other than air differ in their rate of energy absorption, and the roentgen has fallen into disuse in favour of a quantity called the *absorbed dose* (D), which specifies the amount of radiation energy absorbed per unit mass of material. The original unit for this was the rad (radiation absorbed dose) equal to 100 ergs per gram (1 erg $= 10^{-7}$ J). The modern SI unit is the gray (Gy) and 1 Gy $= 1$ J kg$^{-1} = 100$ rad.

For radiation protection purposes, it is useful to define an average absorbed dose for a tissue or organ:

$$D_T = \frac{\varepsilon_T}{m_T} \tag{7.1}$$

where ε_T is the total energy deposited in a mass m_T of tissue or organ, which may range from less than 10 g for the ovaries to over 70 kg for the whole body.

7.3.2   Dose rate

Biological effects depend on the *rate* as well as the total dose delivered to the tissue. This is because mechanisms exist within organisms that enable certain molecules, such as deoxyribonucleic acid (DNA), to be repaired if they have not been too badly damaged. Thus, if a potentially lethal dose is applied at a sufficiently slow rate, it is possible for the damage to be repaired before it has gone too far.

For the same reason, when a single dose is split into smaller doses (*fractionation*), with a time interval separating each application, the biological damage is considerably reduced. For example, 10 Gy given as a single dose may kill almost 100% of a given population of cells in a radiobiological experiment, while two doses of 5 Gy each, given 24 h apart may kill only 40% of the cells. As we shall see in Section 9.6, fractionation is an important consideration in cancer radiotherapy.

7.3.3   Dose distribution and relative biological effectiveness

The biological effects of radiation depend not only on the overall absorbed dose, however it is delivered, but also on the way the energy is distributed within the tissue. As we have seen, this varies with the type of radiation and its energy.

The term *linear energy transfer* (LET) is used to specify the density of deposited energy along the path taken by a charged particle. It is defined as the mean energy deposited per unit path length in the absorbing material, in $keV\mu m^{-1}$. It includes both excitation and ionization and is directly related to stopping power. In water, which is similar in stopping power to tissue, LET varies from less than $1\,keV\mu m^{-1}$ for minimum-ionizing protons or electrons to about $100\,keV\mu m^{-1}$ for protons near the Bragg peak. Heavy charged particles at low or moderate energies constitute high LET radiation – as do neutrons because they lose their energy in tissue mainly to heavy charged particles. Fast electrons are low LET radiation. X-rays and γ rays are also considered to be low LET radiation, because the initial transfer of each photon's energy is to one or more electrons.

In ordinary matter, the spatial distribution of energy is not normally of great significance. In a detector, for example, while there may be some recombination if the ionization density is very high, the output signal mainly reflects the total energy deposited by the radiation and depends only weakly on the way the energy was distributed in the detector material. In living tissue, on the other hand, energy distribution is a major factor. Exposing the brain and bone marrow to radiation has more serious consequences than irradiating surface tissue or bone and so the effects of penetrating radiation are more severe than the same dose of a radiation which is more easily stopped. Also, the local, deposited energy density within a cell strongly influences its survival prospects. In general, highly ionizing (high LET) radiation produces more irreparable damage than the same dose delivered by radiation that is weakly ionizing.

The dependence of the consequences to an organism on the nature of the radiation leads to the concept of *relative biological effectiveness* (RBE). This is a dimensionless quantity which gives the biological response to a given radiation dose relative to that induced by 250-keV X-rays or γ rays. Thus, an absorbed dose of 1 Gy of radiation with an RBE of 2 causes as much damage (e.g. cells killed or rendered inactive) as 2 Gy of X-radiation which, by definition, has an RBE of unity. The general dependence of RBE on LET is shown in Figure 7.1. Initially, it increases relatively slowly and then more rapidly to a peak. It then begins to decrease because, for very high LET, the energy density is greater than that needed to damage the tissue beyond repair and so is 'wasted' to some extent. In practice, RBE is difficult to work with because it can be a complicated function of radiation energy. Instead, a radiation weighting factor w_R is specified, which is obtained by averaging the RBE for a given radiation over a range of energy. Table 7.1 lists values of w_R taken from a 1996 publication of the International Commission on Radiation Protection (ICRP). The ICRP is a body which, since it was established in 1928, has been responsible for assessing information on the biological effects of radiation and recommending international standards pertaining to radiation protection.

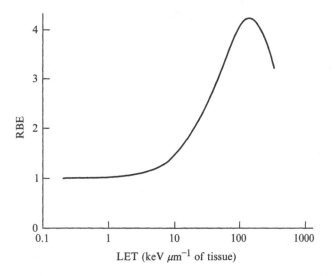

Figure 7.1   An example of the general relationship between the relative biological effectiveness (RBE) of radiation and its linear energy transfer (LET).

Table 7.1   Weighting factors for different radiations[a].

Type of radiation	Energy range	Weighting factor, w_R
Photons, electrons	All energies	1
Neutrons	< 10 keV	5
	10–100 keV	10
	100 keV–2 MeV	20
	2–20 MeV	10
	> 20 MeV	5
Protons	< 20 MeV	5
Alpha particles, fission fragments, heavy nuclei		20

[a] ICRP Publication 74, Annals of the ICRP **26** (3/4), 1996.

7.3.4   Equivalent dose

A quantity called the *equivalent dose* (*H*) is defined to indicate the biological implications of radiation exposure at levels of absorbed dose encountered in normal radiation protection. The equivalent dose H_T in a tissue or organ is given by

$$H_T = w_R \times D_{T,R} \tag{7.2}$$

where $D_{T,R}$ is the average absorbed dose in tissue T from a given type of radiation R. Thus, a 1-Gy dose given by 1-MeV neutrons ($w_R = 20$) is biologically equivalent to a 20-Gy dose of γ radiation ($w_R = 1$). If there are several types of radiation present, the equivalent dose is given by the weighted sum over all the contributions:

$$H_T = \sum_R w_R D_{T,R}. \tag{7.3}$$

The SI unit for equivalent dose is called the sievert (Sv). The old unit is the rem, which is given if the absorbed dose is specified in rads. Thus, $1\,\text{Sv} = 100\,\text{rem}$. The weighting factors w_R are dimensionless, hence, $1\,\text{Sv} = 1\,\text{J kg}^{-1}$.

7.3.5   Effective dose

The biological consequences to a person exposed to radiation depend not only on the nature of the radiation but also on whether or not the whole body is irradiated uniformly. Certain organs and parts of the body are more radiosensitive than others and are given tissue weighting factors w_T, which allow the different sensitivities to damage of the different tissues to be taken into account. Using these factors, another quantity called the *effective dose* (E) is specified, which is the sum of the equivalent doses to different tissues [Equation (7.3)] each weighted by w_T:

$$E = \sum_T w_T H_T. \tag{7.4}$$

Evidently,

$$E = \sum_T w_T \sum_R w_R D_{T,R}. \tag{7.5}$$

Recommended values for tissue weighting factors are given in Table 7.2. The values are chosen so that the calculated effective dose is numerically equal to that of a

Table 7.2   Weighting factors for individual organs[a].

Tissue	Weighting factor, w_T
Gonads	0.20
Red bone marrow	0.12
Colon	0.12
Lung	0.12
Stomach	0.12
Bladder	0.05
Breast	0.05
Liver	0.05
Oesophagus	0.05
Thyroid	0.05
Skin	0.01
Bone surface	0.01
Remainder	0.05

[a] ICRP Publication 74, Annals of the ICRP **26** (3/4), 1996.

uniform equivalent dose given to the whole body. Weighting factors are dimensionless, and the SI unit of effective dose is the sievert (Sv), as is the unit for equivalent dose.

7.4   DAMAGE TO CRITICAL TISSUES

Progressive stages of the biological consequences of exposure to radiation begin with damage to complex molecules and continue through effects at the sub-cellular, cellular and tissue levels and, for sufficiently high doses, to the ultimate survival of the organism and to possible genetic changes affecting entire animal populations. Details of these processes lie outside the scope of this text and we simply present here some illustrative examples of measured effects of radiation exposure on the activity of complex molecules and the survival probability of exposed cells. We note also the importance of certain modifying factors, such as the phase of the cell cycle and the level of tissue oxygenation.

7.4.1   Complex molecules

Proteins are complex macromolecules formed of chains of amino acids. They may have any molecular weight from 10^3 to 10^6. Some act as structural components in cells while others act as organic catalysts (enzymes) of the cell's biochemical processes.

   Proteins need much higher doses to inhibit their function than are required to kill cells and so their destruction is not the critical factor determining the death of a cell. The effects of radiation on a complex biological molecule are illustrated in Figure 7.2. The molecule is the enzyme deoxyribonuclease (DNAase), which acts on deoxyribonucleic acid (DNA) by splitting it in two. Exposing it to radiation decreases its

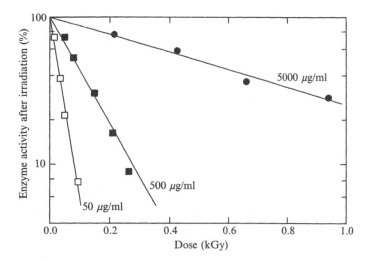

Figure 7.2   Percentage activity of the enzyme DNAase remaining after irradiation with varying doses of radiation for three different concentrations of the enzyme in solution. From Okada (1957).

effectiveness (activity), which can be measured. The graphs, which are for three different concentrations of the enzyme in solution, plot the logarithm of enzyme activity remaining after irradiation as a function of dose. In all cases, the percentage of surviving active molecules decreases with increasing dose as expected. However, the rate of decrease was found to be strongly dependent on the concentration in solution and this is considered to be an example of the importance of indirect chemical damage due to radiation. It is thought that at high concentrations, a more significant fraction of biological damage arises from the radiation acting directly on the enzyme molecules whereas, at low concentrations, an even greater amount of secondary damage is done to the enzyme by free radicals generated as a result of the ionization of water, as described in Section 7.2.2.

The general exponential dependence of 'survival' on dose can be understood in terms of a target model and the statistics of random events. Poisson statistics[1] gives the probability of obtaining a certain number of events n as

$$P(n) = \mu^n e^{-\mu}/n! \qquad (7.6)$$

where μ is the expected mean value. If we assume a simple target model in which a single damaging event or hit renders the molecule inactive, the survival probability is given by the probability $P(0) = e^{-\mu}$ that a molecule will receive no hits. The mean number of hits expected (μ) is proportional to the dose and so the survival probability will decrease exponentially with dose. The data in Figure 7.2 are in accord with this simple model.

7.4.2   Nucleic acids and damage repair

It is generally accepted that DNA is the critical molecule determining a cell's ability to survive exposure to radiation. The DNA molecule consists of two spiral strands twisted together to form the now famous, double-helical structure. During cell division (mitosis), a single DNA molecule splits in two. Each half is used as a template enabling enzymes to construct the matching half, and two DNA molecules identical to the original are the result. The DNA structure, therefore, contains the information necessary for a cell to reproduce itself and, when it is destroyed, the cell cannot survive. On the basis of simple target theory, size suggests that the probability of radiation damage would be much higher for a DNA molecule than for a much smaller protein molecule. Indeed, the size of a lethal dose reflects this expectation – being 1–2 Gy for a mammalian cell compared with several hundred Gy (see Figure 7.2) for a much simpler virus or protein.

Different levels of damage occur to the DNA in a living cell and some of it can be repaired. The evidence for this is found by comparing cell survival curves taken in different conditions. Figure 7.3 shows the surviving fraction of tumour cells irradiated with doses of X-rays or neutrons. When they are assayed for viability, a greater number of functioning cells are found if the assay is carried out after a 5 h

---

[1] See, for example, Barlow (1989), p. 29.

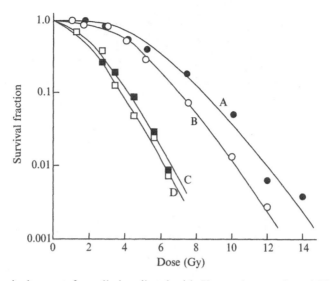

Figure 7.3 Survival curves for cells irradiated with X-rays (curves A and B) and neutrons (curves C and D). The lines are to guide the eye. For curves A and C (closed symbols) there was a delay of 5 h between irradiation and the assay to determine the surviving fraction. In the case of the culture irradiated with X-rays, this delay allowed some recovery from potentially lethal damage to take place, but not for the cells irradiated with neutrons. From Rasey and Nelson (1981).

delay (curve A) compared with a measurement made immediately after exposure (curve B). The delay evidently allows recovery from potentially lethal damage to occur for some of the X-irradiated cells. By contrast, there is little, if any, evidence of recovery of cells irradiated with neutrons (curves C and D). This is consistent with the difference in weighting factor for the two types of radiation and shows the much greater likelihood of irrecoverable, lethal damage from highly ionizing (high LET) radiation.

It should also be noted that the cell survival curves, shown in Figure 7.3, do not follow the exponential fall off with dose predicted by the simple, single-hit, target model. A dependence similar to that seen in Figure 7.3 would be the result if we assume a model in which two or more targets must receive damaging hits before the cell is killed. The probability of survival now is given using Poisson statistics by the sum of two terms:

$$P(0) + P(1) = (1 + \mu)e^{-\mu} \tag{7.7}$$

which is the probability that there is at most only one event or hit. Equation (7.7) has the form shown by the solid curve in Figure 7.4, which exhibits the main characteristics of many cell survival curves. The slope at very low doses is zero, which implies that any damage caused by low doses can always be repaired and does not kill cells. The majority of data on real cells, however, do not support an initial zero slope, suggesting that there are several different damaging events that can kill cells. A model which better describes mammalian cell survival is the multi-target/single-hit model in

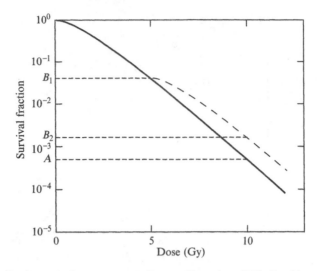

Figure 7.4 Idealized survival curves according to Equation (7.7). In this example, a single dose of 10 Gy reduces survival to level A. The effect of splitting the 10 Gy into two equal fractions separated by some time interval is such that B_1 and B_2 are the levels of survival produced by the first and second 5-Gy dose fractions, respectively.

which a cell can be killed by a single hit if it occurs at a critical place, but requires more than one hit in other places.

Figure 7.4 also shows the effect of splitting the dose in two, which is the simplest form of dose fractionation. In the example shown, a 10-Gy single dose results in a survival fraction of A. If, on the other hand, a single 5-Gy dose is given, the surviving fraction is B_1 and, if this dose is followed by a suitable interval of a few hours, the surviving cells 'forget' the initial exposure. Any non-lethal damage is repaired and the response to the second 5-Gy exposure follows the upper (dashed) curve, leading to a survival value of B_2. The recovery of the cells is indicated by the reappearance of the shoulder between the first and second doses. If no recovery had occurred, there would be no difference in the survival fraction by splitting the dose. In the case of survival curves with little or no shoulder, such as curves C and D in Figure 7.3 given by high LET neutron irradiation, little or no recovery is obtained with a split dose. This is because high LET radiation is efficient at killing cells however small the dose.

7.4.3   Modifying factors

We have already seen how the effect of a dose of radiation on a biological organism is influenced by several physical factors, namely, the size of the dose, the rate at which it is administered, the distribution of energy density (LET) and whether the dose is given wholly or in several fractions. The importance of water in the cell's environment has also been emphasized. There are, in addition, other significant factors which affect a cell's response to radiation. These relate to the oxygen effect, the use of chemicals and drugs and the phase of the cell cycle.

Phase of the cell cycle

As we have noted, a cell proliferates by mitosis, resulting in two daughter cells. After an interval of time, each of these will then undergo further division. This is known as the cell cycle and the time between successive divisions is the cycle time. Experiments show that the greatest damage is caused to cells undergoing mitosis. This explains why cells most sensitive to damage are the rapidly dividing ones (bone marrow, gonads). Least sensitive to radiation are the slow-dividing cells (brain, muscles, kidney, liver). With low LET radiation, there are variations of about a factor of two in the dose required for a given level of cell killing between the different phases of the cell cycle. However, this difference is much less marked for high LET radiation.

The dependence on the phase of the cell cycle significantly modifies the effect of dose fractionation, as is shown in Figure 7.5. Here, the relative surviving fraction of mouse bone-marrow cells is plotted against the time interval T between two 2-Gy doses of γ rays. At zero interval, the surviving fraction is lower than for any other time interval. As T is increased, so does the surviving fraction until it reaches a peak at about 5 h. Some cell repair evidently takes place promptly and, indeed, may not depend on the cell cycle. Further increases in T, however, lead to a dip at 10 h followed by an eventual recovery to a maximum surviving fraction which, in this example, is when the delay interval exceeds about 20 h. This behaviour can be understood if it is assumed that the first dose kills the most sensitive cells, i.e. those in mitosis, leaving a population of partially synchronized cells in more resistant parts of the cell cycle. Later, this population moves to the more sensitive phases and fewer survive if the second dose is applied during this period leading to the dip. Eventually, if enough time is allowed to elapse, the population will move beyond this phase and become more resistant again, resulting in the second rising part of the curve. The cells also tend to become desynchronized and, eventually, the response to the second dose approaches that of a fully recovered survival fraction. This varying response to a fractionated dose can be used to good effect if there are significant differences in cell cycle times between normal and tumour cells.

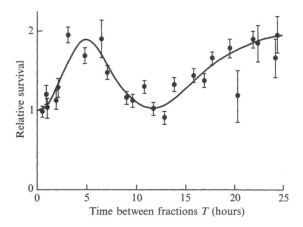

Figure 7.5   Graph showing the survival of mouse bone-marrow cells, irradiated with two 2-Gy doses of γ rays separated in time by the intervals shown, relative to the survival of cells irradiated with a single 4-Gy dose. From Till and McCulloch (1963).

Oxygen effect

The amplification of the chemical action of free radicals due to the presence of oxygen in tissue was described in Section 7.2.2 and referred to as the oxygen effect. Tissue rich in oxygen is much more radiosensitive than when it is irradiated at low levels of oxygen (hypoxic). Oxygen modifies the quantitative amount of damage done but does not alter it qualitatively. The effect is illustrated in Figure 7.6, which shows the survival fraction of cells, exposed to different radiation doses, when in the presence or absence of oxygen. The oxygen enhancement ratio is the ratio of doses required to produce a given effect in hypoxic and oxygen-rich tissue. Indications are that it may be as much as a factor of two to three, as the example in the figure shows. However, there is also a strong LET dependence. The oxygen enhancement ratio decreases as LET exceeds about 10 and approaches unity at very high LET. This is possibly because oxygen may be liberated in some quantity in dense ionization tracks and actually oxygenates and so sensitizes the cell at the instant of irradiation. Also, as noted above, high LET already destroys cells efficiently and the additional damage generated from the oxygen in tissue does not necessarily kill significantly more.

Chemical agents

Certain chemicals are known which alter the degree of cellular damage caused by radiation. Some increase it and others reduce it.

An important group of the first type includes the compounds known as 'hypoxic cell sensitizers'. These are drugs which, by having a strong electron affinity, act like oxygen and increase the sensitivity of hypoxic cells to radiation. They have little effect on well-oxygenated tissues, i.e. normal tissue.

There are also chemical protective agents which reduce the effectiveness of radiation. Examples are cysteine, glutathione and other sulphydryl-containing compounds. Some act by scavenging free $OH^\bullet$ radicals that are responsible for the

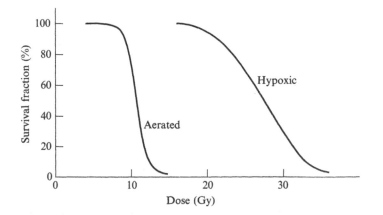

Figure 7.6   Curves illustrating the difference in survival of cells exposed to low LET radiation (e.g. X-rays), when in an environment that is rich in oxygen (aerated) or oxygen-poor (hypoxic).

indirect action of radiation. Others may transfer H atoms to biological free radicals (R•) generated by the radiation damage. The free radicals are thus repaired and converted back into their original form, RH.

Sensitizers and protectors are used in the radiotherapeutic treatment of specific cancers.

7.5 HUMAN EXPOSURE TO RADIATION

Radiation is damaging and we are exposed to it in varying amounts every day of our lives. It is important, therefore, to evaluate doses received and the consequential risks to health in order to be able to establish limits on safe levels for the general public and for radiation workers. Over the years, both the hazards and acceptable dose levels have been continually revised in the light of new research data, with the result that maximum permissible levels of radiation are considerably lower today than they were in the early days of the ICRP. In this section, we consider how to estimate the doses we receive from external and internal sources. In the next section, we indicate how the health risks, which form the basis of the ICRP's current recommendations, are assessed. First, we review the sources and levels of radiation we encounter normally in our daily activities.

7.5.1 Radiation in the environment

The main categories of radiation sources in our natural and working environments are listed in Table 7.3, together with average effective doses delivered annually to the UK population. The data were taken from a 1994 publication of the National Radiological Protection Board, which was set up by the UK Parliament in 1970 to conduct research and provide authoritative advice on radiological protection.

The largest dose we receive is from the natural background. The rest comes from a number of artificial sources, including medical and dental procedures, consumer products, fall-out from weapons testing, and discharges from the nuclear power and other industries. In addition, occupational activities can significantly increase the average for certain groups of people.

Natural sources

Natural background radiation contributes about 85% of the total annual dose of 2600 μSv. Some of it comes from cosmic radiation and some from food and drink. The greatest fraction, and the one which is the most variable, is from sources in the earth's crust.

Cosmic rays are extremely energetic charged particles (mainly protons and α particles), which rain down on the earth from outer space, interact in the upper atmosphere and produce showers of γ rays and electrons, a fraction of which penetrates to the earth's surface. Dose rates vary considerably with latitude and,

Table 7.3 Average annual effective doses to the UK population from various sources, with percentages given in brackets[a].

Source	Annual effective dose (μSv)	Range (μSv)
Natural sources (85%)		
Cosmic rays	260	200–300
Food and drink	300	100–1000
Radon	1300	300–100 000
Terrestrial γ rays	350	100–1000
Subtotal	**2210**	
Artificial sources (15%)		
Medical	370	Large for individuals
Environmental discharges	<1	Up to 150–200
Fallout	5	Up to 15
Misc. (consumer products)	0.4	1–100
Subtotal	**376**	
Occupational (0.3%)	8	Up to 20 mSv to a few individuals
Total (approx.)	**2600**	
Special groups		
In radon-prone areas	5000	
Nuclear industry	1000	
General radiation workers	500	
Medical radiation workers	100	

[a]   Source: National Radiological Protection Board (UK) 1994.

especially, with altitude. In Denver and Mexico City, at about 2000 m above sea level, rates are about three times those at sea level and, in a commercial jet airliner flying at 10 000 m, they are about 150 times higher. Frequent air travellers receive relatively high doses from cosmic rays, increasing their annual total to 3000 μSv. For some air couriers, the annual total could be as high as 8000 μSv.

Terrestrial radiation comes from naturally occurring radioactive isotopes present in all rocks and, therefore, in many building materials. The commonest of the primordial radioisotopes are ^{238}U, ^{232}Th and ^{40}K. Gamma rays from these nuclides and their radioactive daughters irradiate the whole body more or less uniformly. This natural exposure from soil and rocks varies widely over the earth's surface. Among the highest levels are those in Brazil where exposure levels on some black sand beaches may be as high as 100 μSv h^{-1}, which is over a 1000 times normal for the UK and greater than the maximum UK permissible dose rate. Even in the UK there are marked regional variations and annual doses vary from about 100 to 1000 μSv.

Internal irradiation is caused by ingesting and inhaling naturally occurring radio-nuclides of which ^{40}K, ^{87}Rb, ^{210}Pb and $^{226, 228}$Ra are important. About 300 μSv y^{-1} (per year) is from eating and drinking and, therefore, unavoidable. However, by far the most significant internal dose is from breathing in radon gas and its decay products.

Isotopes of radon (^{219}Rn, ^{220}Rn and ^{222}Rn) are part of the natural radioactive decay chains (e.g. see Section 1.5.6). Radon seeps into the atmosphere from the earth's crust or from floors and walls of buildings. Outside, it is dispersed safely, but it accumulates in enclosed spaces, such as mines and poorly ventilated buildings, where levels can be very high. It decays by α emission, and its daughters include radioactive isotopes of lead, bismuth and polonium. These attach themselves to dust particles which, when inhaled, lodge in the lungs and pose a health hazard. Radon inhalation was the cause of many deaths among early uranium miners. Radon doses in the UK vary markedly from one region to another. They are particularly high in the predominantly granite-rock areas of the southwest and in Aberdeenshire. In Cornwall, for example, radon levels are about five times the national average and the radon dose alone accounts for about 6400 μSv out of an annual total for the population there of 7800 μSv.

There are other minor sources of internal radiation from ^{3}H, ^{14}C and ^{7}Be, which are formed by the interaction of cosmic rays in the atmosphere.

Artificial sources

Table 7.3 shows that medical exposure from procedures carried out by doctors and dentists account for the vast majority of doses received by the general population from artificial sources. The average is 370 μSv y^{-1}, but individual doses may vary from zero to a few hundred times the average. Doses have increased somewhat in recent years because of the greater use of tomographic imaging, which gives higher doses than conventional X-radiography. Cancer patients on radiotherapy receive very high (lethal) doses to specific tissues and are a special group. These and other medical uses of radiation are described in Chapter 9.

Other artificial sources contribute only about 1% of the total. These include environmental discharges from the nuclear and other industries, universities and hospitals. The average dose is less than 1 μSv y^{-1}. However, a small number of people, such as heavy consumers of seafood in Cumbria, receive annual doses due to discharges from the nuclear industry of between 150 and 200 μSv. Radioactive fall-out, from the testing of nuclear weapons in the atmosphere, accounts for about 5 μSv y^{-1}, having declined from a high of 140 μSv y^{-1} in the 1960s when weapons testing was at its peak.

Many very minor doses are from consumer products. The β emitters, ^{3}H or ^{147}Pm, are used in luminous paint. Smoke detectors contain the α emitter ^{241}Am, and traces of radioisotopes are used in some dental products to add whiteness and fluorescence to porcelain in teeth and in certain ophthalmic glasses for tinted lenses and eyeglasses. Uranium oxide is used to give the colours red, yellow, brown, black and green to glazes applied to crockery. The most serious exposure due to consumer products is from smoking. Tobacco is an absorber of radon and has above average concentrations of ^{210}Pb and ^{210}Po, which are radioactive daughter products of radon. These cause problems in the lungs, as noted above, where the ^{210}Pb is retained in localized regions or 'hot spots'. The ^{210}Po compounds are generally volatile and not retained. The local effective doses can be high and probably contribute significantly to the causes of smoking-related lung cancer.

Occupational exposure

The average occupational exposure to the whole population is relatively low, but it falls disproportionately on a few special groups who, in the course of their work, receive additional amounts comparable with and even greater than the total national average. About 50 000 people have workplaces in radon-prone areas. Their additional doses are about 5000 μSv y^{-1} on average. About 24 000 aircrew workers average about 2000 μSv y^{-1} from cosmic rays. Nearly 160 000 workers are exposed to artificially produced radiation which is used in their work. Very few of them receive doses greater than 20 mSv y^{-1}. Approximate average doses, in μSv y^{-1}, are: 1000 for the nuclear industry, 500 for general radiation workers and 100 for medical practitioners.

7.5.2   Evaluating the dose

The absorbed dose D depends on the intensity of the radiation, exposure time, the energy of the radiation and the fraction f of the radiation energy that is transferred locally to the tissue. In general, the dose is difficult to evaluate. Any calculation must take into account the energy and tissue dependence of f and time variations in the flux. For heavy charged particles, f is essentially unity. Energy is absorbed according to the stopping power and closely follows the form of the Bragg curve (Figure 5.2) along the path of the radiation. In the case of fast electrons, there is a correction because part of the energy loss is radiated as bremsstrahlung, which may not all be absorbed locally. However, for electron energies less than a few MeV in tissue, this is not a large effect. Loss of photons from a beam of γ rays or X-rays is determined by the mass attenuation coefficient. In tissue, photon attenuation is dominated by Compton scattering over a wide range of energies, in which case only a fraction of the photon's energy is transferred locally to the tissue at the point of interaction. The scattered photon may be absorbed elsewhere or it may escape altogether.

External radioactive source

In addition to the above considerations, the dose received from exposure to an external source emitting radiation in all directions also depends on its distance and any shielding which may be in place. Alpha and β rays are relatively easy to shield and this is normally done to minimize any accidental exposure. Shielding γ radiation has been discussed briefly in Section 5.4.4 (see also Problem 7.12).

A useful, approximate formula for estimating the rate at which a dose is accumulated in tissue at a distance r metres from a radioactive source, emitting $\mathcal{A}$ million photons per second of energy (E_γ), in MeV, is the following:

$$\text{Dose rate } \frac{\mathrm{d}D}{\mathrm{d}t}(\mu\text{Sv h}^{-1}) \approx \frac{\mathcal{A}(\text{MBq}) \times E_\gamma(\text{MeV})}{6 \times [r(\text{m})]^2}. \tag{7.8}$$

The approximate linear dependence on E_γ is valid from less than 0.1 MeV to several MeV.[2] Caesium-137 is widely used as a source of γ rays. The γ-ray energy is 662 keV and for a 10 μCi (0.37 MBq) laboratory source, the estimated dose rate at 2 m is approximately 0.01 μSv h^{-1}. This is safe enough at this distance. However, one should remember that distance is an important way of minimizing dose when handling radioactive material since the dose rate [Equation (7.8)] is inversely proportional to the square of the distance.

Inhaled and ingested radioactivity

In the event of an accident, a major risk to local personnel and to the population at large can arise from breathing air and eating food contaminated by radioactive material released into the environment. There will be severe damage to an individual if an excessive amount is inhaled or ingested. In addition, there will be varying risks of delayed effects to people who take in amounts too small for immediate symptoms to be apparent. Assessing these risks is discussed in the next section.

An internal source of activity $\mathcal{A}$ and radiation energy E_R emits energy at a rate of $\mathcal{A}E_R$ and, if f is the fraction of the energy deposited in an organ or tissue of mass m_T, the dose rate is given by

$$\frac{dD_T}{dt} = \frac{\mathcal{A}E_R f}{m_T} \tag{7.9}$$

As an example, consider the dose we all receive from the naturally occurring radioisotope ^{14}C in our tissue. Our bodies contain about 12 kg of carbon of which a little over 1 part in 10^{12} is ^{14}C. This isotope has a half-life of 5730 years and emits β^- particles (electrons) with an average energy of about 50 keV. There are no γ rays emitted. The ^{14}C activity is about 250 Bq kg^{-1}.

We can assume that all the energy is absorbed locally (i.e. $f = 1$), giving a total rate of energy absorption of $12 \times 250 \times 50 \times 1.6 \times 10^{-16} = 2.4 \times 10^{-11}$ watts. If we take m_T to be 70 kg and $w_R = 1$, this gives an equivalent dose rate of

$$\frac{2.4 \times 10^{-11} \text{ (watts)} \times 3.15 \times 10^7 \text{ (seconds per year)}}{70 \text{ (kg)}} \approx 11 \ \mu\text{Sv y}^{-1}$$

This is very small compared with the total background dose. We receive a much larger dose (over 100 μSv y^{-1}) from the ^{40}K present in our bodies (see Problem 7.7).

In general, internal doses are not simple to estimate. As noted above, some of the emitted energy may not be absorbed and the activity will vary with time, depending on both the radioisotope's half-life and the biological half-life. The latter is the time taken for the body to eliminate half the radioactive material from itself and it can be much less than the nuclear half-life. Furthermore, certain nuclides tend to concentrate in specific parts of the body, e.g. iodine in the thyroid, and so doses to different organs will not be the same.

---

[2] See, for example, Marion and Young (1968), p. 99.

Table 7.4   Effective doses due to several radionuclides if taken internally[a].

Isotope	Principal organ(s) affected	Inhalation dose (μSv/kBq)	Ingestion dose (μSv/kBq)
^{90}Sr	Red bone marrow, lungs	30	28
^{131}I	Thyroid	11	22
^{123}I	Thyroid	0.11	0.21
^{137}Cs	Whole body	6.7	13

[a] Radionuclide and Radiation Protection Handbook, 1998.

Table 7.4 gives some examples of organs affected by several important radio-nuclides and the effective dose (μSv) resulting from an intake of 1 kBq in each case. A few specific examples of doses from ingested radionuclides are considered in the problems at the end of this chapter.

7.6   RISK ASSESSMENT

Detrimental effects of radiation are referred to as 'somatic' if they affect the exposed individual or 'hereditary' if they affect his or her descendents. Somatic effects are of two types:

(a) Deterministic, where there is a threshold below which no detrimental effects are seen and the severity depends on the dose: e.g. damage to blood vessels, induction of cataracts or impairment of fertility.

(b) Stochastic, where there is apparently no threshold and where the severity of the damage is independent of the size of the dose causing it. For example, the most important delayed effect of radiation is cancer. It is assumed (without complete justification) that cancer risk varies linearly with dose and that there is a finite risk even at very low doses. However, with increasing dose, the severity of the cancer itself does not increase, but rather there is a greater probability that it will develop.

Calculating the risk of the induction of a stochastic effect (predominantly cancer) involves risk factors that vary with different parts of the body. The ICRP has issued a list of probabilities for radiation-induced fatal cancer, some of which are given in Table 7.5. From this table, one can obtain an overall risk per sievert of irradiation to the whole body and the fractional risks that different tissues may contribute to it. The latter enable estimates to be made from partial-body irradiation, especially when a radioactive nuclide is ingested and concentrates in a particular organ.

There is a great deal of uncertainty in these risk estimates because they are largely based on extrapolated data, and different sets of published data vary considerably depending on how the extrapolation was done. Also, it is important to note that these are average values intended to give rough estimates for populations at large. The risk to an individual or group of individuals will depend on factors such as age, sex and race. For example, the risk of breast cancer in men is virtually zero but, for women, it is about 4×10^{-3}Sv^{-1}, resulting in an 'average' value of 2×10^{-3}Sv^{-1}.

Table 7.5 Risk factors for radiation-induced fatal cancer[a].

Tissue or organ	Effect	Probability[b] per Sv
Breast	Cancer	2.0×10^{-3}
Red bone marrow	Leukæmia	5.0×10^{-3}
Lung	Cancer	8.5×10^{-3}
Thyroid	Cancer	8.0×10^{-4}
Bone surfaces	Cancer	5.0×10^{-4}
Other tissues	Cancer	3.4×10^{-2}
Whole body, all cancer effects		5×10^{-2}

[a] ICRP Publication 74, Annals of the ICRP **26** (3/4), 1996.
[b] Values relate to a population of equal numbers of both sexes and a wide range of ages.

The total overall risk probability is 5×10^{-2} Sv^{-1} which means that, on average, 5 in 100 people exposed to an effective dose of 1 Sv will contract a fatal cancer as a result at some point in their lives. This estimate is likely to be most realistic if the equivalent dose is distributed more or less uniformly throughout the body. The tissue weighting factors, given in Table 7.2, allow for non-fatal damage and hereditary conditions as well as fatal cancer, and estimates based on the effective (whole body) dose or the equivalent dose to an organ can differ considerably.

As an example, suppose a person ingests a total activity of 10 kBq of ^{131}I, which concentrates in the thyroid. The effective whole-body dose, from Table 7.4, is 2.2×10^{-4} Sv and the equivalent dose to the thyroid (weighting factor 0.05) is 4.4×10^{-3} Sv. Using the equivalent dose, the estimate of the risk of a fatal cancer of the thyroid is $4.4 \times 10^{-3} \times 8 \times 10^{-4} = 3.5 \times 10^{-6}$ or about four out of a million people exposed in this way. However, using the effective dose and the whole-body risk factor gives $2.2 \times 10^{-4} \times 5 \times 10^{-2} = 11 \times 10^{-6}$, which is about three times larger. The difference reflects the fact that the ratio of fatal to non-fatal cancer risk is much less if the dose is to the thyroid than if the whole body is uniformly irradiated. Therefore, calculating the effective dose and using the whole- body risk factor gives an unrealistically large estimate since, in this case, we know that the radioactive iodine was concentrated in the thyroid.

7.6.1 Risk to occupationally exposed workers

Effective-dose limits have been recommended by the ICRP for three categories of people: radiation workers, trainees (who spend part of their time working with radiation) and the general public. They have recently been adopted into UK legislation and are summarized in Table 7.6. The whole-body limit of 20 mSv y^{-1} for a radiation worker is set so that, in practice, an overall occupational safety level is maintained for those workers, which is consistent with that attained in other occupations (see Table 7.7).

Table 7.6   Recommended effective dose limits[a].

Tissue	Radiation workers	Trainees	General public
Whole body (mSv y^{-1})	20	6	1
Skin (mSv per dose)	500	150	50
Eye lens (mSv per dose)	150	50	15
Fœtus (mSv per pregnancy)	1		

[a] The Ionizing Radiations Regulations 1999, No. 3232.

Table 7.7   Annual fatal accident rates in some UK industries and estimated fatal cancer risk for radiation workers.

Industry	Deaths per million workers per year
Deep sea fishing	2500
Coal mining	250
Construction	200
Average radiation worker[a]	50
Textile	25
Clothing and footware	3
All employment	50

[a] See text.

In order to compare risks, Table 7.7 lists annual fatal accident rates in some UK industries and those estimated from the cancer risk to radiation workers. The latter is based on an assumed annual effective dose of about 1 mSv y^{-1} for the special group of workers in the nuclear industry (see Table 7.3) and an overall risk factor of 5×10^{-2} Sv^{-1}. At this level, the average annual risk of death is comparable with other 'safe' occupations.

PROBLEMS 7

7.1 Define the terms: absorbed dose, equivalent dose and effective dose. An equivalent whole-body dose of 2 mSv is received from γ irradiation. Calculate the total energy absorbed in joules (mass of body = 70 kg).

Calculate the absorbed, equivalent and effective doses if 30 μJ of energy is deposited in the liver (mass = 1.7 kg) using (1) γ rays or (2) 10-MeV neutrons.

7.2 During a diagnostic X-ray, a broken leg with a mass of 5 kg receives an equivalent dose of 0.5 mSv. If the X-ray energy is 50 keV, how many X-ray photons were absorbed?

7.3 Equation (7.7) gives the survival probability for an irradiated cell based on a model which allows a cell to repair a single break of a molecular bond in DNA, but not two breaks. What would be the survival probability if cells were irradiated with a dose such that the mean number of bond breakages per DNA molecule is unity? Calculate the survival probabilities if 10 times this dose was given either (1) as a single large dose or (2) as 10 equal fractions with enough time allowed for cell repair to take place between each fraction.

7.4 Show that if a radionuclide, with a physical half-life $t_{1/2}$, is ingested and has a biological half-life $t^b_{1/2}$, the effective half-life $t^e_{1/2}$ for the removal of the radioactive nuclide from the system is given by $t^e_{1/2} = (t^b_{1/2} t_{1/2})/(t^b_{1/2} + t_{1/2})$.

7.5 Show that, in Problem 7.4, the integrated dose delivered to the system is given by

$$D = \frac{t^e_{1/2}}{\ln 2}\left[\frac{dD}{dt}\right]_0, \text{ where } \left[\frac{dD}{dt}\right]_0 \text{ is the initial dose rate.}$$

7.6 You accidentally drink tritiated water and absorb 1 Ci of tritium. By drinking beer you reduce the biological half-life to only 3 days. Using the result of Problem 7.5, calculate the absorbed dose you receive, assuming the tritium is distributed throughout the whole body (mass $M = 70$ kg).

Tritium decay: t $\rightarrow$ ^{3}He $+ \beta^- + \bar{\nu}$, $Q = 18.6$ keV, $t_{1/2} = 12.3$ years. There are no γ rays. Assume the average β energy is $Q/3$.

7.7 On average, 0.27% of the mass of the human body is potassium of which 0.012% is radioactive ^{40}K ($t_{1/2} = 1.25 \times 10^9$ y). Each decay releases an average of 0.5-MeV β and γ radiation, which is absorbed by the body. Calculate the absorbed dose over a lifetime (70 years) from this source of background exposure.

7.8 By mistake, a person holds a $10\,\mu$Ci cm^{-2} source, emitting 5-MeV α particles, in contact with his skin for 30 s. Estimate the skin equivalent dose in sieverts given that the range R' of 5-MeV α particles in tissue is 3.7 mg cm^{-2}.

7.9 Show that the approximate formula for estimating the dose rate for a source of 1-MeV γ rays is given by Equation (7.8) if the linear attenuation coefficient μ in tissue (density $\rho \approx 1$ g cm^{-3}) for 1-MeV γ rays is 0.07 cm^{-1} and you assume an average thickness x of tissue presented to the rays of 20 cm.

7.10 Estimate the strength of a ^{60}Co γ-ray source which would give a dose rate approximately equal to that of the background (2.6 mSv y^{-1}) to a person working at a distance of 1.5 m from the source; ^{60}Co emits two γ rays per disintegration with energies of 1.17 and 1.33 MeV.

7.11 How many hours of flying a week at a height of 10 km would give a person twice the effective weekly dose due to natural sources of background radiation compared with that received by a person at sea level? Cosmic-ray background dose rates are: 0.03 μSv h^{-1} at sea level and 5 μSv h^{-1} at 10 km.

7.12 Estimate the thickness x of a suitable lead shield which would enable a radiation worker to operate 2 m from a 3-Ci ^{137}Cs γ-ray source for up to 8 h a day, 5 days a week and accumulate only 10% of his recommended effective dose limit. ^{137}Cs emits one 662-keV γ ray per disintegration. Lead: density $\rho = 11$ g cm^{-3}; mass attenuation coefficient for 662-keV γ rays $= 0.1$ cm^2 g^{-1}.

7.13 Given that the dose to the whole body (70 kg) from ingesting 1 kBq of ^{137}Cs is 13 μSv (see Table 7.4) and using the result of Problem 7.5, estimate the biological half-life of ^{137}Cs in the system ($\ll$ physical half-life). Assume that the ^{137}Cs is distributed uniformly throughout the body and that the body absorbs 75% of the γ-ray energy (0.662 MeV per disintegration) and 100% of the β-ray energy (about 0.2 MeV per disintegration).

7.14 A standard course of post-operational γ or X-ray therapy, normally given after operations to remove breast cancer, consists of daily fractions of 2 Gy, 5 days a week for 5 weeks. Estimate the risk of breast-cancer induction from this treatment.

7.15 Given that the world population (approximately 6 billion) receives an annual effective dose of 2.6 mSv from background radiation, estimate the incidence of fatal cancers per year due to this source and compare it with the actual incidence of cancer, which accounts for approximately 30% of all deaths. Assume an average human lifespan of 60 years.

7.16 The Chernobyl disaster may have affected about 200 million people in Eastern Europe. Assume that each person ingested 1 μCi of ^{90}Sr, which principally affects the bone marrow, 1 μCi of ^{131}I, which concentrates in the thyroid, and 2 μCi of ^{137}Cs. Calculate the effective dose each person receives and estimate the number of resulting cancer deaths.

7.17 As a result of an accident, it is found that a radiation worker's bones and bone marrow are permanently contaminated with ^{239}Pu to an activity level of 0.5 Bq g^{-1}. Calculate the absorbed and equivalent doses per year due to the contamination, assuming that all the decay energy E is absorbed. ^{239}Pu has a long half-life (2.4 $\times$ 10^4 years) and emits an α particle with a decay energy of 5.25 MeV.

   Is the accident likely to affect his ability to continue to be employed as a radiation worker?

7.18 Within a given year, a radiation worker received a low-energy (<10 keV) neutron whole-body dose of 2 mGy, and ingested 100 kBq of ^{90}Sr. How many MBq of ^{131}I could he be allowed to ingest as treatment for a thyroid problem if he is to avoid exceeding the recommended effective-dose limit?

8

Industrial and Analytic Applications

8.1 INTRODUCTION

In this chapter, we begin to address the main focus of Part II of this book which is to explore and discuss ways whereby the atomic nucleus has come to affect our lives and our technological society. Later chapters deal specifically with the major areas of nuclear medicine, energy from fission and fusion, and nuclear astrophysics. Here, we give a brief introduction to the prolific range of uses to which nuclear technology is applied in industry, other sciences and the arts. The aim is not to attempt a comprehensive coverage, which would be impossible in a text of this nature, but rather to illustrate, with sufficient examples, the diversity of ways in which nuclear physics and nuclear physics methods are exploited today.

We begin in Section 8.2 with an outline of a number of general applications with examples taken mainly from industry. Later sections contain brief descriptions of several specific, analytical procedures for determining the compositions of a wide range of materials and objects. We conclude the chapter with a short discussion of the way we assign significance to a measurement when the signal is comparable with or masked by background.

8.2 INDUSTRIAL USES

8.2.1 Tracing

Many applications of nuclear physics exploit the high sensitivity by which nuclear radiation can be detected and one of the most wide ranging of these is radioactive tracing. The tracer principle was introduced by George de Hevesy in the 1940s for which he was awarded the Nobel prize. Radioactive nuclides in extremely small, 'tracer' amounts are added to non-radioactive substances of interest. Their pathways into a complex system are then followed by detecting the radioactivity as it appears in different locations. Often, the radioactive tracer chosen is an isotope of the main element being studied because the behaviour of the tracer mimics that of the ordinary stable nuclide to which it is virtually chemically identical. The great advantage of tracing is that, with a suitable choice of radioactive isotope, it can usually be done with minimal disruption to the system, unlike chemical methods which require relatively large numbers of atoms and generally involve some disturbance. Tracing is widely used as a research tool in areas as diverse as medicine, chemistry, engineering, agriculture, metallurgy, geology, zoology and criminology. We will cite a few examples to illustrate the power of the method.

The earliest applications of tracing were in diagnostic nuclear medicine where the importance of having minimal effect on the physiological function under investigation was paramount. The first specific use, carried out in the 1940s, was to study iodine kinetics of the thyroid. Radioactive iodine was given to a patient and the buildup of its activity in the thyroid was recorded in a radiation detector placed nearby. The radioiodine had a half-life short enough that the patient received only a small dose of radiation. For the first time, it became possible to measure metabolic function directly in a living person. Nowadays, radioactive tracing has become an indispensable and sophisticated diagnostic tool in medicine. Further details of advances in this area, particularly of imaging techniques, are described in the next chapter.

Tracing also finds broad application in farming and agriculture. Many elements are required by farm animals in trace amounts and it is important to discover how effectively they are taken up in certain compounds. By administering vitamins, labelled with a radioisotope, it was possible to measure their rate of transfer, for example, to milk in cows and from a hen to the contents of its egg. Isotopic tracer studies enabled livestock feeders to improve efficiency by optimizing the calcium-to-phosphorus ratio in the diet and by reducing or eliminating certain substances which were found adversely to affect absorption of important nutritional elements. In agriculture, many studies have been carried out simply and effectively using radioactive-labelled compounds. For example, the protection of trees from leaf-eating insect larvae is an important problem, which was investigated using pesticides labelled with radioactive ^{35}S. The rate of accumulation of the pesticide in the leaves was measured by detecting the buildup of activity and without otherwise interfering with the plant in any way. The simplicity of the method enabled many different pesticides and methods for applying them to be compared cheaply and efficiently. Similarly, the migration rate of labelled fertilizer into soil was established by measuring how quickly the activity spread from the point of application. The rate was found to be much greater than had previously been supposed, an observation which led to simplifications in the way fertilizers were spread on the land and with considerable saving in cost.

The volume of liquid in a complex, closed system can be determined straightforwardly with a tracer. A measured volume of known activity containing the tracer is introduced and allowed to mix thoroughly. The dilution factor is determined by measuring the activity of an extracted sample and comparing it with that of a similar volume of the original material. The dilution factor multiplied by the original volume gives the unknown volume. An example of its use might be to measure the volume of blood in a human or animal.

The wear on a machine component is rapidly and sensitively measured by activating a thin layer on its surface. An example is an internal combustion engine where measurements on the cylinder wall or a piston ring may be required. A radioactive layer of known thickness and extent can be created in the surface by particle bombardment. The activity accumulating in the lubricating oil as the engine is run measures the wear taking place and enables the relative effectiveness of different types of oil for minimizing wear to be assessed easily and accurately. A similar method has been used to obtain information on tyre wear by introducing the radio-active isotope ^{32}P into the outer layer of the tread. Emitted γ rays were counted in a detector placed in the frame above the tyre and the decrease in counting rate recorded as the vehicle was driven under different conditions of load, speed, road type, etc.

The integrity of manufacturing processes can be monitored by using a tracer. For example, mercury vapour is a health hazard in plants manufacturing fluorescent lamps. By mixing a small amount of ^{197}Hg with the mercury, the level of vapour in the air of the working environment can be continually monitored by passing air samples over a Geiger counter.

8.2.2 Gauging

The combination of a radioactive source and radiation detector offers a simple, non-contact way of measuring and gauging dimensions and properties of materials. Nuclear radiation is slowed down and/or attenuated as it interacts with matter and, thus, the thickness of a piece of material, for example, can be determined by measuring the amount by which β particles or γ rays are attenuated or α particles lose energy in passing through it. The amount of material traversed can be calculated from the measurement by using known data on the interaction of the radiation in the material under study. Alternatively, the method can be calibrated by measuring the effect with known thicknesses of material. Thicknesses can be measured ranging from very thin films, such as ink on paper, to thick coal seams or slabs of hot steel.

Thickness gauging is routinely used in controlling the manufacture of films and sheeting. With a suitable source and detector, it is possible to detect a deviation from the required thickness in a short time and then feed back an appropriate correction signal to adjust the rollers determining the thickness of the film or sheet. The time required to make a measurement depends on the counting rate, the precision with which control is required and the sensitivity of the measurement to a given change (see Problem 8.1).

Levels of liquids or materials in closed containers can be determined by placing a source and detector at the same height on opposite sides. A drastic change in count rate is noted when the level moves above the line between the source and detector. This technique is particularly useful when it is undesirable to open a container because it contains hazardous substances. For example, fragments of fuel rods from nuclear reactor waste are loaded into drums and shipped to a reprocessing site for treatment. On arrival it is important to know if the nature and distribution of the contents are as they should be. The height of the material in the container is determined as described above and, from the weight, the average density can be established. A refinement of the absorption technique measures the differential attenuation of more than one γ-ray energy through the material. As described in Section 5.4, the γ-ray attenuation coefficient varies with Z in a way that is energy dependent. Therefore, γ rays of different energies will be absorbed by different amounts by different materials. This can be calculated and compared with the measurement on the container, thus providing a valuable check that the actual contents are as expected.

8.2.3   Material modification

Radiation alters the nature of materials and there are many examples where it is used in industry to improve the quality of manufactured goods or reduce production costs. Products range from computer disks, tyres, cables and plastics to hot water pipes. Under irradiation from penetrating electrons or γ rays, showers of low-energy electrons are liberated in the material. These may either break molecular bonds or create chemically active sites which introduce new forms of chemical bonding, such as cross-linking of long-chain organic molecules (polymers). Radiation is used to cure special kinds of coatings such as inks or paints on metal, paper or plastic surfaces. The irradiations are often carried out with small accelerators. Electron beams up to 750 keV are applied to thin films, and an exposure of only a few seconds is enough to cross-link the polymers and cure the coating. The advantages are that it may be less damaging than heat curing and it can be accomplished quickly, thereby improving efficiency by reducing the overall time needed for the coating process.

A related use of radiation in industry is to create materials with special properties. An example is the production of wood – plastic composites. The process involves injecting a liquid containing small organic molecules (monomers) into porous wood and exposing the material to radiation. This causes the monomers to link together into a network of larger molecules or polymers. The resulting plastic wood is hard and durable. Radiation improves the temperature stability of polythene which, without radiation-induced cross-links, melts at temperatures between 100° and 120°C. One of the most successful industrial applications of this is the manufacture of heat-shrink materials. A cross-linked polythene film will exhibit rubber-like properties when it is taken above its crystalline melting point. However, if it is held in a stretched or shaped form after being heated above this point, it will stay in this shape provided that it is not released until after it has cooled down. If it is then reheated, it reverts back to the shape it had before it was stretched. This property, which is called the memory effect, is given to heat-shrinkable wrapping films and shrinkable connectors for pipes, splices and tubes. Heat-shrinkable tubes have even been made delicate enough to connect human blood vessels.

A well-established, commercial application of charged-particle beams from small accelerators is ion implantation. Particles implanted into metals form alloys in a thin surface layer and produce dramatic improvements in hardness and resistance to wear and corrosion. Ion species used include boron, nitrogen, carbon, titanium, chromium and tantalum. A typical implantation dose is around $2 \times 10^{17} \text{cm}^{-2}$ and requires high beam currents. Energies typically vary between 50 and 200 keV. Traditional hardening processes require high temperatures, which can cause deformation. Ion implantation, on the other hand, is a relatively cold process. The implanted layer is integrated into the substrate whose properties may be 'engineered' to have the desired qualities and yet be quite different from those of the bulk material. Examples of industrial applications are: surface treatments of prostheses (hip and knee joints), ion implantation into high-speed ball bearings to protect against aqueous corrosion in jet engines, hardening of metal-cutting tools and reduction of wear on plastic moulding tools, which are expensive to produce.

For many years, particle-beam implanters have been used to produce special embedded layers in silicon for manufacturing semiconductor devices. The most commonly used ions for this are those of boron, phosphorus and arsenic. The depth of implant is usually about 1 μm and is precisely determined by controlling the ion energy. Doping levels vary from 10^{10} to 10^{15} cm^{-2}, which are about two orders of magnitude lower than those needed for altering the surface properties of metals.

8.2.4 Sterilization

Currently, one of the main uses of radiation is the sterilization of a large variety of single-use products, particularly medical products such as hypodermic needles, rubber gloves, surgical packs, dressings, sutures and many others. It is a simple procedure. Sealed packages, placed on a conveyor belt, are carried through a shielded vault where they pass through an electron beam at a speed regulated to achieve a given degree of sterilization. The energy absorbed increases the temperature no more than about 10°C, whereas conventional thermal sterilization would require treatment for about 15 min at 120°C. It is important to note that the irradiated materials do not become radioactive. Charged particles and, especially, neutrons can cause nuclear reactions leading to radioactive products, but low-energy electrons and γ rays, used for the purposes described here, cannot.

The technique exploits the ability of radiation to kill pathogenic microorganisms, which is almost certainly related to the damage done specifically to the DNA molecule. Microorganisms contain relatively little DNA compared with a human cell and so require a much greater dose to render them inactive. Figure 8.1 shows plots of survival curves for several different microorganisms, illustrating the wide range of different sensitivities encountered. Some exhibit a simple exponential fall off with dose in accordance with the single-hit target theory (see Section 7.4). A shouldered curve is indicative of a more complex mechanism. Also, as was discussed briefly in Section 7.2, radiation is found to be less effective in a dry rather than an aqueous state and also in an oxygen-poor environment compared with one rich in oxygen.

The statistical nature of the damage mechanism means that one can only reduce the number of microorganisms rather than eliminate them completely. Doses are specified which will achieve a certain level of reduction for a given microorganism and set of conditions. They have to be established empirically.

The choice of dose depends on the level of reduction required and is most critical for medical products. In Scandinavia, where a reduction level of 10^{-8} was recommended in some cases, scientists determined that this would need a dose as high as 45 kGy. Other countries specified different levels. In the USA, a level of 10^{-6} is normally

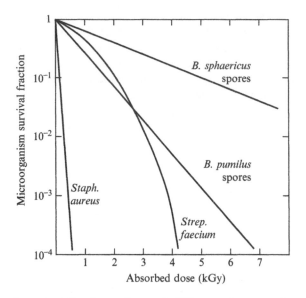

Figure 8.1   Plots of survival fractions of several different microorganisms as a function of absorbed dose. Data from A. Tallentire (private communication).

adopted. However, this can be as high as 10^{-3} when the medical product is used non-invasively and, therefore, poses a lower risk of infection.

8.2.5   Food preservation

The action of radiation to inhibit cell division can substantially prolong the shelf-life of many kinds of food produce by inhibiting sprouting, reducing the number of bacteria and microorganisms, killing insects and preventing their eggs and larvae from hatching. The lowering of enzyme activity by radiation delays growth and the ripening of fruit and vegetables. Irradiation is an effective alternative to conventional food treatment processes such as canning, deep freezing and smoking or the application of chemical preservatives and insecticides, which is not without hazard.

Treatment doses cover a very wide range, depending on the desired effect. A few tens of grays are usually sufficient to inhibit sprouting. Most insects, larvae and eggs are killed or sterilized by a dose of 600 Gy. The highest doses are required for eliminating pathogenic microorganisms from dried herbs and spices and for irradiating deep-frozen meat or fish. This is because radiation is less effective at low temperatures and, as noted above, in the absence of water. These treatment doses are usually in the range 5–10 kGy.

Much research has been carried out to ascertain if there is any degradation of nutritional value in irradiated food or whether any radiation-induced changes might be carcinogenic or toxic. So far, there are no indications of toxicological or nutritional deficiencies, certainly for foodstuffs irradiated to an average dose of 10 kGy. However, consumers remain apprehensive and their concern is the chief limitation on the general application of food irradiation in many countries.

8.2.6   Other applications

Many other uses of radiation and radioactive sources could be cited. For example, radioactive sources are used in compact power supplies and for radiography. Radioactive decay produces heat, which can be converted into electric power. The amount of power is not large. For example, 1 g of ^{238}Pu, which has a half-life of 88 years and decays by α emission, generates about 0.6 W of heat. If this is used as the heat source for a thermoelectric converter, the result is a small source of electric power. Its advantages are simplicity and reliability. The device has no moving parts and will provide power continually for a time determined only by the decay time of the radioactive material. It is ideally suited for use in situations where maintenance is impossible or undesirable, such as in a space probe or in a human body for powering a cardiac pacemaker.

In γ radiography, a small source of γ rays is placed on one side of an object and an image is produced on a piece of film placed on the other side. The principle is closely related to the more familiar X-radiography (see Chapter 9), but a γ-ray source is cheaper and sometimes more convenient. For example, the peculiar shapes and thicknesses of many metal castings sometimes do not permit standard X-ray methods to be used for testing their integrity because either the geometry is unsuitable or the attenuation of X-rays too severe. In these cases, a high-energy γ-ray emitter, approximating a point source, is placed inside the casting and the film is attached to the outside. Gamma radiography is particularly useful if the measurement has to be made in a remote location, e.g. the inspection of welds in an installed pipeline.

Electron and γ radiations are being used for several novel purposes such as the treatment of solid sludge or noxious flue components or the preservation of art objects.

Sludge is potentially beneficial as a fertilizer. However, untreated sludge contains large amounts of bacteria and bio-organisms, which through uptake in the food chain could be harmful to animals and humans. One of the methods being tried to reduce microbiological levels is irradiation.

Sulphur oxide and nitrogen oxide emissions from the burning of coal and oil have been blamed for acid rain which seriously degrades large amounts of forest in North America and Europe and has serious economic and environmental consequences. Facilities to remove these emissions are being built and research carried out to find better effective methods for such reductions. The use of radiation is one method being studied and electron accelerators in the range of a few 100 keV are well suited for this purpose. If ammonia is added to the gases, radiation converts the oxides into the particulate products ammonium sulphate and nitrates of ammonia. These compounds are salts used in fertilizers and, therefore, have economic value.

Ancient objects made of or containing organic material are often destroyed by microbiological activity. The decay may be stopped by exposing the object to radiation and killing the microorganisms. The dose needs to be carefully controlled, however, because the radiation itself can cause damage to some objects.

8.3   NEUTRON ACTIVATION ANALYSIS

The ability to detect radioactivity with great sensitivity is put to good use in an important technique for identifying and measuring trace amounts of specific nuclides in materials. Many stable nuclides have high cross sections for thermal-neutron capture (n, γ), which converts them into radioactive isotopes. For example, gold has only one stable isotope ^{197}Au and when it captures a neutron, it is transformed into radioactive ^{198}Au, which has a half-life of 2.7 days. About 95% of the time, ^{198}Au β^- decays to an excited state of ^{198}Pt, which quickly de-excites to its ground state by emitting a 412-keV γ ray. This energy and the 2.7-day decay half-life are unique to ^{198}Au. In fact, every radioisotope can be identified by measuring its half-life and the type and energy of the emitted radiation. Therefore, if a sample is bombarded with neutrons and we find activity induced in it, we can deduce what elements must be present in the sample by measuring the emitted radiation. This technique for identifying elemental constituents in a sample is called neutron activation analysis (NAA). Activation analysis can be done with beams of charged particles, but neutrons are more commonly used because they are more penetrating and, therefore, better for analysing thicker samples.

The NAA technique has two important advantages over chemical analytical methods. First, it is non-destructive. Apart from the induced activity, which eventually decays away, the specimen is essentially unaffected by the process. This is very important if, for example, the object is a valuable painting, historic document or a piece of forensic evidence which might be needed later in a criminal investigation. Second, the sensitivity, at least for some elements, is much greater than is possible with chemical analysis. As we shall see, quantities as small as 10^{-13} g can be detected. Most elements can be detected with NAA, but with widely varying sensitivities because the capture cross sections vary by many orders of magnitude as do the half-lives of the radioactive nuclides produced. Nevertheless, of all the naturally occurring elements, only 12 cannot be detected within reasonable limits. These are the eight lightest elements (hydrogen to oxygen) and phosphorus, sulphur, thallium and bismuth.

The activity induced in a sample depends on a number of factors. Consider N atoms of the element to be detected exposed to a constant neutron flux Φ. If σ is the neutron-capture cross section leading to the radioactive product, the rate of production [from Equation (1.22)] is $N\sigma\Phi$. The activity $\mathcal{A}$ of the sample increases with bombarding time t according to Equation (1.29). Thus,

$$\mathcal{A} = \lambda n(t) = \Phi\sigma N(1 - e^{-\lambda t}) \tag{8.1}$$

where λ is the decay constant of the radioisotope created by the nuclear reaction, and $n(t)$ is the number of radioactive atoms present at time t. If the bombarding time is short ($\lambda t \ll 1$), the activity is proportional to t. As t is increased, the activity also continues to increase, but at a slower rate. It eventually levels off and approaches an equilibrium value when the rate of decay is equal to the rate of production. The calculation is more complicated if, for example, the flux is not constant or if the measured activity is from the daughter of the product of the irradiation. In these cases, quantitative information is usually obtained by comparing the activity induced in an unknown specimen with that of a known standard irradiated under the same conditions.

Ideally, the half-life should be such that the activity builds up in a convenient time, but not be significantly less than the time taken to transport the sample from where it is irradiated to the counting system measuring its activity. It is also most useful when the decay involves the emission of a γ ray. Decay γ rays, as we have noted, have discrete energies E_γ characteristic of the element emitting them, and the use of high-resolution (germanium) γ-ray detectors enables scientists to analyse complex γ spectra containing a number of activated species. It is best if E_γ is of the order of 100 keV, high enough that self-absorption in the specimen is small yet low enough that a good fraction of the detected γ rays appear in a full-energy peak in the spectrum.

We will illustrate the method with an example. About 96% of indium consists of the isotope ^{115}In. It has a thermal-neutron capture cross section of about 170 b, leading to a metastable state in the radioisotope ^{116}In, which β decays to ^{116}Sn with a half-life of 54.3 min ($\lambda = 2.13 \times 10^{-4}$ s^{-1}). The relevant parts of the decay scheme, sketched in Figure 8.2, show four β-decay branches to several excited states of ^{116}Sn all of which emit γ rays. The particular transition from the first excited state at 1.294 MeV to the ground state occurs in about 80% of all the β decays and, therefore, is a suitable transition for measuring the ^{116}In activity.

We suppose that a sample containing indium is irradiated for a time t_1 and, after a time t_2, is placed in front of a detector and counted for a time t_3. The number of

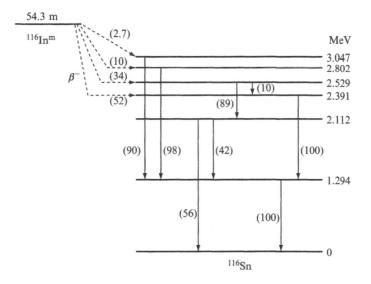

Figure 8.2 Energy-level scheme showing the β decay of a metastable state in indium ^{116}In^m to excited states of ^{116}Sn and their γ decays. The numbers in parentheses give approximate percentages for the four β-decay branches and for the main γ decays of each of the excited states of ^{116}Sn, which form part of the decay scheme.

radioactive nuclei immediately after irradiation $n(t_1)$ is obtained from Equation (8.1) with $t = t_1$. The fraction of these remaining after time t_2 has elapsed is $e^{-\lambda t_2}$ and the fraction of active nuclei which decay in time t_3 is given by $(1 - e^{-\lambda t_3})$. Hence, we obtain for the number of counts recorded by the detector:

$$N_c = n(t_1)e^{-\lambda t_2} (1 - e^{\lambda t_3})f\varepsilon \qquad (8.2)$$

where f is the fraction of decays resulting in the emission of γ rays, and ε is the efficiency for detecting them.

For simplicity, we assume that t_2 is small (i.e. $e^{-\lambda t_2} \approx 1$) and that t_1 and t_3 are the same and equal to the decay half-life $t_{1/2}$. Thus, $e^{-\lambda t_1} = e^{-\lambda t_3} = 0.5$ [since $\exp(-\lambda t_{1/2}) = 0.5$] and we can write $N_c = 0.25\,\sigma\Phi Nf\varepsilon/\lambda$. If the sample contains $m = 10^{-12}$ g of indium, we have $N = 0.96\,mN_A/A = 5.03 \times 10^9$ ^{115}In atoms in the sample and, if $\Phi = 10^{13}$ cm^{-2} s^{-1} and $f\varepsilon = 0.1$, the number counted in just under 1 h is

$$N_c = \frac{0.25 \times 10^{13} \times 170 \times 10^{-24} \times 5.03 \times 10^9 \times 0.1}{2.13 \times 10^{-4}} = 1004 \text{ counts.}$$

Taking a detection limit (see Section 8.7.2) as that which gives about 100 counts h^{-1}, our example leads to an estimate of about 10^{-13} g for the minimum amount of indium detectable under these conditions. This is a favourable case, but there are many other elements which are detectable with NAA to very small fractions of a microgram.

Neutron activation analysis is very well suited for detecting trace quantities of specific elements, and applications are found in a number of fields. In forensic investigations, for example, a suspect may be linked to a crime by analysing a speck of material found on his clothing. If the speck has exactly the same composition as material from the crime scene it is strong evidence that it came from that site. Even minute quantities of barium and antimony, contained in residues remaining on the hand after firing a weapon, are well above the NAA detection threshold. As an example from archaeology, studies of ancient pottery show which pieces have a common origin and help researchers to trace migration routes of ancient cultures and, as a piece of historical research, NAA applied to a strand of Napoleon's hair indicated the presence of arsenic in sufficient quantities to demonstrate that the ex-emperor possibly did not die of natural causes but was poisoned while in exile on St. Helena.

A variation of the NAA technique is being developed for contraband detection. The amount of manpower required to inspect thoroughly even a small fraction of incoming cargo for drugs, explosives or other contraband is prohibitive. Now, a new system using pulsed, fast-neutron analysis allows automatic inspection of loaded containers and trucks. A collimated beam of fast neutrons scanned across the container excites nuclei of a number of common elements found in bulk materials. The primary signals of interest are γ-ray emissions following inelastic scattering of fast neutrons from carbon, nitrogen and oxygen. By measuring the time difference between the emission of a pulse of neutrons from a neutron source and the detection of a γ ray, it is possible to determine the location of the nucleus emitting the γ ray along the line of the neutron beam. This information can be correlated with the scan pattern of the beam across the face of the container, thus enabling an image of the positions of the interactions to be reconstructed, while, at the same time, measuring the γ-ray energies which identify the emitting elements. Observed ratios of elements are used to identify different substances. A high carbon-to-oxygen ratio, for example, is characteristic of drugs.

8.4 RUTHERFORD BACKSCATTERING

A well-established technique for analysing surfaces and thin films is based on measuring the energies of ions scattered at large angles. In a collision, a certain amount of energy is transferred to the recoiling particle, and the scattered ion emerges with an energy that depends on the angle of scatter and the mass of the scattering nucleus. The energy transfer is greatest when the scattering is at 180°.

The energy of an ion, of mass m, elastically scattered at 180° by an initially stationary nucleus of mass $M > m$ is given non-relativistically by Equation (5.26) as

$$E(180°) = E_0 \left(\frac{M - m}{M + m}\right)^2.$$

(8.3)

Note that, if $m > M$, there is no backscattering and both particles proceed in the forward direction.

The energy E_0 immediately before the scattering depends on how far the ion has penetrated into the target. Thus, for a given ion and incident energy, the measured scattered energy depends on M and also on the amount of energy lost by the ion in its passage through the target material both before and after it suffers the nuclear scattering.

The probability of scattering depends on the cross section. If the bombarding energy is below the Coulomb barrier, the differential scattering cross section is given by the well-known Rutherford formula [Equation (1.30)] and the technique is called Rutherford backscattering (RBS). In the laboratory (lab) system, Equation (1.30) takes the form:

$$\frac{d\sigma_R}{d\Omega} = 1.296 \left(\frac{zZ}{E_0}\right)^2 \left[\operatorname{cosec}^4 \frac{\psi}{2} - \left(\frac{m}{M}\right)^2 + \cdots\right] \quad \text{mb sr}^{-1}$$

(8.4)

where ψ is the scattering angle in the lab system, (z, m) and (Z, M) are the atomic numbers and masses of the incident ion and target nucleus, respectively, and E_0 is in units of MeV. The next term in the expression in square brackets is of order $(m/M)^4$. As $m/M \to 0$, the lab and c-m systems become equivalent and Equation (8.4), with ψ replaced by θ, takes the same form as Equation (1.30). For light ions (H or He) of a few MeV, the cross section at $\psi = 180°$ is typically less than 1 b sr^{-1}. The dependence on Z^2 favours heavy elements.

Calculated RBS spectra are shown in Figure 8.3 for a monoenergetic ion beam backscattered from samples of different thicknesses, each containing equal amounts by weight (mass per unit area) of three elements uniformly distributed in the sample. Heavier elements give higher scattered energies and yields, according to

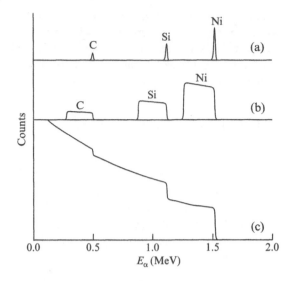

Figure 8.3   Calculated RBS spectra of 2-MeV α particles scattered at large angles from samples, each containing equal amounts by weight of C, Si and Ni. The sample thickness increases progressively from (a) to (c). From Breeze *et al.* (1992), p. 8.

Equations (8.3) and (8.4). The width of each peak increases as the sample thickness increases, reflecting the extra energy loss incurred by ions that penetrate deeper into the sample. Thus, the energy spectrum for a given atomic mass is related directly to its depth profile in the sample. The rise in yield at lower energies in Figure 8.3(c) is related to the energy dependence of $d\sigma_R/d\Omega$. A lower energy corresponds to deeper penetration into the sample and, therefore, to a lower energy E_0 at which scattering takes place. The scattering probability varies inversely as E_0^2 and so the scattered intensity will increase with depth of penetration. The area of each peak is proportional to (a) the number of atoms in the sample giving rise to it and (b) the cross section. Therefore, if the detector efficiency is known, the amounts of the different atoms can be determined from the observed yields normalized to a given amount of integrated beam current. Note that RBS identifies nuclei according to atomic mass and, therefore, cannot distinguish between isobars of different elements.

The ability to measure a given mass with RBS depends on the mass number, the proximity of neighbouring masses, detector resolution and sample thickness. For ions incident with energy E_0 on a thin sample, the energy separation of two adjacent masses ΔE is obtained by substituting appropriate mass numbers into Equation (8.3). For 5-MeV incident α particles, this gives $\Delta E \approx 1.8\,\text{keV}$ for $A = 200$, which is much less than the typical energy resolution of about 15 keV for a high-quality, silicon, surface-barrier detector. Near $A = 40$, however, ΔE is greater than 30 keV, which is certainly resolvable if the sample is not too thick.

In principle, RBS is more generally applicable than NAA because cross sections do not vary greatly from one nuclide to another as they do in NAA. However, the sensitivity of RBS is seriously degraded if the signal of interest has to be detected in the presence of a large background of scattering from heavier masses. Also, the scattered energy depends on the scattering angle. Therefore, to obtain a given energy resolution, the range of scattering angles accepted into the detector must not be too large and this places a limit on the detection efficiency. The best case is the detection of a heavy nucleus in a film of lighter elements. For example, consider a 1-μA beam of 5-MeV α particles backscattered from gold into a detector subtending a solid angle of 0.1 sr and spanning an acceptable angular range of about 20° near to 180°. The Rutherford cross section is 1.29 b sr^{-1} and the counting rate would be about 440 counts h^{-1} from 10^{-10} g cm^{-2} of ^{197}Au. In the absence of contaminants with similar masses, this could easily be measured above the background and implies a good level of sensitivity. However, it is an ideal case and RBS does not generally offer such high sensitivity for trace element detection.

A number of low-energy, ion-beam facilities exist which use RBS as an analytic technique. As we have seen, scattered ions emerge from the surface of a material with energies related directly to the mass of the scattering atom and its depth below the surface. The effect can be seen in Figure 8.4, which shows a spectrum of 100-keV H$^+$ ions scattered from a structure consisting of three thin germanium layers implanted into a silicon wafer forming part of an electronic device. The signals from silicon, germanium and oxygen are clearly identified. The oxygen signal is from a surface oxide layer formed when the sample was exposed to air. As noted above, an ion scattered below the surface loses an extra amount of energy, which is given by the average stopping power of the ion multiplied by the distance it travels in the medium (see Problem 8.10). Thus, the energy loss is related directly to the depth of the scattering atom and, in favourable cases, a depth resolution of a single atomic layer can be achieved. The example, given in Figure 8.4, is from a structure in which adjacent germanium-rich layers are separated by about 5 nm.

In a refinement of the method, in which the scattering angle is varied and measured accurately, it is possible to determine details of surface structure as well as obtain

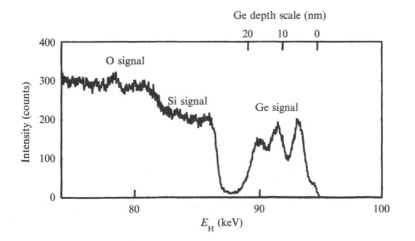

Figure 8.4 Spectrum of 100-keV H$^+$ ions scattered from a structure of three layers of germanium implanted into a silicon wafer, plotted as a function of the scattered energy. The outer surface was oxidized. The computed upper scale enables the germanium signal to be interpreted as a depth profile. Adapted from data taken using the MEIS facility, CLRC Daresbury Laboratory.

information on the atomic masses and depth profiles of elemental constituents. Obtaining crystallographic surface-structure information requires careful adjustment of the incident and scattered directions. When the ion beam is aligned with a crystal axis, the surface atoms shadow deeper atoms from the beam. This alignment, there-fore, makes the technique surface specific. Also, at certain scattering angles, ions scattered from the second and lower layers will have their outward paths blocked by first-layer atoms. The variation in scattered ion intensity with angle thus relates to the geometrical arrangement of surface atoms. A complete analysis of a surface structural arrangement of atoms involves comparing experimental and simulated results for different scattering geometries. An example of a particular effect is shown in Figure 8.5. A schematic arrangement of the scattering geometry and arrangement of surface and near-surface atoms in a simple, regular lattice is shown in Figure 8.5(a). It is often the case that the distance between atomic layers in the surface of a crystalline substance is different from that of the bulk material. In the example shown, the outer layer of atoms is shown in two positions: one corresponding to the average bulk position and one where it is shifted (relaxed) inwards slightly relative to the second layer. Figure 8.5(b) shows the effect of this surface relaxation on the blocking spectra of scattered ions, which exhibit dips corresponding to the angles where the paths from inner atoms are blocked. The scattering geometry [Figure 8.5(a)] shows that, if the upper layer is not shifted relative to the bulk material, scattering from the second (and lower) layers would be blocked at 101°, 108° and 135°. A weaker, blocking angle should also occur at 117° for ions scattered from the third layer.

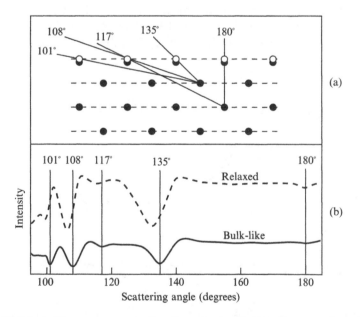

Figure 8.5   (a) Schematic arrangement of surface atoms showing the upper layer in a normal position (open circles) and in an inward-shifted, 'relaxed' position (closed circles). (b) Different blocking patterns of scattered ions due to inward surface-layer relaxation (dashed line) and normal bulk material (solid line). The scattering angle is measured relative to the normal to the crystal surface. Source: MEIS facility, CLRC Daresbury Laboratory.

Dips at these angles appear in the solid curve in Figure 8.5(b). The dashed curve shows the expected shift of blocking features to lower scattering angles if the upper layer is relaxed inwards. The amount of relaxation can be found from the magnitude of the shift, either by calculation or computer simulation. By appropriate choice of scattering geometry, atomic displacements as small as 0.03 Å can be measured.

8.5 PARTICLE-INDUCED X-RAY EMISSION

A sensitive analytical technique for determining the elemental composition of materials is particle-induced X-ray emission (PIXE).The basic principle is similar to that of NAA. Reactions induced in a target bombarded by particles cause it to emit characteristic radiation which identifies elements contained in it. In PIXE, the incident particles are ions which, through their interaction with matter, excite atoms by causing ionization of inner (K or L) atomic shells. The excited atoms lose their energy when the inner-shell vacancies are filled by electrons from outer shells with the subsequent emission of either X-rays or Auger electrons (see Section 5.4.1). A K_α X-ray is emitted if an L-shell electron fills a K-shell vacancy. A more energetic K_β or K_γ X-ray is emitted when the electron transition to the K shell is from an M or N shell, respectively. Similarly, L_α, L_β, etc., X-rays arise when L-shell vacancies are filled by electrons from M, N and higher shells. The K and L shell energies vary, to a good approximation, as the square of the atomic number Z and, therefore, are characteristic of the element. By measuring X-ray energies we can determine what elements are in a sample.

In a standard experimental arrangement, shown schematically in Figure 8.6, the ion beam is directed on to a target, generally in the form of a thin film. The target may be held inside a vacuum chamber (as shown in the figure) or, alternatively, the beam may be brought out through a thin window into air or a helium atmosphere for external irradiations. The X-rays pass through a thin ($\approx 8 \, \mu g \, cm^{-2}$) beryllium window into a detector, which is usually a cooled, silicon semiconductor detector with an energy resolution for X-rays of about 150 eV. The detector efficiency is poor for low-energy X-rays (less than a few keV) because they are attenuated in the beryllium window and in the sample itself. This attenuation and the detector energy resolution limit the use of PIXE to the identification and measurement of elements heavier than sodium.

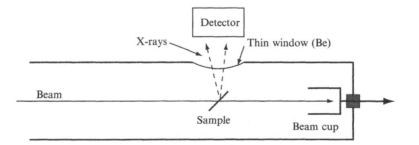

Figure 8.6 Schematic arrangement for carrying out PIXE measurements on a thin sample supported inside an evacuated tube. An ion beam passes through the sample and is stopped in a beam cup (Faraday cup), which enables the beam current to be monitored. X-rays from the sample are recorded in an external detector after passing through a thin beryllium window.

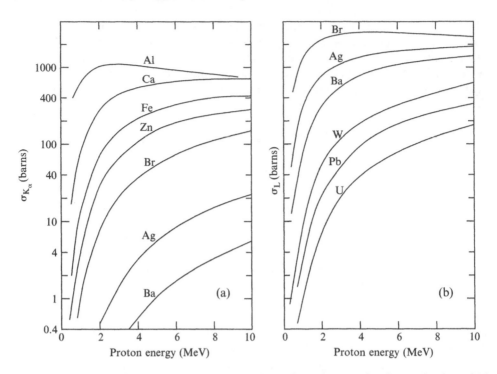

Figure 8.7   Cross sections versus proton energy for various elements for the production of (a) K_α X-rays and (b) all L X-rays. The vertical scales for (a) and (b) are the same. From Cahill (1980), pp. 219 and 220.

Cross sections are large, since they are determined by atomic rather than nuclear dimensions, but they are very dependent on both Z and ion energy. Representative K and L X-ray production cross sections are shown in Figure 8.7 for a range of elements as a function of energy of bombarding protons. The excitation of K_α X-rays, which is as high as 1000 b for low Z elements, decreases with atomic number to become small for Z greater than about 50. Exciting the L shell is much more likely for the heavier elements and, by observing both K and L X-rays, it is possible to cover a wide range of the periodic table in a single run. A standard measurement of a few minutes gives elemental concentrations to an accuracy of about 10%, which can be improved by comparing results for the unknown sample with those for a similar sample prepared with known concentrations. A consequence of the high cross sections is that analyses can often be carried out with very low beam currents (<1 nA), which cause little damage to the sample. This is important if the item is delicate or has historical or artistic significance.

If the X-ray production cross section σ_X is known, the rate of production of X-rays can be calculated from Equation (1.23):

$$R_X = I\sigma_X N dx \tag{8.5}$$

where I is the ion-beam intensity in particles per second and N is the density of target atoms in the sample of thickness dx. If we express the amount of material m in the sample in units of mass per unit area, we can write

$$Ndx = mN_A/A \qquad (8.6)$$

where A is the mass number of the target atoms and N_A is Avogadro's number.

The number N_X of X-rays counted in a run is equal to the X-ray production rate R_X multiplied by the exposure time t and the overall efficiency ε for detecting the X-rays. The efficiency includes the geometric solid angle of the detector as well as effects due to attenuation of X-rays in the sample and beryllium foil. Combining these quantities gives the expression

$$N_X = I\sigma_X \frac{mN_A}{A} t\varepsilon. \qquad (8.7)$$

We will estimate a count for an exposure of a target to 500 pA of protons for 300 s ($It = 9 \times 10^{11}$ protons) to detect trace amounts of antimony ($A = 121$) with an overall detection efficiency $\varepsilon = 10^{-2}$. If the target contains $10^{-9}\,\mathrm{g\,cm^{-2}}$ of antimony ($\approx 5 \times 10^{12}$ atoms cm^{-2}) and $\sigma_X = 1000\,\mathrm{b}$, we obtain, by substituting into Equation (8.7):

$$N_X = 9 \times 10^{11} \times 1000 \times 10^{-24} \times 5 \times 10^{12} \times 10^{-2} = 45 \text{ counts (in 5 min)}$$

This will give a measurable peak if the level of background is low enough (see Section 8.7.2).

The sensitivity of PIXE depends on the element to be detected and the background coming from the sample and the substrate. The continuum background in a PIXE energy spectrum is generally low. Bremsstrahlung radiation from the primary ion beam is negligible. The only significant contribution to bremsstrahlung comes from secondary electrons, with energies up to a few keV, which result from collisions between incident particles and electrons in the sample material [Equation (5.1)]. The low background, combined with the high cross section, implies an excellent sensitivity for the technique. Fractional sensitivities are usually quoted and PIXE offers analytical sensitivities of less than 0.1 part per million at best for elements heavier than sodium and varying by no more than a factor of three to four across the periodic table, depending on the energy and type of bombarding ion. In a 1 mg cm^{-2} sample, this sensitivity implies a minimum detectable amount of about $10^{-10}\,\mathrm{g\,cm^{-2}}$ which, as our example above (for $10^{-9}\,\mathrm{g\,cm^{-2}}$) shows, could give an observable peak in about 50 min. At this level, PIXE is comparable with NAA and has the advantage that it is more universally applicable since its sensitivity varies relatively slowly with Z. The highest sensitivity obtained with NAA is limited to isotopes with high thermal-neutron capture cross sections and which produce radioactive daughters with suitable half-lives and decay radiation.

A disadvantage of PIXE is the possibility of X-ray ambiguity. For example, the lead L_α and arsenic K_α X-ray lines differ by only 10 eV and are unresolvable with a Si detector. However, there are often several X-ray lines associated with the presence of a given element and, in such cases, the ambiguity can usually be resolved. The PIXE technique is not so good for samples much thicker than about 1 mg cm^{-2}. Results from thick samples are difficult to interpret because of particle energy loss and X-ray attenuation in the sample. Both of these need to be known accurately, but they depend on sample composition and are not easy to calculate. In this regard, NAA is better because thermal neutrons do not change energy in the sample and decay γ rays are generally higher in energy than X-rays and more penetrating. It should be noted that PIXE identifies the atomic number Z of an atom and does not give information about the relative amounts of different isotopes of an element.

An ion beam can be focused to a very small diameter, which makes it possible to perform spot analyses on small details of an object being investigated. A common technique is to scan a finely focused, microprobe beam across the sample surface and correlate each detected X-ray with the position of the beam spot. In this way, one can map the spatial variation of different elements in a specimen to a resolution of a fraction of a micrometre.

Even if the beam is brought out through a thin window for an external bombardment, the spot size is preserved to a few tens of micrometres making it still possible to perform spot analyses of a delicate object with minimal disturbance. Figure 8.8 shows PIXE spectra taken using an external microprobe beam to analyse ink spots on two manuscripts written by Galileo. They were taken as part of a recent investigation attempting to establish the chronology of Galileo's writings. There are clear differences between the two spectra shown in the figure, most notably in the relative amounts of lead to the other elements, which indicate that those particular manuscripts were most probably written in quite different periods.

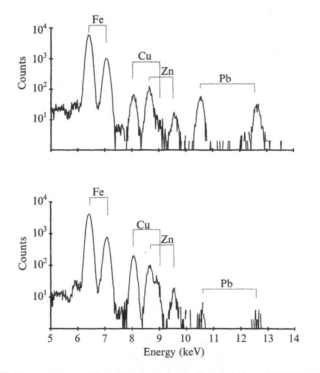

Figure 8.8   PIXE X-ray spectra of medium to heavy elements from 3-MeV proton bombardment of ink spots on two different manuscripts written by Galileo. Courtesy F. Lucarelli and P. Mando (private communication).

8.6   ACCELERATOR MASS SPECTROMETRY

Accelerator mass spectrometry (AMS) is an ultrasensitive, analytic tool in which atoms are counted directly instead of being induced to emit radiation or scatter bombarding particles. It combines a number of techniques originally developed for nuclear and atomic physics research and achieves high sensitivity by maintaining detection efficiency while suppressing background to a very low level. It was developed in 1977 as a way of directly detecting long-lived, cosmogonic radionuclides, such as ^{10}Be, ^{14}C, ^{26}Al, ^{36}Cl and ^{129}I, which occur at very low concentrations in the presence of much larger quantities of their stable isotopes. Cosmogonic radionuclides are produced by the action of cosmic rays in the atmosphere and upper layers of the earth's crust. Rainwater washes them into the land and seas where they migrate and enter into the biosphere by a variety of pathways. The ability to measure them quantitatively has important applications in areas as diverse as archaeology and the arts, geology, environmental science and medicine.

The initial impetus for AMS was to detect ^{14}C, which for many years has been used to date fossils and objects of archaeological and historical significance (see Section 1.5.7). Living organic matter contains about 1 atom of ^{14}C per 10^{12} atoms of ^{12}C or about 5×10^{10} ^{14}C atoms per gram of carbon. After a plant or animal dies, the amount of ^{14}C decays away exponentially with a half-life of 5730 years and so, by comparing the ^{14}C/^{12}C ratio with that of living organic matter, we can find out how long the organism has been dead. The amount of ^{14}C in a fossil can be measured by detecting the β activity, but the counting rate is low and, to obtain a measured precision of 1% or better in a reasonable time, gram quantities of carbon are needed, especially if the specimen is very old (>20 000 years). Counting the activity is clearly inefficient because only a tiny fraction of the ^{14}C nuclei present decay during the counting period. If the ^{14}C atoms could be identified and counted in a spectrometer with an efficiency of about 1%, the sample size could be reduced by a huge factor.

Conventional mass spectrometry is unable to detect isotope ratios at the levels required because of the presence of isobars and molecular species of similar mass. For example, ^{12}CH$_2$, ^{13}CH and ^{14}N all have nearly the same mass as ^{14}C and cannot be resolved from it in a conventional mass spectrometer unless it is operated at ultrahigh mass resolution which, inevitably, means very low efficiency. AMS uses the same basic principles as conventional mass spectrometry, but includes several additional stages that are designed to remove mass ambiguities and reduce background to low and even negligible levels, without sacrificing efficiency.

The AMS method is best described with reference to Figure 8.9, which shows the schematic layout of a commonly used arrangement based on a tandem Van de Graaff accelerator. As we noted in Section 6.8.1, negative ions are extracted from the ion source as the input beam for a tandem accelerator and even at this early stage, certain isobars, which would contaminate an AMS measurement (e.g. ^{14}N in a ^{14}C measurement), are eliminated because they do not form stable negative ions. The low-energy ions next enter an analysing magnet where they are deflected according to their radii of curvature. The radius of curvature of a particle of mass m, speed v, energy E and charge q in a uniform magnetic field B is given by Equation (6.4):

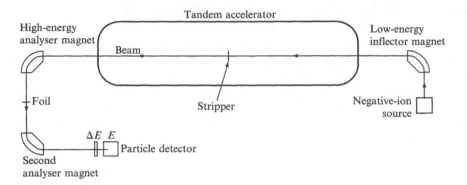

Figure 8.9 Schematic layout of the main components of an accelerator mass spectrometry system based on a tandem Van de Graaff accelerator.

$$r = \frac{mv}{Bq} = \frac{\sqrt{2mE}}{Bq}.$$ (8.8)

Thus, since all the ions at this stage have the same energy and charge state ($q = -e$),[1] this magnet is used to select ions of a given mass and direct them into the accelerator. Molecular isobars are present at this stage but, after the particles have been accelerated to the terminal, they pass through a stripper where several electrons are removed. If more than three electrons are removed, any molecule becomes unstable and breaks up into fragments. Positively charged particles emerging from the stripper region then receive a further stage of acceleration, and a dipole magnet at the high-energy end analyses the beam for a second time and effectively removes the molecular fragments which, in general, will not have the same radius of curvature as the atomic ions of the nuclide of interest. Some arrangements include a second stripper and additional magnetic deflection before the detection system to provide a further level of background suppression.

The detection system, consisting (as shown in Figure. 8.9) of a thin, passing detector (ΔE) and a stopping detector (E), is used for particle identification (see Section 6.7.1). For isobars with the same energy and, therefore, speed, the ΔE signal is a measure of the stopping power which is proportional to z^2 where z is the atomic number of the detected particle [see Equation (5.2)]. Therefore, any particles, other than the ones of interest, which manage to reach this point with a given energy will generate different signals in the passing detector and can be rejected. The result is a highly sensitive technique with extremely low background. It is usual to calibrate an AMS system by comparing an unknown sample with one of known isotope ratio.

Table 8.1 lists some key parameters of AMS, which indicate the sensitivity of the technique for detecting trace amounts of certain cosmogonic nuclides in different sample materials. Different nuclides form negative ions with different probabilities and so the system efficiency, which is the fraction of the number of atoms of the nuclide

---

[1] With very few exceptions, doubly charged negative ions, with two elecrons attached to a neutral atom or molecule, are unstable.

Table 8.1   AMS measurement capabilities.

Nuclide	Half-life (years)	Stable isobar	Sample material	System efficiency (%)	Sensitivity[a]
^{3}H	12.3	^{3}He	TiH_2	1	10^{-14}
^{10}Be	1.5×10^6	^{10}B	BeO	0.1	10^{-15}
^{14}C	5730	^{14}N[b]	C	2	2×10^{-15}
^{26}Al	7.4×10^5	^{26}Mg[b]	Al_2O_3	0.01–0.1	10^{-15}
^{36}Cl	3.01×10^5	^{36}S	AgCl	1	2×10^{-15}
^{41}Ca	1.0×10^5	^{41}K	CaH_2	0.01–0.1	10^{-15}
^{129}I	1.6×10^7	^{129}Xe[b]	AgI	1	10^{-14}

[a]   Limited by background, which is usually from tails of isobars or neighbouring stable isotopes. The ^{14}C background is from sample contamination by modern carbon.
[b]   Does not form a stable negative ion.

in the sample reaching the detector in a sufficiently long measurement, varies with each nuclide accordingly. Although the background in the final detector is small, it is not zero and the numbers in the final column of the table reflect this. These numbers indicate the AMS sensitivity achievable (limited by background) as the minimum amount of nuclide detectable relative to the predominant stable isotope.

As an example of the power of AMS, consider a 1-mg sample of AgCl in which the ^{36}Cl/Cl ratio is 10^{-14}. At this ratio, there are about 40 000 atoms of ^{36}Cl in the sample and, if they are detected with a system efficiency of 1% (see Table 8.1), a total of 400 ions are counted during the run, which might last an hour before the sample is used up in the ion source. With a half-life of 300 000 years, only one of the ^{36}Cl atoms in the sample is likely to decay in 11 years. Measuring the activity in this case is completely impractical.

AMS is most commonly used in special situations where both a high sensitivity and a high degree of background reduction are required. The detection of ^{14}C for dating purposes has been our main illustrative example and it is the most widely-studied, cosmogonic radioisotope. Some famous AMS measurements of ^{14}C have indicated that the Turin shroud is medieval (1260–1390 AD), that various Dead Sea scrolls are consistent with known dates (ranging from 400 to 700 AD) and that the Ice Man found in the Tyrolean Alps lived in the late Neolithic and not the Bronze Age.

Similar levels of sensitivity and background reduction have been achieved with other isotopes. For example, ^{36}Cl is formed by cosmic-ray interactions with argon in the upper atmosphere and is taken up into water which carries it into the ground. Hydrologists use ^{36}Cl to trace pathways of water into underground reservoirs and to date ancient groundwaters and salt deposits, which can be millions of years old. Such studies are important for assessing the stability of potential long-term depositories for storing waste radioactive material from nuclear reactors. Recently, AMS has been extended to measure trace amounts of transuranic nuclei. These nuclei are heavier than uranium and are by-products of the nuclear power industry. As a result of discharges from nuclear installations over the years, they appear in the environment in minute quantities and need to be actively monitored. It is now possible with AMS to measure trace amounts of a number of these nuclei to levels much lower than can be achieved by conventional, low-level activity counting.

There is a growing number of biomedical applications of AMS, using long-lived radioisotopes as tracers. Two examples among many are the study of human metabolism by using ^{14}C-labelled compounds and the use of ^{26}Al as a tracer to measure the response of biological systems and humans to aluminium, which is a known neurotoxin. With AMS, doses to patients or human volunteers are well below permissible levels and tissue samples are very small.

8.7   SIGNIFICANCE OF LOW-LEVEL COUNTING

It is often the case that we try to detect a signal that is very weak. Indeed, a given measurement may yield a null result, but this could be an important finding if, for example, we are attempting to confirm that some material is free from contamination or that a certain process does not occur. However, even a null result needs to be qualified because uncertainties associated with any measurement mean that we cannot be sure that repeating it or measuring for a longer time would not give a finite result. Accordingly, we present the measurement as an upper limit, the value of which depends on the degree of confidence we wish to attach to it, i.e. we say that our result has a certain probability of being less than a particular value. The analysis of a null measurement depends very much on whether or not there is any background present which could mask the signal and in the discussion which follows, we treat these two situations separately.

8.7.1   Null measurements with zero background

If no counts are recorded in a certain period of time T, we suppose that the process does occur at some mean rate λ and then calculate the likelihood that we would not observe a count during the measurement time. This probability is given by Poisson statistics [Equation (7.6)] as

$$P_0(T, \lambda) = \exp(-\lambda T) \tag{8.9}$$

which is the probability of recording no events in the time interval T, if the mean number of events in this time interval is λT. This probability is a function of λ and has the same form as the probability distribution of λ, given that no counts are observed in time T. With this interpretation, we can calculate the probability that λ is less than some chosen value λ_0 as

$$P_0(T, \lambda \leq \lambda_0) = \int_0^{\lambda_0} P_0(T, \lambda)\mathrm{d}\lambda / \int_0^{\infty} P_0(T, \lambda)\mathrm{d}\lambda = 1 - \exp(-\lambda_0 T) \tag{8.10}$$

where the integral in the denominator is required because $P_0(T, \lambda)$ is not normalized. $P_0(T, \lambda \leq \lambda_0)$ is the probability that, in a null experiment with no background (i.e. no counts in time interval T), the true event rate is less than λ_0. It is often referred to as the *confidence level* CL because, the higher the value of $P_0(T, \lambda \leq \lambda_0)$, the more confident we can be that λ does not exceed λ_0. If we require a confidence level of 90%, say, we set $\mathrm{CL} = P_0(T, \lambda \leq \lambda_0) = 0.9$ and, after rearranging Equation (8.10), obtain

$$\lambda_0 = -\frac{1}{T}\ln(1 - \mathrm{CL}) = \frac{2.3}{T}. \tag{8.11}$$

So, for example, if no counts are recorded in $1000\,\mathrm{s}$, we can be 90% confident that the true count rate is less than $0.0023\,\mathrm{s}^{-1}$.

8.7.2   Low-level counting with finite background

In a counting experiment where there is no background, even a single count is significant. However, if there is background, a measured count may or may not be significant, depending on the extent to which it exceeds the background. This leads to the concept of a *critical limit* L_c, which enables us to say, with a certain degree of confidence, that a measurement is or is not significant. If the measurement falls below L_c, it is deemed insignificant and an upper limit is quoted. Alternatively, we can say, again with a specified probability or degree of confidence, that the measurement would have yielded a non-zero result if the expected count due to the process under investigation exceeds the background by a certain amount. This is called the *detection limit*. The upper limit and detection limits are not the same, as we explain below, and their values depend on the degree of confidence we attach to them.

Critical limit and upper limit

Suppose we take a measurement on a sample that has no activity. We expect to obtain a count C consistent with background B, i.e. $C \approx B$. The count S due to the sample is obtained by subtraction i.e. $S = C - B \approx 0$ and if we were to repeat the measurement many times, we would obtain a probability distribution in S, centred about zero and with a standard deviation σ_0, as illustrated in Figure 8.10(a). In our example, the counts C and B are subject to statistical uncertainty and, according to Poisson statistics, they have standard deviations equal to the square roots of C and B, respectively. Following the rules for the propagation of error, we obtain the standard deviation of S to be

$$\sigma_S = \sqrt{C + B} \tag{8.12}$$

and, because $C \approx B$, we obtain

$$\sigma_S \rightarrow \sigma_0 = \sqrt{2B}. \tag{8.13}$$

If we approximate this distribution with a Gaussian form:

$$P_0(S) = K \exp(-S^2/2\sigma_0^2) \tag{8.14}$$

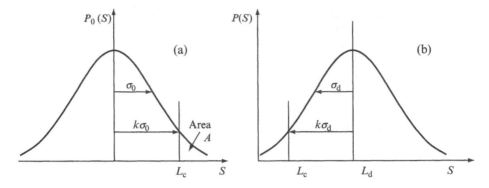

Figure 8.10   Graphs illustrating the concepts of critical limit (L_c) and detection limit (L_d). Probability distributions are shown as a function of net count S for the situations (a) where no activity is present and (b) where the activity would be expected to give a net count L_d. The corresponding standard deviations are σ_0 and σ_d.

where K is a normalization constant, we can use it to obtain the probability that any single measurement will lie within certain limits.

If we now measure an unknown sample, which may be active, we say that the sample activity is not zero if S exceeds a certain value. This is the critical limit L_c and we can state that the measurement is significant (i.e. non-zero) with an increasing degree of confidence the larger the value chosen for L_c. The critical limit is usually quoted as a multiple of the standard deviation, i.e.

$$L_c = k\sigma_0. \tag{8.15}$$

The area A of the distribution $P_0(S)$ above L_c determines the level of confidence. The greater the value of L_c the smaller is the area A and the more confident we can be that a count greater than L_c will be significant and not consistent with zero. For example, if we take $k = 1$, A is approximately 0.15 for a Gaussian distribution and we are, therefore, 85% confident that a count greater than L_c is significant. If we want to be 95% confident, we set $k = 1.645$ which gives $A \approx 0.05$.

If S is significant ($> L_c$), we quote its value with uncertainty limits: $S \pm k\sigma_S$ where k is determined by how confident we wish to be that S lies within the specified range. If $S < 0$ the result is consistent with zero and the upper limit is L_c. If $0 < S < L_c$, the measurement is not significant (by definition). but we specify an upper limit L_u to its value by assuming that S is representative of a probability distribution with a mean value equal to S and standard deviation σ_S as given above [Equation (8.12)]. Accordingly, if we set

$$L_u = S + k\sigma_S \tag{8.16}$$

and take $k = 1.645$ (for a confidence level of 95%), we can state that there is only a 5% chance of the actual value of S being greater than L_u.

Detection limit

The detection limit (L_d) gives an answer to the question: what would the strength of the signal S have to be for us to be confident of detecting it in a given measurement? It should not be confused with the critical limit because $L_d = L_c$ would imply that the mean value of S is equal to L_c. Repeated measurements of S would be distributed symmetrically about L_c and there could only be a 50% chance that our measurement would be significant. Our level of confidence, therefore, depends on the fraction of the measured distribution which lies above L_c. The situation is illustrated in Figure 8.10(b), where we have taken the mean value of S to be the detection limit and σ_d is the standard deviation of the distribution whose mean value is L_d. The chance of not detecting the signal is given by the area of the distribution below L_c. Thus, we can write

$$L_d = L_c + k\sigma_d = k\sigma_0 + k\sigma_d \qquad (8.17)$$

where k is determined by the confidence level we choose to associate with our statements about L_c and L_d.

In our simple example, where we record a count C_d at the detection limit and, independently, a background count B, the signal is equal to $C_d - B \approx L_d$. We can write the variance (the square of the standard deviation) of the distribution in Figure 8.10(b) from Poisson statistics as

$$\sigma_d^2 = C_d + B = (C_d - B) + 2B = L_d + \sigma_0^2. \qquad (8.18)$$

Substituting this into Equation (8.17) gives $L_d = k\sigma_0 + k\sqrt{L_d + \sigma_0^2}$, which can be manipulated to give

$$L_d = k^2 + 2k\sigma_0. \qquad (8.19)$$

As an example, suppose we record $C = 85$ counts from a sample in 5 min and a background count $B = 72$, also in 5 min. The count due to the sample $(S = C - B)$ is 13 counts with a standard deviation $\sigma_S = \sqrt{85 + 72} = 12.5$ counts. We also have $\sigma_0 = \sqrt{2B} = 12$ counts and the critical limit (at 95% confidence) is $1.645\sigma_0 = 19.7$ counts. This is greater than S, so our measurement of the activity of the sample is not significant. From Equation (8.16), we quote an upper limit of $13 + 1.645\sigma_S = 33.6$ counts, i.e. the activity is less than 0.112 Bq (95% confidence).

The minimum detectable activity is obtained from the detection limit L_d. Setting $k = 1.645$ in Equation (8.19) gives $L_d = 2.71 + 3.29\sigma_0 = 42.2$ counts or a minimum detectable activity of 0.141 Bq at the 95% confidence level.

PROBLEMS 8

8.1 Using a monitoring system consisting of a source of photons and a detector, the thickness of a foil is regulated during manufacture by observing the attenuation of the photon beam passing through it. The counting rate with no foil in place is accurately known to be $1000\,\text{s}^{-1}$ and the photon energy is chosen so that the attenuation is exactly 50% at the desired thickness $x_{1/2}$. (a) Obtain the attenuation coefficient μ in terms of $x_{1/2}$, using Equation (5.16). (b) Calculate the counting rate if the thickness increases by 5%. Hence or otherwise, estimate how long it would take to determine that a thickness change of 5% had occurred (to one standard deviation).

8.2 Calculate the mass of ^{210}Po required to generate $10\,\text{W}$ of electric power using a thermoelectric converter that operates with an efficiency of 15%. ^{210}Po has a half-life of 138 days and emits an α particle with decay energy $Q_\alpha = 5.4\,\text{MeV}$.

8.3 A copper foil weighing $100\,\text{mg}$ is irradiated in a neutron flux of $3 \times 10^{12}\,\text{cm}^{-2}\,\text{s}^{-1}$ for $90\,\text{h}$. Calculate the activity (in mCi) of ^{64}Cu (half-life $= 12.7\,\text{h}$) in the sample $8\,\text{h}$ after removal from the reactor. Natural copper consists of 69% ^{63}Cu and 31% ^{65}Cu; ^{63}Cu has a neutron-capture cross section of $4.5\,\text{b}$.

8.4 Irradiating a certain sample of material leads to a radioactive product which has a half-life $t_{1/2}$. If the sample is irradiated for a time $t_{1/2}$ and then immediately placed in front of a detector and counted for a further time $t_{1/2}$, 1000 counts are recorded. How many counts would be recorded if the irradiation and counting times were (i) $0.1t_{1/2}$ or (ii) $10t_{1/2}$?

8.5 A 100-mg foil of gold (^{197}Au) is placed in a thermal-neutron flux for $12\,\text{h}$. It is then set in front of a detector, which has an efficiency of 2×10^{-2} for detecting 412-keV γ rays. If the count rate in the detector, $6\,\text{h}$ after irradiation, is $10\,\text{s}^{-1}$, calculate the neutron flux near the foil.

The thermal-neutron capture cross section for ^{197}Au is $99\,\text{b}$. The half-life of ^{198}Au is 2.7 days and 95% of the time, its decay results in the emission of a 412-keV γ ray.

8.6 A 5-MeV α-particle source is used in Rutherford backscattering analysis. Calculate the energies of backscattered α particles from nuclei with $A = 20$, 50 and 100. What detector energy resolution (in keV) for α particles would be needed to resolve $\Delta A = 1$ for each of the above mass regions?

8.7 You are asked to design an α-particle source consisting of a deposit of ^{210}Po on a supporting disc. If the energy spread of α particles emerging normal to the surface is to be about $20\,\text{keV}$, calculate the area activity of the source in $\text{Ci}\,\text{cm}^{-2}$. Data (see also Problem 8.2): $(-\mathrm{d}E/\rho\mathrm{d}x) = 217\,\text{MeV}\,\text{g}^{-1}\,\text{cm}^2$ for 5.4-MeV α particles in ^{210}Po.

8.8 In a modification of the backscattering technique for analysing thin plastic films, two detectors are used. One is at large angles (near $180°$) and records α particles backscattered from carbon. A second detector is at forward angles (near $0°$) and records protons recoiling from head-on collisions of incident α particles with hydrogen nuclei in the film. Calculate the forward/backward energy ratio in the two detectors.

8.9 A thin target is bombarded with a 10-nA beam of 10-MeV α particles in a backscattering experiment to determine the level of lead contamination. After $5\,\text{min}$, 30 counts corresponding to scattering from lead in the target are recorded in a detector placed near $180°$, which subtends a solid angle of 5×10^{-2} steradian at the target. Calculate the level of lead contamination (mass per unit area). The atomic mass of lead $M_A = 207.2$.

8.10 The depth profile given at the top of Figure 8.4 indicates that the germanium implanted into the silicon wafer is at three different depths separated by about 5 nm. Estimate the average stopping power $(-dE/dx)$ (in keV μm^{-1}) for the H^+ ions in the material. Atomic weight of germanium $= 72.6$.

8.11 In a PIXE experiment, a $0.2 \, mg \, cm^{-2}$ film containing 5 parts per million by weight of an element of mass number 100 is bombarded with a 200-nA beam of protons for 10 min. The cross section for exciting the L shell of the element is 800 b and the probability of the excited atom emitting an L X-ray is 50%. Calculate the number of counts recorded if the overall detection efficiency $\varepsilon = 5 \times 10^{-3}$.

8.12 In a PIXE measurement, a thin target containing 10^{-11} g cm^{-2} of an element of mass number 120 is bombarded with a proton beam of intensity $0.5 \, \mu A$ in a direction perpendicular to the target. The X-ray detection efficiency of the system $\varepsilon = 1\%$. If the count rate is $0.6 \, s^{-1}$ and the background is negligible, calculate the cross section in barns for the production of X-rays and its accuracy (to one standard deviation) in a measurement lasting 25 min.

8.13 In an AMS measurement of a carbon sample, 1000 counts due to transmitted ^{14}C ions are recorded in 5 min. A beam of $10 \, \mu A$ is measured when the system is set to transmit $^{12}C^{3+}$ ions. Calculate the atomic ratio of $^{14}C/^{12}C$ in the sample assuming that the transmissions of ^{14}C and ^{12}C ions through the system are the same. What mass of ^{12}C was in the sample if it is totally consumed in half an hour? Assume a constant rate of consumption during this period and a system efficiency ε of 2%.

8.14 Using ^{26}Al as a tracer, you set out to test the hypothesis that if aluminium is ingested orally, only a fraction $f = 0.01\%$ is absorbed into the blood stream. A certain amount of ^{26}Al is ingested by a volunteer and, after a period of time such that any ^{26}Al entering into the blood stream is well mixed, 5 ml of blood are extracted and converted into an AMS sample of Al_2O_3. Estimate the activity of ^{26}Al the volunteer should ingest in order that the hypothesis could be tested to an accuracy of $\pm 10\%$ (1 standard deviation). A sample of blood taken before the experiment yielded no counts due to ^{26}Al. Assume that the human body contains 5.5 l of blood, that the system efficiency $\varepsilon = 0.1\%$ and that the background is negligible. Half-life of $^{26}Al = 7.4 \times 10^5$ y.

8.15 A counting system has a background counting rate of $5 \times 10^{-3} \, s^{-1}$. Calculate, with a 95% level of confidence, the critical limit and detection limit for measuring an unknown sample in a measurement lasting 1 h.

8.16 You are asked to design a low-level counting system capable of detecting a minimum activity of 5×10^{-4} Bq in a 20-h measurement with 95% confidence. How low must the background rate be?

8.17 In an attempt to detect a certain superheavy nucleus, a ^{208}Pb target of mass $m = 1\,\text{mg cm}^{-2}$ is bombarded with a beam of krypton ions with an intensity $I = 2 \times 10^{12}$ ions s^{-1}. The efficiency ε of the system for detecting any superheavy nuclei that are produced is 20%. If no counts are recorded in 4 weeks of observation, calculate an upper limit for the production cross section at 90% confidence.

9

Nuclear Medicine

9.1 INTRODUCTION

There has always been a close connection between discoveries in atomic and nuclear physics and their use in medicine. In his first publication following the discovery of X-rays, Roentgen included an X-ray image of the skeletal structure of his wife's hand and, shortly after the discovery of radioactivity, the biologically damaging effects of radiation were dramatically demonstrated by Pierre Curie who induced a radiation burn on his arm. Radium, shortly after it had been isolated by Marie and Pierre Curie in 1898, was used in the first treatment of cancer by the direct method of inserting it into the malignant tissue, a technique still in common use today.

Since that time, developments in experimental techniques and nuclear instrumentation have stimulated parallel developments in medical procedures, particularly in the areas pioneered by those early scientists, of non-invasive imaging and cancer therapy. Radioisotopes, created using accelerators and reactors, can be attached to compounds and pharmaceuticals and used as tracers to obtain functional imaging of, for example, blood flow to the brain or the heart, renal excretion of the kidneys and, as mentioned in the last chapter, the activity of the thyroid. Early X-ray images were projections on to a screen or film, but the availability of massive computer power and the development of *computed tomography* (CT) have enabled remarkably detailed two- and three-dimensional images to be obtained of any part of the body. A combination of the use of positron-emitting nuclides and CT led to the development of an imaging technique called *positron emission tomography* (PET). Another powerful method for producing detailed images of the body, called *magnetic resonance imaging* (MRI), detects changes in magnetization due to nuclear magnetic moments in an external magnetic field. Both PET and MRI have improved in efficiency to the point where images can be created sufficiently rapidly that functional behaviour and dynamic processes can be recorded in real time. In the treatment of cancer with radiation, photons of varying energies have been used for many years. Some of the more recent advances have been made using charged-particle beams and potential advantages of new procedures using neutrons are also being investigated.

The field of nuclear medicine is vast and we can touch only briefly on certain aspects of it, concentrating mainly on the physics and nuclear physics principles being exploited. In Section 9.2, we review conventional projection imaging for standard X-radiography and the use of a widely used γ-ray detection system for obtaining projected images of radioactive nuclides taken internally. Section 9.3 provides a brief account of the way CT enables depth information to be obtained from projected images and then Sections 9.4 and 9.5 describe how CT is exploited by the specific techniques of PET and MRI. Finally, some of the main principles of radiation therapy for cancer treatment are presented in Section 9.6, together with some mention of recent developments using neutrons and charged particles.

9.2   PROJECTION IMAGING: X-RADIOGRAPHY AND THE GAMMA CAMERA

Projected medical images are produced by using either external radiation (X-rays or γ rays) from a source located outside the body or internal radiation emitted by radio-active substances inside the body. In conventional external radiography, rays which have penetrated an object are detected with a piece of film or an array of detectors. Internal radiation is detected by using a specially developed instrument called a gamma camera. Below, we outline the basic principles underlying these two techniques.

9.2.1   Imaging with external radiation

In an arrangement, shown schematically in Figure 9.1, radiation from an external point source is directed on to one side of an object to be examined and recorded in two dimensions on an image plane positioned as close as possible to the object on the other side. The image depicts variations in the attenuation of rays travelling along different paths.

Consider a particular ray passing through an absorbing region to an image point on a screen at P. We will assume, for simplicity, that the source is a long way from the object so that the rays passing through the region can be considered to be parallel. According to Equation (5.15), the fractional attenuation of intensity I due to a small element dl of the path is given by

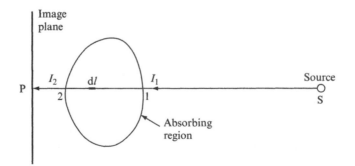

Figure 9.1   Principle of projection imaging using an external source of radiation. A ray from the source is shown passing through a region of absorbing material and arriving at point P on the image plane. I_1 and I_2 are the incident and outgoing intensities at the points 1 and 2, respectively.

$$\frac{\mathrm{d}I}{I} = -\mu \mathrm{d}l \tag{9.1}$$

where μ is the attenuation coefficient and is a function of position. The coefficient depends on γ-ray energy and increases with the density of the absorbing material and as a power of the atomic number Z (see Figure 5.6).

When integrated, Equation (9.1) gives

$$\ln\left(\frac{I_1}{I_2}\right) = \int_1^2 \mu \mathrm{d}l \tag{9.2}$$

where I_1 and I_2 are the intensities immediately before and after the traversed region. Thus, each point on the projected image contains information in the form of a line integral of the attenuation coefficient through the object. The full image reveals transverse variations of this integral in two dimensions only. It contains no depth information along the path of the radiation.

In medical work, X-radiography is used mainly to reveal anatomical structure corresponding to variations in μ within the body. The attenuation coefficient is much greater for bones and organs containing higher Z elements, such as calcium ($Z = 20$), than it is for water and soft tissue and, for this reason, X-radiography is excellent for revealing details of the skeletal system. However, it is not so good for differentiating between different types of soft tissue, e.g. for identifying tumours, particularly in the brain where the tissue is shielded and can be obscured by the mass of bone in the skull. As described in the next subsection, more revealing images for these kinds of studies can be obtained by detecting radiation generated from within the organ or tissue itself.

9.2.2 Imaging with internal radiation

When a radioactive isotope is introduced into the body, its eventual distribution will depend on how it is introduced, the form in which it is taken and how the body deals with it. The ability to be able to image this distribution quantitatively, therefore, gives information not only about anatomical structure but also about the physiological function of the body. Moreover, because of the high sensitivity with which radioactive tracers can be detected, these procedures are non-invasive and can be done at dose levels that pose no significant risk to the patient. Perhaps the best-known example, mentioned in the last chapter, is the use of radioactive iodine, which is taken up preferentially into the thyroid. Measuring the intensity of the radiation and imaging the distribution of the radioiodine gives information about the metabolic activity of the thyroid as well as its size and shape.

Imaging an internal-source distribution requires a different technique to that of imaging with external radiation because internal radiation is emitted in all directions from different parts of the object and, therefore, simply recording a signal in a detector does not indicate the origin of the radiation. The simplest way to locate a source inside an object is to place a collimator in front of a detector and scan the assembly across the object, as illustrated in Figure 9.2(a). Radiation is detected when

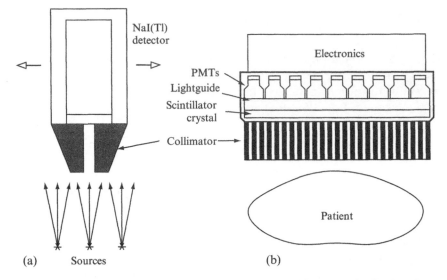

Figure 9.2   Principle of γ-ray imaging. In (a), a single, collimated scintillation detector is scanned across a distribution of γ-ray sources. In (b) is shown a schematic view of a gamma camera consisting of a multichannel collimator in front of a large piece of scintillator material, which is viewed through a lightguide with an array of photomultiplier tubes (PMTs).

the emitting region comes in front of the collimator. However, this method is inefficient. It is adequate for measuring a small, localized region, such as the thyroid, but not for mapping an extended area, and special, large-area detector arrays or gamma cameras have been developed which will produce a complete, two-dimensional image simultaneously.

A simplified sketch indicating the main components of a gamma camera is shown in Figure 9.2(b). The detector unit consists of a large piece of sodium iodide scintillator crystal (approximately 50 cm diameter and 1 cm thick). An array of photomultiplier tubes (PMTs), optically coupled to the rear surface, detects light emitted when a γ ray is converted in the crystal. This light, originating from a point within the scintillator, gives rise to measurable signals from several PMTs close to that point. The relative intensities of these signals depend on the position of the point of origin and, using a computer, they can be analysed to locate the source of the scintillation light to within a few millimetres. A multi-way collimator, commonly consisting of a honeycomb arrangement of narrow channels, restricts the direction of the γ rays entering the detector to a small angular range and, therefore, the spatial distribution of the recorded light from the scintillator is a projected image of the distribution of the radioactive isotope in the object (patient). The overall spatial resolution is typically 8–12 mm, depending on the geometry of the collimator channels.

Figure 9.3(a) shows an image taken with a gamma camera of the brain of a patient who had been injected with a compound labelled with the long-lived metastable isotope $^{99}Tc^m$. Darker areas indicate greater concentrations of the isotope. Normally, systems in the human body actively protect the brain from any impurities in the blood. However, the blood supply to a tumour can by-pass this blood–brain barrier and carry the labelled compound into the diseased area. The image in the figure

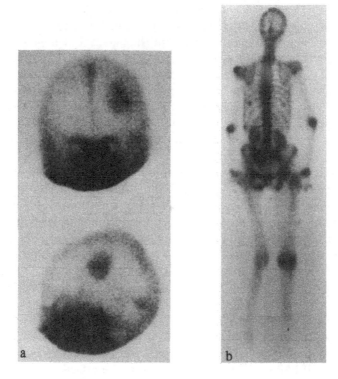

Figure 9.3 Gamma-camera images of (a) a human brain (upper: rear view; lower: side view) and (b) a human skeleton. The radioisotope $^{99}Tc^m$ was carried selectively to tumour sites, which are revealed as dark areas on the images Courtesy H. Sharma, Manchester Royal Infirmary, Manchester.

clearly indicates the presence of such a tumour. Another example, shown in Figure 9.3(b), is a whole-body image of a patient who has been given the drug methylene diphosphonate labelled with $^{99}Tc^m$. This drug moves preferentially to points in the skeleton affected by a form of bone cancer and several of these sites are identified here.

This latter illustration is one of a growing number of examples whereby pharmaceuticals, known to accumulate in particular regions or organs in the body, are used to transport radioactive nuclides to those sites for diagnostic or therapeutic purposes. Today, there are radiopharmaceuticals which can be used to examine the state and physiological function of almost any organ in the body. These include studies of the rate at which bone is remodelled, blood flow to the heart and brain tissue, and the functioning and efficiency of various organs such as the liver, kidneys, thyroid and lungs. Specific antibodies have been labelled and used as tracers. For example, with labelled antibodies to fibrin, blood clots can be identified as they form. Fibrin is a major constituent of blood clots, and antibodies move rapidly to the sites where the fibrin is being produced carrying the radionuclide with them.

The most effective radionuclides for diagnostic use with a gamma camera are those which emit a single γ ray, no β particles and have a half-life that is long enough for the image to be produced, but not so long that the patient receives unnecessary radiation exposure. An activity of 20 MBq (about 500 μCi) is usually enough for a typical imaging period of 15–20 min. The γ-ray energy should not be so low that absorption and scattering of γ rays in the tissue is large but, at the same time, it should not be so high that there is significant penetration into the material of the collimator system, which would blur the image. Also, lower-energy γ rays are better because they are more likely to have all their energy converted in the detector via the photoelectric effect and then it is possible to reject lower-energy signals, from Compton-scattered events in the tissue, which constitute an undesirable background. Energies between 100 and 200 keV are optimally suited to the technique. In general, the half-life should be no more than a few hours, although, as we shall see later in Section 9.4, it can be much less.

As an example, we will compare the suitability of the radioisotopes ^{131}I and ^{123}I, both of which have been used as tracers for imaging the thyroid. Iodine-131 is a fission product and has been available for a long time. It has a half-life of 8 days and β decays to ^{131}Xe, which emits a 364-keV γ ray with about an 81% probability. Iodine-123 became available more recently as an isotope produced in a cyclotron. It decays by electron capture to ^{123}Te, which emits a 159-keV γ ray with a similar (83%) probability. Iodine-123 has a half-life of 13 h, which is much better matched to the requirements for diagnostic imaging than the much longer-lived ^{131}I. Furthermore, the energy of the γ rays from ^{123}I is close to ideal. Even more important is the fact that ^{123}I decays by electron capture and, therefore, emits no β particles, unlike ^{131}I which emits β particles with an average energy of about 200 keV. The β particles from ^{131}I decay would be absorbed locally and contribute a good fraction of the effective radiation dose to the thyroid, which is about 22 μSv per kBq ingested (see Table 7.4). Since about 10–20 MBq are needed for imaging, the effective dose will be several hundred millisieverts. As we shall see in Section 9.6, this dose has a therapeutic application. By comparison, the effective dose from ^{123}I is much smaller at 0.21 μSv per kBq ingested. Clearly, ^{123}I is the preferred isotope for diagnostic imaging. However, its half-life does imply the need for ready access to its point of production.

The radioisotope most widely used in nuclear medicine is technetium-99m, because it has many ideal characteristics and also because it is easy and cheap to use. Technetium-99m is the daughter product of ^{99}Mo, which itself is obtained either as a fission fragment or via the capture of a neutron by ^{98}Mo. Technetium-99m is produced in its first-excited (metastable) state, which decays, with a half-life of 6 h, emitting a 140-keV γ ray about 90% of the time (see Section 1.5.5). Beta emission is very weak and contributes a negligible amount to any radiation dose. The isotope is supplied from a generator consisting of ^{99}Mo adsorbed on to alumina and held in a container, which allows liquid introduced at the top to be collected at the bottom. After a few hours, the ^{99}Tc^m daughter builds up to a useful level and is easily extracted by passing a saline solution through the column. A generator will last for about a week before needing to be replaced.

9.3   COMPUTED TOMOGRAPHY

The projected X-ray and γ-ray images described above have good spatial resolution in two dimensions, but lack resolution in depth. A consequence of this is that the attenuation coefficient μ, for a region being examined of about 1 cm in diameter, must differ from its surroundings by at least 10% or there will be insufficient contrast for it to be seen. A revolutionary advance took place in 1971 when a scanning technique was developed which enabled images of a living brain to be generated as a series of two-dimensional (2-D) sections, each with a contrast resolution in μ of about 1% for regions as small as a few millimetres in diameter. The new technique overcame the two major deficiencies of projection imaging, namely, poor discrimination in μ and the confusion arising from the overlapping images of objects from different depths projected on to a 2-D plane. Modern versions of this scanner have a spot resolution of about 1 mm and a contrast resolution of 0.5%. With the most advanced systems, it is even possible to acquire images in real time and capture motion as rapid as the beating of a heart. The technique is called computed tomography (CT) and it is capable of producing images by using radiation either from an external (X-ray or γ-ray) source or from ingested or inhaled radioisotopes.

The basic principle of CT is based on the fact that all the information necessary to construct an image of a chosen 2-D slice of tissue is contained in a complete set of one-dimensional projected images that cover all possible directions within the plane of the slice. For example, let us assume we have a slice of tissue in the xy plane and we wish to obtain a 2-D image of the variation of the linear attenuation coefficient $\mu(x, y)$ within the slice. As explained in Section 9.2, a projected image of the slice contains information on the function $\mu(x, y)$ in the form of a set of line integrals of μ taken through the region in a particular direction. A projection, taken at a different angle in the plane of the slice, gives a different representation of $\mu(x, y)$ in the form of a different set of line integrals. Once a complete set of line integrals through $\mu(x, y)$ has been obtained, a 2-D function $\mu_B(x, y)$, called a back-projected function, is constructed using computers. The value of the function at each point P is generated by integrating over angle in the xy plane all line integrals which pass through P. The procedure is described below and illustrated in Figure 9.4.

Figure 9.4 shows a 2-D slice lying in the xy plane. Consider the line integral:

$$\int_A^B \mu(x, y)\mathrm{d}l$$

for the line AB, making an angle ϕ with the x-axis and passing through the point P, with coordinates (ξ, η). The summation over all directions of lines passing through P and lying in the xy plane defines the back-projected function:

$$\mu_B(\xi, \eta) = \int_0^\pi \mathrm{d}\phi \int_A^B \mu(x, y)\mathrm{d}l.$$

This expression can be rewritten in a form which allows one to obtain the attenuation coefficient $\mu(x, y)$ from a knowledge of $\mu_B(\xi, \eta)$. We shall omit the derivation. (For the interested reader it is given below.) Here we only quote the result:

$$\mu_B(\xi, \eta) = \int \mathrm{d}A \frac{\mu(x, y)}{r} \tag{9.3}$$

where $\mathrm{d}A$ is an element of area of the slice, the integration is over the whole area of the slice and $r = \sqrt{(x - \xi)^2 + (y - \eta)^2}$ is the distance of the point (x, y) from P. Thus, the

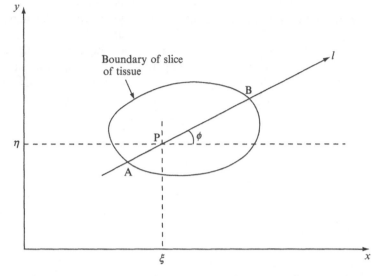

Figure 9.4   Schematic representation of a 2-D region of tissue within which the linear attenuation coefficient is given by the function $\mu(x, y)$. The line AB, making an angle ϕ with the x-axis and passing through the point P, with co-ordinates (ξ, η), is the path of the line integral considered in the text. The value of a back-projected function $\mu_B(\xi, \eta)$ is given by the sum of all line integrals passing through P at different angles ϕ in the xy plane. To obtain a complete knowledge of the function $\mu_B(\xi, \eta)$ requires such a summation to be carried out for each value of (ξ, η) within the 2-D slice.

value of μ_B at each point P, generated by this procedure, contains contributions from all $\mu(x, y)$, but weighted by $1/r$, i.e. by the inverse of the distance of the point (x, y) from P.

For interested readers we give here the derivation of Equation (9.3), though it will not be required later. Introducing polar co-ordinates (r, ϕ), centred at P, we can write our original expression for $\mu_B(\xi, \eta)$ as

$$\mu_B(\xi, \eta) = \int_0^\pi d\phi \int_A^B r\, dl \frac{\mu(x, y)}{r}.$$

In the line integral, we split the range of integration AB into two segments AP and PB. Taking the point P as the origin from which l is measured, it follows that $dl = dr$ for points (x, y) lying on the segment PB and $dl = -dr$ for points lying on AP. Hence,

$$\mu_B(\xi, \eta) = \int_0^\pi d\phi \left\{ \int_P^B r\, dl \frac{\mu(x, y)}{r} + \int_A^P r\, dl \frac{\mu(x, y)}{r} \right\}$$

$$= \int_0^\pi d\phi \int_P^B r\, dr \frac{\mu(x, y)}{r} + \int_\pi^{2\pi} d\phi \int_P^A r\, dr \frac{\mu(x, y)}{r}$$

which at once reduces to Equation (9.3), since $r\, dr\, d\phi$ is the element of area dA expressed in polar co-ordinates.

The integral defining $\mu_B(\xi, \eta)$ in Equation (9.3) is called the convolution of $\mu(x, y)$ with $1/r$. There exists a well-defined mathematical technique, called deconvolution, for recovering the function $\mu(x, y)$ from the back-projected function $\mu_B(\xi, \eta)$. In practice, this is done using powerful computers. We shall not present the deconvolution procedure here and refer the interested reader to more advanced texts on medical imaging.[1]

In a typical modern CT scanner, the patient is surrounded by several hundred detectors, and an X-ray source, emitting a narrow fan of beams, is moved round the patient as illustrated in Figure 9.5. The X-ray source is shown in two positions S_1 and S_2. Each position of the source and that of a detector in the ring define a line through the patient, and a recorded count rate enables a line integral to be computed from Equation (9.2). For example, in Figure 9.5, the source, positioned successively at S_1 and S_2, gives line integrals through P to detectors D_1 and D_2, respectively. As the source is moved round the full angular range, a large set of line integrals is obtained for a complete 2-D section through the patient. This set contains all the information needed to construct an image of $\mu(x, y)$ for that section.

Detector instrumentation and computer power have improved greatly since the technique was pioneered and a complete image can now be obtained in a few seconds. Figure 9.6 is an example of a CT scan through the human trunk showing a cross section of the spine and details of various organs. In real systems, account has to be taken of averaging due to finite resolution. However, as can be seen in the figure, an

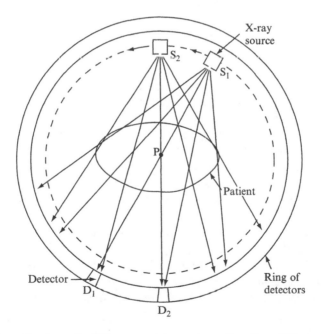

Figure 9.5 Schematic view of a CT scanner. A source emits a fan of X-rays and those that pass through the patient are recorded in an outer ring of detectors. Scans at different angles are obtained by rotating the X-ray source about the patient.

---

[1] See, for example, Hobbie (1997), Chapter 17, and a recent review article by Hendee (1999).

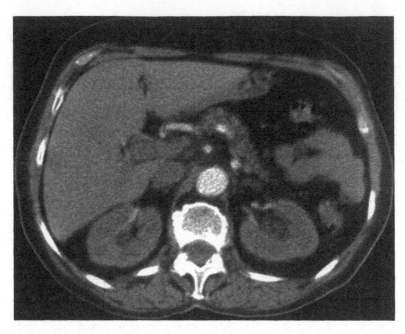

Figure 9.6   X-ray CT image of a section through a human trunk. Bone is indicated by the light areas where attenuation is relatively high compared with that of soft tissue.

image with good definition and contrast can be obtained. One disadvantage is that the patient receives a greater radiation dose with a CT scan than with conventional X-radiography.

Computed tomography is also used to reconstruct from projected images the distribution of internal radionuclides, emitting single γ rays. The method is called single-photon emission computed tomography (SPECT). Instrumentation usually consists of one or more gamma cameras, which can be rotated round the patient and so generate the required set of projected images for CT reconstruction. However, the efficiency for detecting radiation emitted from a given point depends on the distance of the radiating source from the camera. Also, the radiation may be significantly attenuated by intervening tissue before reaching the detector. The latter effect is normally treated approximately by assuming an average linear attenuation coefficient for the γ rays. Both these effects mean that there are inevitable inaccuracies in the measurement of the line integral of the radionuclide distribution, which can give rise to artefacts in the reconstructed image. This limitation means that, while SPECT imaging using single, γ-emitting tracers is useful for clinical diagnostic work, it is not suitable for accurate quantitative measurement.

9.4   POSITRON EMISSION TOMOGRAPHY

For imaging with SPECT, it is necessary to collimate the detectors in order to define the lines taken through an object by γ rays emitted from the internal distribution of radioactive material. This introduces position-dependent artefacts, as we noted at the end of the last section, but, more importantly, it reduces the efficiency by which the radiation can be detected and this increases the time required to produce an image or the amount of radioactive material needing to be taken internally to give an image in a given time. In positron emission tomography (PET), a compound labelled with a positron-emitting radioisotope is introduced into the patient in a manner similar to that used for SPECT imaging. However, in this case, the detectors need not be collimated, which increases the overall detection efficiency and, thereby, reduces the radiation exposure to the patient.

The PET technique is based on the fact that positron annihilation usually occurs when the positron is essentially at rest, resulting in two 511-keV γ rays being emitted close to 180° relative to each other (see Section 5.4.3). This means that if two 511-keV γ rays are detected in coincidence by two detectors D_1 and D_2, as indicated in Figure 9.7, the γ rays must have been emitted from a point (e.g. P), which lies on the line connecting D_1 and D_2. Note that this statement does not require the detectors to be collimated. Also, the emitting nuclide must lie close to this line, since the positron typically has an energy less than 0.5 MeV and travels no more than a few millimetres before coming to rest. Thus, the coincidence count rate is a measure of the integral of the positron-source activity within the patient along the line segment AB, after effects of attenuation have been taken into account.

A ring of detectors, as shown in Figure 9.7, defines a plane or slice through the patient and records coincidences from pairs of γ rays along all possible directions within the plane. A complete set of data from all the different combinations of

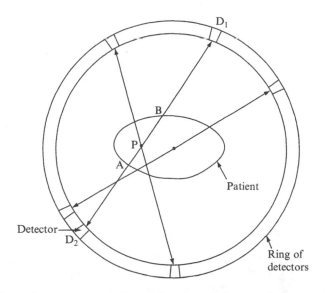

Figure 9.7 Schematic arrangement for performing positron emission tomography. Each positron, from a β^+ emitter inside the patient, results in two 511-keV γ rays emitted in opposite directions. These may trigger two detectors in the ring simultaneously. For example, detectors D_1 and D_2 will record coincident γ rays from β^+ emitters distributed along the line segment AB through the patient. The coincident count rate, therefore, is a measure of the line integral of the β^+ source distribution along AB. Other detector pairs measure different line integrals through the patient.

detector pairs in the ring, therefore, contains all the information necessary to generate a full set of line integrals of the source distribution $S(x, y)$ within the slice. An image of $S(x, y)$ can then be reconstructed from this set using CT, as described in the previous section. The best PET systems give images with a spatial resolution of 5–10 mm, limited mainly by the physical size of the detectors and the finite distance travelled by the positrons before undergoing annihilation.

The most frequently used radionuclides for PET are ^{11}C, ^{13}N, ^{15}O and ^{18}F. The first three are isotopes of elements found in bio-organic compounds and, therefore, are ideal tracers for medical studies. All are commonly produced in small, dedicated cyclotrons, using beams of protons or deuterons. Some of the reactions used to generate them are listed in Table 9.1, together with their half-lives and common diagnostic uses. Medical applications of PET for clinical research and diagnosis are well established. In oncology, PET is used to detect and grade primary tumours before surgery and to assess their response to therapy. Applications in neurology include the localization of seizure sites in epilepsy and the early detection of Alzheimer's disease. Two examples of its use to detect tumours are shown in Figure 9.8. Before being scanned, patients ingest radioactive glucose (taken as ^{18}F-deoxyglucose), which is found to accumulate selectively in tissue, such as malignant tissue, where the metabolic rate is abnormally high. The localized activity shows as dark regions and several tumours are clearly indicated on the PET images.

These images also show that the labelled tracer reaches the brain and a rapidly growing application of PET is its use for mapping brain function. Oxygen and glucose are examples of substances which can cross the blood–brain barrier and by using radioactive ^{15}O or ^{18}F-deoxyglucose, it is possible to trace these substances in the brain using PET. The images highlight areas of increased activity where the blood flow and metabolic rate are relatively high. In this way, areas of abnormal activity in a diseased brain can be identified by comparing with a PET image of a normal brain. The procedures are sensitive enough to reveal regions in the brain that are activated

Table 9.1   Common radionuclides used in PET.

Nuclide (half-life)	Production reactions	Common beam energies (MeV)	Diagnostic uses
^{11}C (20.4 m)	^{14}N $(p, \alpha)^{11}$C	14	Dopamine binding (brain) Heart metabolism Amino acid metabolism (cancer detection)
^{13}N (10.0 m)	^{16}O $(p, \alpha)^{13}$N ^{13}C $(p, n)^{13}$N	20 8	Heart blood flow Protein synthesis
^{15}O (2.0 m)	^{14}N $(d, n)^{15}$O ^{15}N $(p, n)^{15}$O ^{16}O $(p, pn)^{15}$O	8 8 29	Brain blood flow Oxygen metabolism Blood volume
^{18}F (109.8 m)	^{18}O $(p, n)^{18}$F ^{20}Ne $(d, \alpha)^{18}$F	14 14	Glucose metabolism (all tissues) Dopamine synthesis (brain)

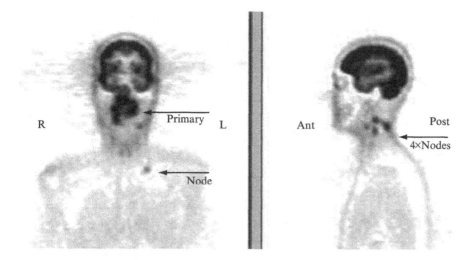

Figure 9.8 PET images of the human head and chest region. The patient ingested ^{18}F-labelled glucose before the PET scan. A primary tumour is shown in the nasal region in the front view (left). Several smaller tumours (nodes) are evident in the side view (right). Courtesy D. Hastings, Christie Hospital, Manchester.

as a result of external stimuli such as audio or visual impressions. The ability to be able to obtain such information, non-invasively, is stimulating much new research on correlations between mental processes and special functions of the brain.

9.5 MAGNETIC RESONANCE IMAGING

Magnetic resonance imaging (MRI) is a tomographic technique, which differs from PET and X-ray CT in that it uses a property of stable nuclei, rather than potentially hazardous radiation, to obtain information about the interior of an object. Whereas X-ray CT probes electron density and PET determines the distribution of internal radioactive material, MRI measures the magnetization due to nuclear magnetic moments in the object being scanned. Hydrogen occurs in great abundance in tissue and for this and other practical reasons, it is the behaviour of proton magnetic moments in an applied field that is usually measured in MRI.[2] During its development, MRI was used mainly to reveal anatomical structure and it is especially useful for differentiating between soft tissues, a task which is difficult to do with X-ray CT. More recently, the technique has been extended to kinetic studies and it is now capable of showing details of localized activity inside the brain and other parts of the body.

In MRI, the way protons interact with externally applied magnetic fields causes a signal to be generated in an external receiver, from which the spatial distribution of hydrogen atoms in the sample can be reconstructed. Also, by measuring the time

---

[2] Note that a nuclear magnetic moment is associated with the spin of the nucleus, and MRI cannot be used to study the distribution of even–even nuclei, which have zero spin.

variation of the recorded signal, it is possible to obtain information about variations in the local chemical environment in which the protons exist. A complete, rigorous derivation of MRI lies well beyond the scope of this text and we present here a brief outline only of the main physics principles that form the basis of the technique. References are given in the bibliography.

9.5.1   Principles of MRI

As noted in Section 1.3.1, the proton is a spin-1/2 particle which has a magnetic moment μ_p. In a magnetic field $\mathbf{B}$, there is an interaction energy $(-\mu_p \cdot \mathbf{B})$ and the proton has two energy states $\pm \mu_p B$ corresponding to its two magnetic substates $m_s = \pm 1/2$. The energy difference $\Delta E = 2\mu_p B = hf$, where f is the frequency of a photon, which will cause a transition between the two states. This is called the Larmor frequency and it corresponds to the classical precession frequency of a magnetic moment about an axis parallel to a magnetic field. The energy splitting is typically very small. For a proton in a field of 1 T, $\Delta E = 1.76 \times 10^{-7} \text{eV}$ and the Larmor frequency is 42.6 MHz. As we shall see, the fact that f depends on the magnetic field is of crucial importance in MRI.

If we have a collection of protons at a temperature T in a magnetic field, the energy splitting ΔE gives rise to unequal populations in the two states, according to the Boltzmann factor $\exp(-\Delta E/kT)$, and a stable equilibrium value of magnetization $\mathbf{M}$ (net magnetic moment per unit volume), pointing in the direction of $\mathbf{B}$, will be reached after a period of time. This time is known as the spin–lattice or longitudinal relaxation time T_1. Immediately after the field is applied, both states are equally populated and $\mathbf{M} = 0$, but energy can be exchanged directly between the proton and its surroundings and this allows the population ratio to change towards the thermal equilibrium value. This spin–lattice relaxation process is quite weak and a typical time for an energy exchange to take place is in the range of milliseconds to seconds.

Nothing is detected if the system is in equilibrium with $\mathbf{M}$ parallel to $\mathbf{B}$. However, a signal is generated if $\mathbf{M}$ has a component $M_\perp$ in the plane orthogonal to $\mathbf{B}$. Suppose $\mathbf{M}$ has been rotated into this plane, as shown in Figure 9.9. There is angular momentum associated with the magnetization and, therefore, $\mathbf{M}$ will precess about

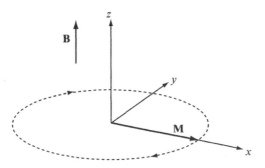

Figure 9.9   Precession of the magnetization $\mathbf{M}$ in the xy plane due to a torque exerted by an external magnetic field $\mathbf{B}$ along the z-axis. $\mathbf{M}$ precesses at the Larmor frequency and electro-magnetic energy is radiated preferentially in the xy plane as it does so.

B under the influence of a torque $\mathbf{M} \times \mathbf{B}$. The precession frequency is equal to the Larmor frequency (see Problem 9.10) and as **M** precesses, the external field due to **M** will vary with time at this frequency and induce an e.m.f. in a pickup coil surrounding the object. This is the MRI signal. Its strength is a maximum when **M** lies in the plane of rotation (and $M_\perp$ is maximum) and is zero when **M** is parallel to the z axis at $\theta = 0°$ or $180°$ (and $M_\perp$ is zero). The strength of the signal is a measure of the number of precessing protons in the sample.

The direction of **M** relative to **B** can be changed by applying an alternating magnetic field to the sample at the Larmor frequency and, if this is done in a particular way, **M** can be rotated through a predetermined angle. Consider the equilibrium situation with **M** pointing initially in the z direction along the main static field **B**. A r.f. field, of amplitude B_{ex}, is then applied at right angles to **B**. This alternating field is equivalent to the superposition of two rotating fields, each of constant magnitude $B_{ex}/2$, one rotating clockwise and the other anticlockwise in the xy plane at the Larmor frequency. Thus, as **M** precesses about **B**, one component of the r.f. field rotates in phase with it. This component applies a steady torque on **M** and causes it to precess about another axis perpendicular to **B**, i.e. in addition to precessing about **B**, **M** also precesses about an axis which is rotating at the Larmor frequency f in the horizontal plane. The ensuing motion is complicated, but, if it is viewed from a frame of reference which itself is rotating at frequency f about **B**, as shown in Figure 9.10, we see **M** precessing about the x' axis at an angular velocity proportional to B_{ex}.

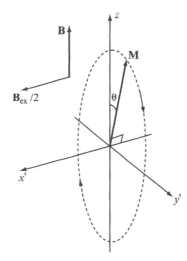

Figure 9.10   View taken from a frame of reference rotating at the Larmor frequency about the z-axis, which is parallel to the main DC magnetic field **B** (see Figure 9.9). An additional magnetic field $B_{ex}/2$ is applied along a direction (x', say) in the plane orthogonal to **B**. This field is rotating about the z-axis at the Larmor frequency and, therefore, is in phase with magnetization **M** as it precesses about **B**. This rotating field component causes **M**, initially parallel to **B**, to precess about the x'-axis (as shown) at a rate which depends on its magnitude $B_{ex}/2$.

The angle of rotation θ depends on the strength of $\mathbf{B}_{ex}$ and the time for which it is applied (see Problem 9.11). It can be controlled and is usually set to be either 90° or 180°. This is the excitation phase of the MRI measurement. It is important to note that the frequency of the external r.f. field must match the Larmor frequency of the protons to be excited. This is the resonance condition and it is used in the following way to excite selectively proton magnetic moments in a chosen slice of tissue.

9.5.2   Excitation of a selected region

First the object is placed in the fixed field $\mathbf{B}$ pointing along the z axis. A second static field $\mathbf{B}_{zz}$ is turned on which is parallel to $\mathbf{B}$ and is z dependent. This means that the Larmor frequency, which is proportional to the magnetic field, will vary along the z axis. Thus, when the r.f. field $\mathbf{B}_{ex}$ is applied with a narrow band of frequencies about f_{ex}, the only protons to be resonantly excited will be those within a narrow slice, of thickness dz at the particular values of z corresponding to the narrow band of frequencies. The principle is illustrated in Figure 9.11.

The exciting field $\mathbf{B}_{ex}$ is applied until the magnetization in the slice has been rotated through either 90° or 180° depending on what measurements are to be taken. Both $\mathbf{B}_{ex}$ and $\mathbf{B}_{zz}$ are then turned off.

9.5.3   Readout and MRI image formation

In order to construct a spatial image of the magnetization $\mathbf{M}(x, y)$ in the excited region, the recorded MRI signal must be encoded with information that enables its different points of origin to be identified. There are a number of ways this can be done and we shall outline, as an example, one method, used in early applications of

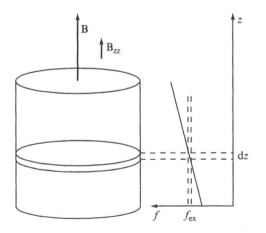

Figure 9.11   Excitation of a selected slice of tissue. The combination of the applied magnetic fields $\mathbf{B}$ and $\mathbf{B}_{zz}$ has a z dependence and, therefore, the Larmor frequency f also varies with z as shown in the graph on the right-hand side. Protons within a narrow slice, thickness dz, will resonate at frequencies within a narrow band about f_{ex}.

MRI, which is based on the CT reconstruction principles outlined in Section 9.3.[3] In this method, encoding the spatial information is achieved by turning on a static magnetic field $\mathbf{B}_{z\phi}$ during the time the signal from the excited region is being recorded. This field is also parallel to $\mathbf{B}$, but has a strength which varies linearly along a chosen direction ϕ with respect to the x-axis, as shown in Figure 9.12. Thus, the precession frequency of protons in the slice will vary along that direction and the recorded signal will be made up of a spectrum of different precession frequencies. The result of this is that the strength of any frequency component of the total recorded signal is a measure of the integral of the magnetization $\mathbf{M}(x, y)$ along the line corresponding to that frequency. Therefore, by analysing the total signal as a function of frequency, we obtain a set of quantities which are measures of the line integrals through the slice in a direction perpendicular to ϕ. By varying ϕ in repeated measurements, a full set of line integrals through the slice is then obtained, from which an image of $\mathbf{M}(x, y)$ can be reconstructed using CT.

9.5.4 Time variations of the signal

We now consider how the amplitudes of the signal recorded from different parts of the slice vary with time.

If $\mathbf{B}$ were uniform throughout the excited region, all the protons would precess at the same frequency and remain in phase. Spin–lattice relaxation processes would gradually cause the system to revert to equilibrium with $\mathbf{M}$ parallel to $\mathbf{B}$ and, therefore, the radiated signal would decrease with time at a rate depending on T_1. However, there are always small irregularities in the field experienced by the protons either because of inhomogeneity of the external field $\mathbf{B}$ or because of local internal microscopic fields associated with the magnetic moments of neigbouring nuclear and

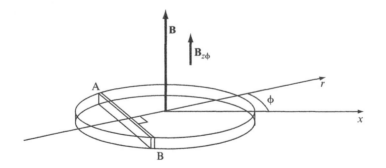

Figure 9.12   Diagram illustrating readout from a slice of tissue generating a MRI signal. During readout, a field $\mathbf{B}_{z\phi}$ is applied parallel to the main magnetic field $\mathbf{B}$. Its strength varies linearly along the r-axis, which is at an angle ϕ to the x-axis. Protons in the line element AB, which is perpendicular to r, experience the same magnetic field and, therefore, all contribute to the MRI signal at a given frequency corresponding to that field. This frequency will depend on the position of the line element along the r-axis.

---

[3] There are many more techniques available for MRI than for CT and we refer the reader to Hobbie (1997), Chapter 17, and to Hendee (1999) for further details and references.

atomic spins in the material. These spatial variations mean that protons in different places precess at slightly different rates, and the signal decays because the magnitude of $M_\perp$ decreases as the individual moments lose phase coherence with each other. This is illustrated in Figure 9.13, which shows precession in the xy plane observed (as in Figure 9.10) from a frame of reference rotating about **B** at the average precession frequency. The directions of the individual nuclear moments become increasingly spread out in phase angle with time. The net value of $M_\perp$, therefore, decreases with a characteristic decay time and eventually reaches zero. This decay time is called the spin–spin or transverse relaxation time T_2. It is normally much shorter than T_1 and typical values are in the range of one to a few milliseconds. Thus, if the excitation field $\mathbf{B}_{ex}$ is applied to rotate **M** through 90°, dephasing will cause the signal to decay with the characteristic time T_2, which can be measured. If $T_2 \ll T_1$, this will occur long before the spin–lattice interaction has caused **M** to reset back to equilibrium in the z direction. However, both T_1 and T_2 can be measured in the following way.

If, after a time τ say, the excitation field $\mathbf{B}_{ex}$ is reapplied long enough to rotate all the precessing proton magnetic moments through 180° about $\mathbf{B}_{ex}$, the moments will continue to precess with slightly different frequencies, as before, but will now move back into phase with each other. The process is shown schematically in Figure 9.14. The signal builds up again to a maximum after a time 2τ, in what is called a *spin echo*, and then decays again according to T_2. This cycle of re-excitation and readout can be done repeatedly and the spin-echo signal will recover each time, as shown in Figure 9.15, but with a gradually reducing amplitude as energy exchanges due to spin–lattice relaxation cause an increasing number of proton magnetic moments to revert to their equilibrium condition of alignment with **B** at a rate depending on the spin-lattice relaxation time constant T_1.

Thus, it is possible to measure not only the magnetization **M** from the signal strength, but also the relaxation times T_1 and T_2 from the time dependence of the signal. All three quantities vary spatially within the body. Images can be made of them and each gives different useful information. For example, relaxation times are often different for tumour tissue compared with normal tissue, and tumours can be identified by comparing images where the time allowed for relaxation has been

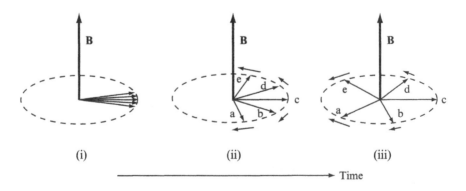

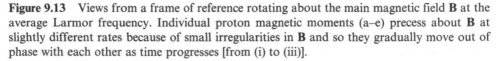

Figure 9.13   Views from a frame of reference rotating about the main magnetic field **B** at the average Larmor frequency. Individual proton magnetic moments (a–e) precess about **B** at slightly different rates because of small irregularities in **B** and so they gradually move out of phase with each other as time progresses [from (i) to (iii)].

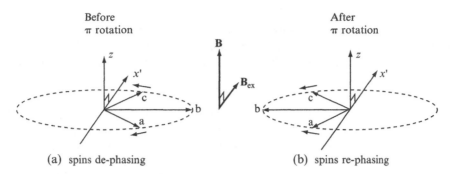

Figure 9.14 Principle of the spin echo. Illustration (a) is viewed from a rotating frame (as in Figure 9.13) and shows three representative, precessing proton magnetic moments moving apart in phase. An excitation field $\mathbf{B}_{ex}$ (parallel to x') is applied to rotate all the magnetic moments by 180° about x', as shown, until they reach the configuration shown in (b). $\mathbf{B}_{ex}$ is then turned off and the moments continue to precess, as before, but now move in phase towards each other.

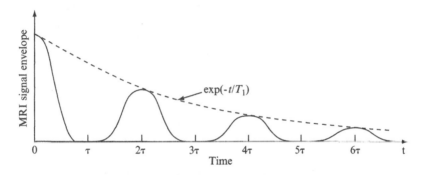

Figure 9.15 Time variation of a sequence of spin-echo signals. Following an initial rotation of the magnetization $\mathbf{M}$ on to the xy plane (see Figure 9.10) at time $t = 0$, the amplitude of the signal decays at a rate depending on the spin–spin relaxation time T_2. After a time τ, $\mathbf{M}$ is rotated through 180° (see Figure 9.14), which causes the moments to move back in phase and the signal grows to a new peak, called a 'spin echo', at time 2τ. If the 180° excitation is reapplied at 3τ, 5τ, ..., spin echoes recur at 4τ, 6τ, ..., but with decreasing intensity due to spin–lattice relaxation realigning the proton magnetic moments in the equilibrium direction.

varied. There are many different excitation–readout sequences used by MRI practitioners, other than the one outlined above, each enabling a different combination of measurable quantities to be obtained related to the nature of the scanned object. This flexibility of use, and the sensitivity of the technique, has established MRI as one of the most sophisticated tools used today for clinical diagnostic work and medical research.

9.5.5 Functional MRI

Relatively recent developments in MRI have been directed to studying the way the brain functions. The method, called functional magnetic resonance imaging (fMRI), is based on the fact that a change in blood flow to a region of tissue alters the magnetic susceptibility in that area. Magnetic susceptibility is proportional to the net magnetization acquired by a substance when it is exposed to a magnetic field. If the susceptibility is changed in the tissue, the magnetization induced by the external field changes and, therefore, any local magnetic field gradients due to the induced magnetization also change. As we have noted above, local field gradients directly affect the spin–spin relaxation time T_2.

The magnetic susceptibility of blood depends on the oxygenated state of blood hæmoglobin, since deoxyhæmoglobin is paramagnetic, whereas oxyhæmoglobin is diamagnetic with a susceptibility similar to that of normal tissue. An increase in nerve cell activity causes more blood to flow locally, which increases the amount of oxygenated blood and decreases the concentration of deoxyhæmoglobin. The magnetic susceptibility of blood in the activated region, therefore, becomes closer to that of normal tissue. This reduces inhomogeneities in the degree of magnetic susceptibility and, therefore, in the associated microscopic magnetic field variations throughout the region. This leads to a small but measurable increase in T_2 which can be detected in the MRI signal. The technique is referred to as BOLD (blood oxygenation level dependent) imaging because it depends on the level of local oxygen concentration in the blood. If the stimulus is applied cyclically, the activated area will exhibit a corresponding pattern of increased and decreased activity, and very small effects can be detected because the MRI signal-to-noise ratio can be improved by averaging the measurement over many cycles. An advantage of MRI over PET and X-ray CT is that it is completely non-invasive, involves no external radiation dose and so can be repeated many times on the same subject.

Examples of fMRI images are shown in Figure 9.16 and the coloured plate facing p. 3. Data for Figure 9.16 were obtained as a series of images of brain cortex while the

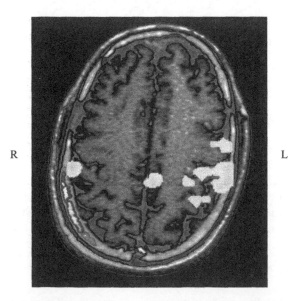

Figure 9.16   fMRI image of a slice through a human head showing details of the brain cortex. The white regions in the brain indicate areas of increased activity resulting from tactile stimuli. Courtesy P. Meltzer, Vanderbilt University, USA.

right-hand finger of the subject was being stroked with a brush. The light patches indicate increased activity in the cortical and motor areas in both hemispheres. However, the activation is more extensive in the hemisphere opposite to the stimulated finger. The coloured plate shows a section through the skull of a volunteer, who was cyclically given visual stimuli and periods of rest in 30-s intervals. The red patches indicate the activated areas responsible for processing the sensation of sight.

9.6 RADIATION THERAPY

As noted in Section 9.1, the use of radiation to treat cancer dates back to the days when the destructive effects of radiation were first being realized. By damaging DNA, radiation disables the ability of an affected cell to reproduce (see Section 7.4.2) and so can be used to remove unwanted tissue. Today, radiation therapy is a standard procedure, which is used in conjunction with other treatments designed to combat cancer (surgery and chemotherapy). Of overriding concern is the need to minimize the negative effects of radiation exposure, which are the damage and destruction of healthy tissue and the possible risk of inducing a new cancer in the irradiated areas. Unfortunately, many cancers and tumours are less oxygenated and, therefore, are often more radiation resistant than normal tissue. This can be alleviated to some extent by using drugs, reducing oxygen to all the tissue in the affected area or, as we shall describe, by fractionating the dose or by using high LET radiation for which the oxygen-enhancement ratio is low (see Section 7.4.3).

Different types of radiation are used for therapy. Most treatments employ photon or β emitters because they are cheaper and more readily available than alternatives which require sources of neutrons or charged particles. However, the latter do have particular applications and we shall mention some recent developments of their use.

9.6.1 Photons and electrons

In most cancer therapy facilities, photons are generated using a small, linear accelerator capable of accelerating electrons up to an energy of about 6 MeV. The electron beam strikes a target which emits X-ray photons and bremsstrahlung radiation, and a beam of these photons is directed on to the region to be irradiated.

Normally, the accelerator system is mounted on a movable gantry above the patient, and the photon beam is passed through a collimator whose shape is adjusted to match the tumour profile as closely as possible. The collimator restricts lateral exposure satisfactorily but, inevitably, normal healthy tissue is irradiated in front of and behind the tumour. To reduce this dosage to normal tissue, the tumour is usually irradiated from different directions. The principle is simple to understand and is illustrated in Figure 9.17. The tumour is always irradiated and receives a lethal dose whereas, although more normal tissue is exposed, the dose to it can be maintained at a sub-lethal level. Note that photons are attenuated exponentially with depth and, therefore, the dose to the tissue also generally decreases with depth. Even if a tumour is irradiated from different directions, it is generally better to use penetrating

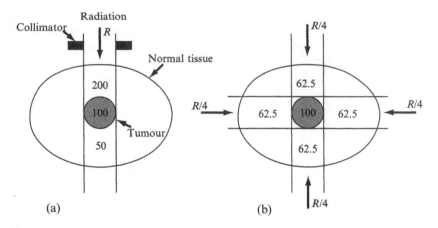

Figure 9.17   Effect of irradiating a tumour from different directions. In (a), an amount R of collimated radiation is directed from one direction only. The dose decreases approximately exponentially with depth of penetration and, therefore, is greater in front of the tumour than at the tumour itself. The numbers are for illustrative purposes only and indicate doses delivered to tissue before and beyond the tumour that are double or half that given to the tumour itself, which is arbitrarily set to be 100. In (b), radiation is delivered in four smaller amounts ($R/4$) from four different directions, as shown. The tumour receives the same dose as in (a), i.e. 100 units, but, although more tissue is exposed, its average dose is reduced to below that delivered to the tumour.

radiation to treat deep-seated cancers than radiation which is more easily absorbed (see Problem 9.12).

   In some instances, the electron beam itself is applied directly to treat a cancerous area. Electrons have lower penetration ($\simeq 0.5\,\text{cm/MeV}$) than photons of the same energy and, therefore, are better suited for treating surface tissue without affecting deeper and more critical anatomical regions.

   Photon therapy, or treatment using any low LET radiation, is usually found to be more effective if the dose is given as a series of small doses rather than as a single large one. There are thought to be two main reasons for this: one is that the survival probability of normal tissue is improved and the second is that the chance of killing tumour cells is increased due to an effect known as tumour reoxygenation.

   As was explained in Section 7.4.2, the survival curve of a cell often exhibits a shoulder at low doses because of the cell's ability to repair the damage done to DNA, if it is not too great. If there is a shoulder, fractionating the dose effectively changes the slope of the curve for a large accumulated dose, as shown in Figure 9.18, which illustrates a situation in which a dose is given as a series of equal fractions. If enough time is allowed between exposures for all sub-lethal damage to be repaired, the shoulder is reproduced for each dose fraction. The net survival curve, shown dashed, exhibits no shoulder, but is shallower than if the accumulated dose were given in a single exposure (curve A). In short, lowering the dose rate for normal, oxygenated tissue reduces the biological effect of the total dose.

   Fractionation has a different effect on a tumour. Tumours often consist of rapidly multiplying cells which outgrow their blood supply. This leads to a tumour having a structure, as shown in Figure 9.19, with a hypoxic core surrounded by a region which

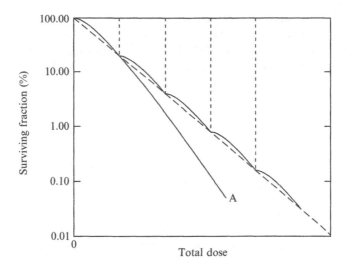

Figure 9.18 Results of an ideal fractionation experiment. Curve A represents the survival curve of cells which can repair sub-lethal damage but are subject to a single exposure. The dashed curve is obtained if the total dose is given as a series of equal fractions with enough time being allowed between each exposure for damage repair to take place.

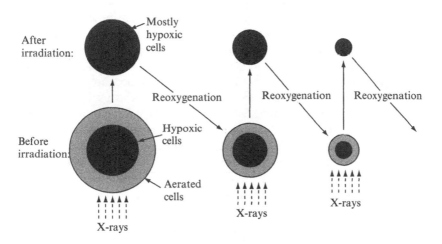

Figure 9.19 Effect of reoxygenation on the reduction of tumour volume irradiated in a series of dose fractions. Radiation selectively destroys the more radiosensitive aerated cells (rich in oxygen) after which, during reoxygenation, cells in the hypoxic core (low in oxygen) receive blood and become aerated. Successive irradiations repeat the cycle.

is relatively rich in oxygen. If such a tumour is exposed to a single dose, cells in the oxygenated surface region are killed while those in the radio-resistant core survive. However, if the dose is fractionated, then, as each smaller dose kills some of the aerated cells in the surface, blood then becomes available to supply and reoxygenate the outer layer of the core before the next dose fraction is applied. The next fraction kills this newly reoxygenated layer and the process continues, as illustrated in Figure 9.19, causing the tumour to shrink in size until it is destroyed altogether. For the treatment to be most effective, the time between the application of each dose fraction should match the time needed for tumour reoxygenation to take place.

9.6.2   Radionuclides

Many localized cancers are treated with radiation emitted by radioactive nuclides. These are introduced into the region to be irradiated either surgically, e.g. on the tip of a needle or wire, or by injection or ingestion with the radionuclide attached to a compound which is taken up selectively into the cancerous tissue. The ideal radio-nuclide for therapy is a pure β emitter, since β particles are less penetrating than γ rays and, therefore, deliver their dose locally. This is opposite to the requirements for diagnostic use where the radiation must emerge to trigger a detector and should deposit as little energy as possible in the tissue.

Since the early 1940s, the treatment of hyperthyroidism and thyroid cancer has been based on the specific uptake of iodine by thyroid tissues and it is the most common example of targeting a radioactive substance to a specific organ. A calculated amount of radioactive iodine administered to the patient accumulates in the thyroid and delivers the required dose during the time it remains there.

An effective, modern method of treatment is immunobiology in which an antibody is used to deliver a radionuclide to a tumour. An example is antiferritin which will target sites producing ferritin, an iron-storage protein. Ferritin is secreted preferentially by a broad range of malignancies relative to normal tissue and, thus, the labelled antibody, antiferritin, will carry the radionuclide selectively into the diseased regions. Iodine-131 was used in most of the early studies to label antibodies, because experience over the years in using radioactive iodine to treat thyroid cancer was useful in estimating the required doses. However, ^{131}I emits 364-keV γ rays as well as β particles (0.2 MeV) and a dose necessary to treat a cancer incurs a large, general exposure to the patient who normally has to be hospitalized. As the technique became better established, ^{131}I was replaced with ^{90}Y, which is almost a pure β emitter of relatively high average energy (0.9 MeV). Compared with ^{131}I, ^{90}Y delivers a much higher relative dose to the tumour than to normal tissue and treatment can be given on an outpatient basis.

9.6.3   Neutron therapy

Neutrons constitute penetrating radiation, which is also high LET and, therefore, should be more effective than photons for treating radio-resistant tumours. However, the results of several studies indicated that improvements in treatment effectiveness were less marked than had been hoped, mainly because is is difficult to collimate a neutron beam satisfactorily to achieve an optimum dose distribution between the tumour and normal tissue. Also, neutron therapy requires a special facility near a nuclear reactor or accelerator and so is much more costly than treatment with photons or electrons. Consequently, after the initial surge of interest in the 1940s and 1950s, the number of facilities delivering neutron therapy declined.

More recently, however, there has been renewed interest in the use of slow (thermal) neutrons for therapy as a result of developments which promise to improve radically the dose distribution. The new approach is known as boron neutron cancer therapy (BNCT). In BNCT, the element boron is enriched in tumour cells by attaching it to a compound that specifically targets tumour sites. The presence of boron increases the number of nuclear interactions in any boron-loaded tissue exposed to slow neutrons because ^{10}B has a very high, thermal-neutron reaction cross section (see Figure 6.10). Also, the resulting compound nucleus ^{11}B breaks up into two fragments (^{7}Li and an α particle), which share an energy release of 2.79 or 2.31 MeV (see Section 6.6.1). These fragments deposit their energy in a very short distance in a cell and, therefore, cells containing ^{10}B are preferentially and efficiently killed. The technique of immunobiology, referred to above, promises to improve the effectiveness of BNCT by using boron-carrying antibodies which selectively bind at sites characteristic of malignant tissue. Results are encouraging and research into this area is continuing.

9.6.4 Heavy charged particles

In contrast to photons and neutrons, protons and heavier charged particles exhibit an inverted dose profile with depth due to their interaction with matter becoming stronger as they slow down. Figure 9.20 illustrates qualitatively how the resulting absorbed dose (energy per unit mass) delivered to tissue varies for different types of radiation. Photons and neutrons are attenuated exponentially in matter and the doses they deliver exhibit a general decrease with depth of penetration. Doses due to charged particles (e.g. nitrogen ions), on the other hand, exhibit the opposite behaviour. As explained in Section 5.2.3, the rate of energy loss of a fast charged particle passing through matter increases with increasing penetration to a maximum, called the Bragg peak, and then falls off sharply close to the end of the range (see Figure 5.2). The position of the Bragg peak depends on particle energy, and can be

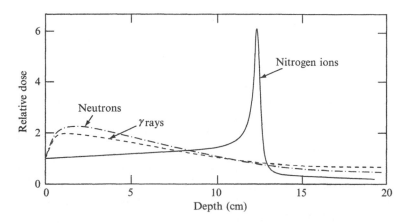

Figure 9.20 Composition of relative dose–depth distributions in tissue for ^{60}Co γ rays, neutrons and nitrogen ions. The doses delivered by neutrons and γ rays exhibit a mainly exponential decay with depth, whereas nitrogen ions have an inverted profile, typical of charged particles, with a high relative dose near the end of their range.

shifted to any required depth. Arranging that it lies within the volume of a tumour, will lead to a significant improvement in the ratio of tumour dose to that of surrounding tissue. Moreover, compared with high-energy photons and neutrons, charged particles are much easier to collimate and there is less deviation due to scattering as they pass through the tissue.

For these reasons, improved spatial resolution and dose distribution can be achieved with charged particles, and a number of facilities exist which use protons for cancer treatment. However, proton therapy does require access to an accelerator capable of delivering upwards of 60-MeV protons and, therefore, is quite expensive. Nevertheless, results suggest significant advantages for cases where precise targeting of the dose is important. For example, protons are particularly well suited for treating localized tumours in the eye and at the base of the skull, certain prostate and uterus cancers and, in general, any tumour situated near a critical organ for which exposure to radiation should be avoided if at all possible.

More recently, techniques involving active spot scanning and the use of heavy ions, such as carbon, are being developed to give an even better match in three dimensions of delivered dose to tumour volume and with a greater range of biological response than can be achieved with other types of radiation. In most therapy centres, passive collimators and degrader foils are used to match the region irradiated with the beam to that of the tumour. In active scanning, a beam is focused to a small spot and scanned with magnetic deflectors in two orthogonal directions. In this way, the Bragg peak, corresponding to the maximum dose, is scanned within a given two-dimensional slice of the tumour and, by varying the energy and beam intensity, a three-dimensional dose distribution can be tailored and delivered to a tumour volume of any irregular shape.

Much of this work is being developed with heavy ions, such as carbon, because they can be delivered to a particular location with good spatial precision. As described in Section 5.2, charged particles lose energy in matter by colliding with atomic electrons. The heavier the ion, the less it is deflected from its path by these interactions and, therefore, the smaller will be the increase in beam spot size with depth of penetration due to multiple scattering. Heavy ions are also favoured by the fact that the fractional variation in range decreases as the mass of the ion is increased. There are two factors contributing to this. First, the kinetic energy of a heavy ion, in general, must be considerably greater than that of a light ion in order to penetrate a given distance. For example, a 1-GeV carbon ion has approximately the same range as a 50-MeV proton. Second, the fractional energy loss suffered by an ion in a collision with an electron varies inversely as its mass M [see Equation (5.1)], so a carbon ion will suffer a much greater number of collisions N before coming to rest than will a proton with the same range. The range depends on N and the variation in the range depends in part on the statistical variation of this number, which is given approximately by $\sqrt{N}$. Therefore, the fractional variation in the range, which varies as $1/\sqrt{N}$, decreases as M is increased. With a carbon beam, for example, it is possible to work to a depth precision below 1 mm, which allows good control of the dose profile near the boundary between a tumour and normal tissue.

High-energy, heavy charged particles may also offer significant advantages for treating certain types of deep-seated tumour that may be radio-resistant to X-rays to such an extent that they would require a dose which would do unacceptable damage to surrounding sensitive tissue. Indeed, carbon ions may be particularly suitable because of the way the relative biological effectiveness varies along the path as they slow down. At the beginning of the path, as the particle enters the tissue, the energy is high and the rate of energy loss low enough that the effect of the irradiation is comparable with that of low LET radiations such as photons and low-energy β particles. Near the end of the range, the LET and local ionization increase dramatically, as we have noted in Figure 9.20, and even radio-resistant tumours with a high capacity to withstand photon irradiation will become radio-sensitive to heavy ions.

Another feature of high-energy, heavy ions, which is attracting attention, is that they can deposit positron-emitting radionuclides in the tissue, which allow the irradiated region to be imaged using PET. For example, as a beam of ^{12}C ions passes through tissue, a small fraction of them undergo nuclear reactions which remove one or two neutrons converting them into ^{11}C or ^{10}C nuclei, respectively. The velocity of these residual nuclei is not changed a great deal by these neutron-transfer reactions from that of the ^{12}C ions and they stop close to where the primary beam stops. Both ^{11}C and ^{10}C are positron emitters with half-lives of 20.4 m in and 19 s, respectively, and they generate annihilation radiation that can be detected and imaged as described in Section 9.4. In this way, it is possible to realize a dream of radiation therapists, which is to be able to monitor directly how the beam is being delivered to a tumour without incurring any additional radiation dose to the patient they are treating.

PROBLEMS 9

9.1   **Table 9.2**   Mass attenuation coefficients μ_m (cm^2 g^{-1}) for γ rays or X-rays.

Material	Density (g cm^{-3})	E_γ (keV)				
		60	140	159	364	511
Tissue	1.0	0.20	0.15	0.15	0.11	0.097
Bone	1.8	0.32	0.16	0.15	0.10	0.090
Copper	8.9	1.5	0.25	0.20	0.098	0.081
NaI(Tl)	3.67	5.7	0.66	0.52	0.13	0.090

(a)   Using data listed in Table 9.2, calculate the percentage transmission of γ rays of energies 60 and 511 keV through a person represented by 20 cm of tissue.
(b)   What would be the additional reductions in transmitted intensities if the person had swallowed a copper coin, 1 mm thick?

9.2

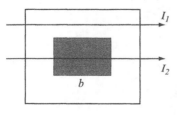

The sketch is a schematic representation of a piece of bone surrounded by tissue irradiated with 60-keV γ rays. If the ratio of transmitted γ-ray intensities I_1 (through tissue) to I_2 (through tissue-plus-bone) is 2, calculate the thickness of the bone using the information in Table 9.2.

9.3 Using the information given in Table 7.4 and Section 9.2.2, estimate the ratio of doses to the thyroid of using either ^{131}I or ^{123}I for imaging purposes. Imaging is done using a gamma camera, which contains a piece of NaI(Tl) scintillator (thickness $x = 1$ cm). Assume that either radioisotope is taken up quickly into the thyroid and that the time to acquire an image is much less than the half-life. Also assume, for simplicity, that detection efficiency is proportional to the probability of a γ ray being attenuated in the NaI(Tl) scintillator.

9.4

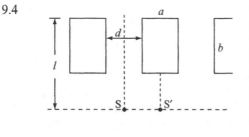

The diagram shows a section through a gamma-camera collimator along a line joining the centres of two adjacent channels. Dimensions are such that rays from a point S, directly below a channel, cannot pass through any other channel – assuming that a photon is stopped if it strikes the collimator. The width d of a channel is small compared with the distance l.

(a) For a given spatial resolution (determined by d), calculate the transmission of the collimator (fraction of rays from the source at S, which pass through the collimator) if the channel is (i) circular, (ii) square or (iii) hexagonal.

(b) Estimate the width a of the wall separating adjacent square channels if $l = 2b$ and if the transmission of rays from S′ (midway between two channels) and S are the same.

9.5 A gamma-camera system is used for both SPECT imaging (using ^{99}Tc^m) and PET imaging of a tumour in the centre of the brain. Obtain an approximate estimate of the ratio of efficiencies of PET and SPECT imaging. Note that the overall efficiency depends on the attenuation of γ rays in the brain and skull, the transmission of rays to the detector and the detector efficiency.

Information: Na(Tl) detector: radius $r = 7.5$ cm, thickness $x = 1$ cm. Brain: diameter $= 18$ cm, skull thickness $= 1$ cm. Distance of tumour to detector $= 20$ cm. Make reasonable assumptions in arriving at your estimate. In SPECT imaging, assume that the collimator has 5 mm square apertures (to give an overall spatial resolution of about 10 mm) and has a transmission (see Problem 9.4) that is independent of the source of radiation in the tumour. In PET imaging (with no collimator), assume that if one of the 511-keV γ rays enters the 15 cm diameter detector, there is a coincidence with the other γ ray, which enters a similar (but larger) detector on the opposite side of the brain.

9.6 The positron-emitting nucleus ^{15}O is produced in the ^{15}N(p, n)^{15}O reaction at a certain energy with a cross section of 50 mb. If a 1 cm thick cell, filled with ^{15}N gas, is bombarded with a 1-μA beam of protons, calculate the minimum gas pressure (at standard temperature) to produce 1 MBq of ^{15}O.

9.7 You require an amount of ^{99}Mo (half-life $t_{1/2}^{Mo} = 66$ h) for use as a generator of metastable technetium ^{99}Tc^m ($t_{1/2}^{Tc} = 6$ h). It should be possible to extract an activity of at least 5 MBq of ^{99}Tc^m every 6 h (assume 100% efficiency for this) and the generator should last 1 week before needing to be replaced. What activity of ^{99}Mo should you order? About 85% of the ^{99}Mo decays populate the metastable state

^{99}Tc^m. Assume that $t^{Mo}_{1/2} \gg t^{Tc}_{1/2}$. You may need to refer back to Sections 1.5.4 and 1.5.5 and Equation (1.17).

9.8 Calculate the Larmor frequency of a proton in a magnetic field of 1 T.

9.9 Show that, in a sample containing N hydrogen atoms per unit volume at temperature T, the difference between the numbers of protons with spins parallel and spins anti-parallel to a magnetic field B is approximately equal to $\mu_p NB/kT$ for reasonable values of B and T, where $\mu_p = 1.41 \times 10^{-26}$ J T^{-1} is the magnetic moment of the proton and k is the Boltzmann constant. Hence, determine M, the magnitude of the magnetization per unit volume of water, at 20°C in a field of 1 T.

9.10 Using the relationship between magnetization and proton magnetic moment from Problem 9.9 and the classical result that torque = rate of change of angular momentum, show that, in general, the magnetization M, due to aligned proton magnetic moments (μ_p), will precess at the Larmor frequency ($f = 2\mu_p B/h$) about a constant magnetic field B.

9.11 Using the result of Problem 9.10, show that the magnetization M, initially aligned with a constant magnetic field B, would precess about a rotating axis, as shown in Figure 9.10, with an angular frequency of $\mu_p B_{ex}/\hbar$, where B_{ex} is the magnitude of an external r.f. magnetic field applied at 90° to B.

Hence, calculate the value of B_{ex}, applied at the Larmor frequency, that would cause M to precess 90° in 10^{-5} s.

9.12 Consider radiotherapy treatment of a tumour in the centre of the brain by irradiating it with γ rays from four directions, as illustrated in Figure 9.17. Assume (for simplicity) that the dose to tissue (brain or tumour) decreases exponentially with depth x as $\exp(-\mu_{en}x)$, where μ_{en} is an energy-deposition coefficient that is related to, but is not the same as, the attenuation coefficient μ defined in relation to Equation (5.16).

Show that it is better to use penetrating radiation for treating deep-seated tumours. Do this by considering two radiation energies, with different values of μ_{en}: 0.2 and 0.1 cm^{-1}, and estimate doses, relative to that given to the tumour, at points A and B midway between the surface of the brain (radius 9 cm) and its centre.

9.13 Fifty per cent of cells in a piece of tissue survive after being irradiated with 10 equal dose fractions, given at intervals sufficiently far apart in time for cell recovery to take place. Calculate the surviving fraction if the total dose were given in a single exposure. Assume that the cell-survival probability is given by Equation (7.7).

10

Power from Fission

10.1 INTRODUCTION

Nuclear fission was first observed in the late 1930s in the same decade as the discovery of the neutron by Chadwick. Scientists studying nuclear phenomena quickly realized that this new particle was ideal for creating hitherto unknown nuclei because it is uncharged and there is no Coulomb barrier to prevent it entering a nucleus and causing a nuclear reaction. In the following years, many experiments were carried out using neutrons to produce new nuclei, including investigations with uranium as the target, which were designed to create nuclei heavier than any that are encountered on earth.

A far-reaching result of this work was the discovery by Hahn and Strassmann in 1939 that one of the products of bombarding uranium with neutrons was barium, an element near the middle of the periodic table. Evidently, the heavy nucleus had divided into two fragments of intermediate mass. It was also found that there was a great release of energy when the nucleus was split in two in this way. A further, crucial observation was that when fission takes place several neutrons are emitted, which can go on to induce further fissions enabling the process to continue in a chain reaction. A chain reaction enables large amounts of energy to be obtained once the reaction has been initiated. However, a chain reaction is unstable. If it is uncontrolled, the reaction rate can grow exponentially and very rapidly and is the basis of the fission bomb. However, as is explained in Section 10.5, it is possible to stabilize the chain reaction, and controlled energy obtained from a fission reactor is the source of all commercial nuclear power used today.

The basic physics which determines why nuclei undergo fission is discussed in Chapters 2 and 4. Here, we are concerned more with the consequences of fission and its exploitation. In Section 10.2, we describe how energy released in fission is shared among the various reaction products and how the cross sections for fission induced by neutrons vary with energy and target nucleus. The key factors, which determine the design of a reactor, are presented in Section 10.3 using, as a simple example, a hypothetical infinite system. How we can then determine the size of a practical, finite reactor is discussed in Section 10.4, which may be omitted without loss of continuity. Some important aspects of reactor operation are discussed in Section 10.5. In Section 10.6, we describe several different types of reactor, which have been and are in use commercially throughout the world. Finally, in Section 10.7, we discuss briefly some aspects dealing with how nuclear power may be used in the future.

10.2   CHARACTERISTICS OF FISSION

10.2.1   Fission and fission products

Neutron-induced fission proceeds in several stages. If we consider ^{235}U as the target nucleus, neutron capture leads to the formation of the compound nucleus ^{236}U in an excited state. Sometimes this nucleus decays by γ emission in a capture reaction, but more often it undergoes fission when some of the excitation energy goes into deforming the nucleus, as shown in Figure 10.1. The behaviour may be likened to that of a liquid drop driven into oscillation by the impact of another (smaller) droplet. Break-up will occur if the amplitude of the oscillation is sufficiently great. The nucleus is like a charged droplet and when the deformation reaches a certain critical point, the repulsive Coulomb force driving the deformation overcomes the attractive nuclear force, and the nucleus breaks apart into two large fragments and several neutrons. Immediately after separation, the two fragments acquire approximately 170 MeV of kinetic energy from the Coulomb potential energy as the electric force continues to drive them apart (see Problem 10.1). This forms the bulk of the prompt release of energy in fission. Energy is also carried by neutrons and γ rays. The latter arise because the fission fragments are usually created with several MeV of excitation energy and they de-excite to their ground states by γ emission.

An example of a typical fission reaction is

$$n + {}^{235}U \rightarrow {}^{236}U^* \rightarrow {}^{147}La + {}^{87}Br + 2n.$$

It is not unique and many different mass partitions for the fission fragments are possible, as is shown in Figure 10.2 for low-energy induced fission of ^{235}U. Fragment mass numbers vary between 70 and 160 with the most probable values being about 96 and 135. Symmetric fission is relatively rare and occurs only once in about 20 000 fissions.

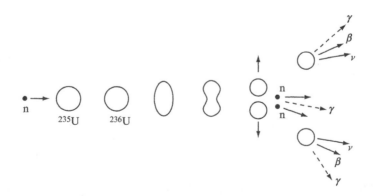

Figure 10.1   Stages in the variation of nuclear shape leading to fission.

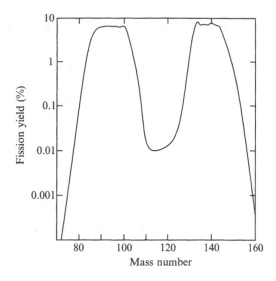

Figure 10.2 Distribution of fission fragment masses from the fission of ^{235}U induced by thermal neutrons.

Many fission fragments are radioactive, because they are formed from a nucleus which has a neutron-to-proton ratio of over 1.5, whereas, in the mass region 70 to 160, the corresponding stable nuclei have lower ratios (1.3–1.4). Thus, the fragments are neutron rich and decay towards stability mainly by a series of β^- emissions although, in a few cases, the decays also include a neutron emission. These decay sequences are called the fission decay chains. The radioactive decay energy from the fission chains is the second release of energy due to fission. It is much less than the prompt energy, but it is a significant amount and is why reactors must continue to be cooled after they have been shut down and why the waste products must be handled with great care and stored safely.

Some of the fission fragments are nuclides that happen to have very large cross sections for neutron capture and, even in small quantities, can remove a significant number of neutrons which otherwise could be used to continue the chain reaction. These nuclides are called *reactor poisons* and we shall see in Section 10.5 that they can have an important effect on reactor operation.

10.2.2 Fission energy budget

The total energy released in fission can be estimated from the binding energy per nucleon which, as shown in Figure 1.1, is about 7.6 MeV u^{-1} for uranium and in the mass region near $A = 117$ it is about 8.5 MeV u^{-1}. Thus, the change in binding energy when fission takes place produces about 0.9 MeV u^{-1}, equivalent to a total of 212 MeV for ^{235}U. Results of a more accurate calculation are summarized in Table 10.1, which also shows how the energy is distributed among the various products of fission.

Table 10.1   Distribution of emitted energy from neutron-induced fission of ^{235}U.

	Emitted energy (MeV)
Prompt energy:	
Fission fragments	168
Fission neutrons	5
γ emission: (photons and internal-conversion electrons)	7
Radioactivity:	
β decay: (electrons)	8
β decay: (neutrinos)	12
γ emission: (photons and internal-conversion electrons)	7
Total	207

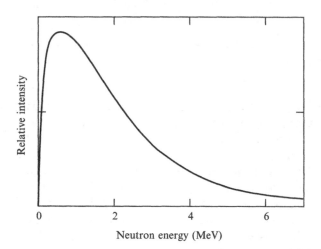

Figure 10.3   Energy spectrum of neutrons emitted in the thermal-neutron induced fission of ^{235}U.

About 87% of the total energy is emitted promptly and, of this, the fission fragments account for over 90%. They are easily stopped and travel only a fraction of a millimetre from their point of origin. Prompt neutrons are emitted with a range of energies as shown in Figure 10.3. Their mean energy is about 2 MeV, and the 5 MeV listed in the table corresponds to an average of 2.5 neutrons per fission each with the mean energy. This energy is also recovered as heat mainly by the neutrons gradually losing their energy by successive scatterings in the reactor.

About 13% of the energy is released in radioactive decay. Energy carried by electrons, γ rays and internal-conversion electrons is absorbed and converted into heat, but the neutrino energy is not recoverable. A further small energy source of about 5 MeV (not listed in Table 10.1) comes from the radiative capture (n, γ) of neutrons which do not cause fission. The emitted radiation is also absorbed. The net result is that about 200 MeV per fission is recoverable for use and, although not exact, it is the generally accepted value.

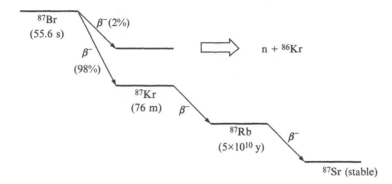

Figure 10.4   Delayed neutron emission following the decay of the fission fragment ^{87}Br; 2% of ^{87}Br decays are to excited states of ^{87}Kr, which have enough energy to decay rapidly by neutron emission to ^{86}Kr.

10.2.3   Delayed neutrons

Over 99% of neutrons are released within about 10^{-14} s of the time fission takes place and are the prompt neutrons. However, as mentioned above, a few are released later as delayed neutrons during some of the fission decay chains. An example of a decay chain, which includes a neutron emission, is shown in Figure 10.4. The fission fragment ^{87}Br decays by β^- emission to ^{87}Kr with a half-life of 55.6 s. Two per cent of the decays are to excited states of ^{87}Kr, which lie above the neutron separation threshold and decay very rapidly by neutron emission to ^{86}Kr. These decays represent a small source of neutrons that are delayed by an average of 80 s, the mean β-decay lifetime of ^{87}Br. About 0.65% of all neutrons emitted (in the case of ^{235}U fission) are delayed. This is only a small fraction of the total but, as we shall see in Section 10.5, it is the delayed neutrons that make it possible for a nuclear reactor to be controlled.

10.2.4   Neutron interactions

There are three principal processes resulting from the interaction of neutrons in a reactor, which need to be treated separately. These are scattering, radiative capture and fission. The total cross section is the sum:

$$\sigma_T = \sigma_s + \sigma_c + \sigma_f \tag{10.1}$$

where the subscripts s, c and f refer to scattering, radiative capture and fission, respectively. The absorption cross section σ_a is the sum of the capture and fission cross sections i.e. $\sigma_a = \sigma_c + \sigma_f$. In the case of non-fissionable nuclei, of course, the absorption and capture cross sections are the same.

Cross sections are often highly dependent on E_n, the energy of the incoming neutron. Figure 10.5 illustrates this for ^{235}U and ^{238}U. Uranium-235 has a large fission cross section, which has a $1/v$ dependence at low energies where it dominates the total. The cross section also exhibits a number of resonances between 1 and

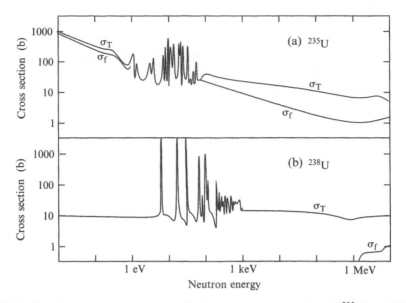

Figure 10.5   Total σ_T and neutron-induced fission cross sections σ_f for (a) ^{235}U and (b) ^{238}U as a function of neutron energy. From Hughes and Schwartz (1958).

100 eV. These are due to the existence of states in the compound nucleus ^{236}U and when the incoming neutron produces ^{236}U at an energy which matches one of these states, the reaction probability is greatly enhanced and gives a high cross section. The scattering cross section is typically 10 b and is relatively independent of energy. Thus, as σ_f falls with increasing E_n, σ_s assumes a greater fraction of the total. At 2 MeV, for example, σ_s is approximately six times greater than σ_f.

The dominant isotope of uranium ^{238}U behaves very differently to ^{235}U. Fission is zero at low neutron energies and only begins to occur for E_n greater than about 1 MeV. The scattering cross section is approximately 8 b over the energy range shown. It represents the largest contribution to σ_T, except in the region between about 10 and several hundred electron volts where there are resonances in the absorption cross section, just as there are in the ^{235}U interaction. However, in this case, absorption leads to neutron capture (n, γ) and not fission. This is important because it can be a significant source of neutron loss in a reactor. It is also an example of a so-called *breeder reaction* as is explained below.

10.2.5   Breeder reactions

Neutron capture by ^{238}U leads to the isotope ^{239}U. This nucleus is radioactive and is the start of a nuclear decay chain that produces first, ^{239}Np and, eventually, ^{239}Pu, which is long lived. The set of processes is

$$n + {}^{238}U \longrightarrow {}^{239}U \ (23\,m) \xrightarrow{\beta^-} {}^{239}Np \ (2.36\,d) \xrightarrow{\beta^-} {}^{239}Pu \ (2.4 \times 10^4\,y)$$

where half-lives are indicated in minutes (m), days (d) and years (y). The importance of this reaction is that ^{239}Pu is fissile, which means it can easily be induced to undergo fission with neutrons of all energies. In this regard, its properties are similar to those of ^{235}U and it can be used in a reactor with slow as well as fast neutrons.

Thorium-232 is another naturally occurring nuclide, which behaves similarly to ^{238}U. It will undergo fission with fast neutrons above about 1.4 MeV, but will capture lower-energy neutrons leading, eventually, to the production of ^{233}U which is also fissionable with slow neutrons. The processes leading to the formation of ^{233}U are

$$\text{n} + {}^{232}\text{Th} \longrightarrow {}^{233}\text{Th} \ (22\,\text{m}) \xrightarrow{\beta^-} {}^{233}\text{Pa}(27\,\text{d}) \xrightarrow{\beta^-} {}^{233}\text{U} \ (1.6 \times 10^5\text{y}).$$

The use of these reactions for breeding new fissile fuel is discussed in Section 10.7.

10.3 THE CHAIN REACTION IN A THERMAL FISSION REACTOR

We have already noted that neutron-induced fission generates extra neutrons which can induce further fissions in the next generation and so on in a chain reaction. The chain reaction is characterized by the *neutron multiplication factor k*, which is defined as the ratio of the number of neutrons in one generation to the number in the preceding generation. If, in a reactor, k is less than unity, the reactor is subcritical, the number of neutrons decreases and the chain reaction dies out. If $k > 1$, the reactor is supercritical and the chain reaction diverges. This is the situation in a fission bomb where growth is at an explosive rate. If k is exactly unity, the reactions proceed at a steady rate and the reactor is said to be critical.

It is not possible to build a bomb or a reactor consisting only of natural uranium because over 99% of it is ^{238}U which absorbs too many neutrons to allow k to exceed unity. It is only if uranium is highly enriched, for example, in ^{235}U, that a chain reaction in uranium can be established with $k > 1$. This is the basis of the *fast* reactor, which is discussed in Section 10.7. The word 'fast' indicates that fission is caused by neutrons that have not been slowed down significantly. However, if we turn our attention again to Figure 10.5, we see that very low-energy (thermal) neutrons are very effective in causing fission in ^{235}U, so much so that its high fission cross section for thermal neutrons more than offsets its low abundance, which is about 0.7% in natural uranium. It makes it possible to achieve criticality in a reactor using natural uranium as fuel, provided that the neutrons have been efficiently moderated to thermal energies (see Section 5.5.2). A reactor operating with thermal neutrons is called a *thermal* reactor and in the remainder of this and the next three sections we shall consider only this type of reactor.

10.3.1 A nuclear power plant

A schematic layout of a typical power plant, based on a thermal fission reactor, is shown in Figure 10.6. The reactor core contains a mixture of fuel, moderator and other materials such as fuel cladding, structural components, etc. It usually consists

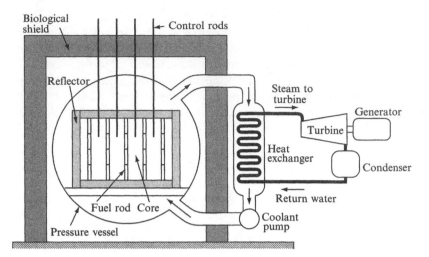

Figure 10.6   Simplified schematic layout of a typical reactor power plant.

of a heterogeneous arrangement of fuel rods, which are separate from but surrounded by moderator material. Heat generated from fission reactions in the core is removed by a circulating gas or liquid coolant.

In most reactors, the fuel is either natural uranium or uranium enriched in ^{235}U. In some designs, the moderator is a solid (e.g. graphite), with the coolant circulated through open channels; in others, water (both light H_2O and/or heavy D_2O) is used as both moderator and coolant. Sometimes the water moderator is circulated through the core acting also as coolant or it may be piped separately as coolant in a number of loops through the fuel matrix sitting in a static pool of moderator. The core is surrounded by a *reflector* made of material that has a low absorption and high scattering probability for neutrons. Its purpose is to reflect back into the core some of the neutrons which leak out and would otherwise be lost to the system. Control rods, containing neutron-absorbing elements (boron or cadmium) are often mounted above or below the reactor. They are driven in and out of the core in order to maintain the overall k factor at unity, once the desired level of operating power has been reached.

The whole assembly is encased in a pressure vessel. As we shall see later, some pressure vessels must be capable of withstanding internal pressures up to 160 bar. A biological shield, normally several feet of concrete, surrounds the entire system. Its purpose is to attenuate the intense γ and neutron radiations to levels that are safe for humans outside the plant.

The coolant is pumped through the core inside the pressure vessel and through heat exchangers outside, where steam is generated and used to drive turbines for generating electric power. There may be four to eight separate cooling loops and heat exchangers so that one can be shut down and serviced while the reactor remains operating.

Many different types of reactor have been designed and built. Some of those that are used commercially are described briefly in Section 10.6, and in Section 10.7 we consider the breeder reactor and an accelerator-driven system as possible reactors for providing energy in the future.

10.3.2   The neutron cycle in a thermal reactor

If we wish to determine the multiplication factor k for a particular reactor assembly, we must take into account everything that can happen to a neutron in the cycle of events leading from one generation to the next. Calculating what happens inside any realistic reactor is very complicated and sophisticated computer codes exist for this purpose. Here, we shall illustrate our discussion of the general principles with the example of a homogeneous reactor in which ^{235}U is the fissile nucleus.

In a homogeneous reactor, the fuel and moderator are intimately mixed, which means that we can assume that the neutron flux is the same for both components. This is a major simplification. In a heterogeneous reactor, where the fuel is distributed in finite-sized lumps surrounded by moderator, neither the flux nor the neutron energy spectrum can be assumed to be the same for fuel and moderator and we shall see at the end of this section that this has a significant effect on k.

Relevant data, which we shall use in several examples, are listed in Table 10.2. The quantity v is the average number of neutrons emitted when a fissile nucleus (in this case, ^{235}U) undergoes fission. Data are given for neutrons moving at 2200 m s^{-1}, this being the most probable thermal speed at 20°C. Numbers will be different in a hot reactor and this is one of many reasons why k will be temperature dependent.

A reactor is a mixture of different nuclei in different proportions, and the basic equations for evaluating the factors in k are those which give average reaction rates, cross sections, etc., for such mixtures. The average cross section for a mixture is the weighted mean of the constituent microscopic cross sections σ_i:

$$\sigma = \sum w_i \sigma_i = \sum N_i \sigma_i / \sum N_i \qquad (10.2)$$

where N_i is the number density of component i of the mixture. For example, a mixture of uranium enriched to 1.6% in ^{235}U has an absorption cross section (using data from Table 10.2) given by

$$\sigma_a = 0.016 \times 680 + 0.984 \times 2.72 = 13.56 \text{ b.} \qquad (10.3)$$

The macroscopic cross section for a mixture is defined as

$$\Sigma = N\sigma = \sum N_i \sigma_i = \sum \Sigma_i \qquad (10.4)$$

Table 10.2   Uranium data for reactor calculations.

Nucleus	Density (g cm^{-3})	σ_f (b)	σ_c (b)	σ_a (b)	σ_s (b)	v
^{235}U	18.7	579	101	680	10	2.42
^{238}U	18.9	–	2.72	2.72	8.3	–
^{nat}U	18.9	4.17	3.43	7.60	8.3	–

where $N = \sum N_i$ is the number density of all atoms in the mixture and $\Sigma_i = N_i\sigma_i$ is the macroscopic cross section of component i in the mixture. We see from Equation (10.2) that the σ in Equation (10.4) is the average microscopic cross section of a mixture and that the macroscopic cross section Σ is simply the sum of the macroscopic cross sections Σ_i of the components.

There are several factors that combine to determine the probability that a neutron produced in a reactor will survive to initiate another fission in the chain reaction. These factors are basic to the design of a reactor and we consider each of them in turn below.

(1) For each thermal neutron absorbed in the fuel, a number eta (η) of fast neutrons are produced. This is less than the average number v of neutrons produced in a fission reaction because not all neutrons absorbed in the fuel cause fission. Indeed, both ^{235}U and ^{238}U have cross sections which do not result in fission (see Table 10.2). Thus, η is equal to v multiplied by the fraction of the total absorption cross section in the fuel that leads to fission. For example, if the neutron is absorbed by pure ^{235}U, we have

$$\eta = v \times \left[\frac{\sigma_f}{\sigma_f + \sigma_c} \right] = 2.06. \tag{10.5}$$

For natural uranium fuel, which contains 0.72% ^{235}U, the value of η for the mixture is

$$\eta = v \times \left[\frac{0.72\sigma_f(235)}{0.72\sigma_a(235) + 99.28\sigma_c(238)} \right] = 1.328. \tag{10.6}$$

We can vary η by changing the enrichment: e.g. with an enrichment of 1.6%, which we shall take as our design example, η is equal to 1.654.

(2) Some fast neutrons cause fission before they are slowed down. This increases the number of neutrons by what is called the *fast fission factor* ε. However, many thermal reactors are *dilute* reactors in which the ratio of moderator to fuel atoms is large and, in a dilute reactor, there is very little fission caused by fast neutrons because they most probably interact first with moderator atoms and quickly lose energy. Consequently, in all our examples, we shall take the fast fission factor to be unity.

(3) As we have noted, fast neutrons lose energy mainly by scattering off moderator atoms. As they slow down, a fraction of them will be captured by ^{238}U, particularly in the resonance region between about 10 and 100 eV (see Figure 10.5). The probability that a neutron will *avoid* resonance capture by ^{238}U is the *resonance escape probability* p. It is an important factor, which depends on the rate at which neutrons are slowed down and on the number N_{238} of ^{238}U atoms relative to moderator atoms. It is not simple to calculate and an approximate expression for a homogeneous system is[1]

$$p = \exp\left[-\frac{2.73}{\langle \xi \rangle} \left\{ \frac{N_{238}}{\Sigma_s} \right\}^{0.514} \right] \tag{10.7}$$

---

[1] For example, see Bennet and Thomson (1989), section 4.5.

where Σ_s is the total macroscopic scattering cross section, assumed to be independent of neutron energy. The logarithmic energy decrement ξ for a single substance was defined in Equation (5.30). The quantity $\langle \xi \rangle$ in Equation (10.7) is the mean logarithmic energy decrement for a mixture, with each constituent value ξ_i weighted according to its scattering probability. Thus,

$$\langle \xi \rangle = \sum N_i \sigma_{s_i} \xi_i \Big/ \sum N_i \sigma_{s_i}. \tag{10.8}$$

In a dilute reactor, both Σ_s and $\langle \xi \rangle$ are approximately the same as the corresponding values for the pure moderator (see Table 10.3 below).

As expected, p increases with $\langle \xi \rangle$ because the larger the value of $\langle \xi \rangle$ the greater is the rate at which the neutrons are slowed down; p also increases with moderator-to-fuel ratio because this increases the probability of the neutron being scattered by a moderator atom relative to its interacting with a ^{238}U nucleus. For example, in a reactor containing 1.6% enriched uranium in graphite with a moderator-to-fuel ratio of 600, Equations (10.7) and (10.8) give $p = 0.749$ (see Problem 10.3).

(4) Once they have been thermalized, the neutrons diffuse as they continue to be scattered in the reactor. Some are absorbed by the moderator and various structural components, but a fraction is absorbed back into the fuel. This fraction is called the *thermal utilization factor f*. It is given by

$$f = \frac{\Sigma_a(F)}{\Sigma_a(C)} = \frac{\Sigma_a(F)}{\Sigma_a(F) + \Sigma_a(M) + \cdots} = \frac{N_F \sigma_a(F)}{N_F \sigma_a(F) + N_M \sigma_a(M) + \cdots} \tag{10.9}$$

where C, F, M $\cdots$ refer to the core, fuel, moderator, etc. In our design example of 1.6% enriched uranium with 1 atom of uranium per 600 atoms of carbon, we shall consider absorption due to fuel and moderator only in the core. After substituting $\sigma_a(F) = 13.56\,\text{b}$ from Equation (10.3), $\sigma_a(M) = 0.0045\,\text{b}$ (from Table 10.3) and $N_M/N_F = 600$ into Equation (10.9), we obtain $f = 0.834$.

So far, we have forgotten about neutrons which may leak out of a real finite reactor. A fraction l_f of these may be fast neutrons and there will be a further fractional loss l_s of thermal neutrons. So the fraction of neutrons that do not leak out of the reactor is $(1 - l_f)(1 - l_s)$.

We now have all the terms we need to determine how many neutrons per single, original neutron survive to continue the cycle, i.e. we can write the neutron multiplication factor as:

$$k = \eta \varepsilon p f (1 - l_f)(1 - l_s). \tag{10.10}$$

The corresponding formula for an infinite reactor, for which there is no leakage, is

$$k_\infty = \eta \varepsilon p f \tag{10.11}$$

which is called the four-factor formula. Combining the factors for the example we have been using throughout this section (1.6% enriched U with a graphite/fuel ratio of 600), we obtain

$$k_\infty = 1.654 \times 1 \times 0.749 \times 0.834 = 1.033. \qquad (10.12)$$

We have a successful design and our reactor could be made to run at a critical level ($k = 1$), provided that we lose no more than 3.3% of the neutrons by leakage.

10.3.3   Moderator

It is clear from the above analysis that the choice of moderator is vitally important for the design of a successful thermal reactor. Neutrons must be slowed down as rapidly as possible to minimize their capture by ^{238}U in the resonance region and the moderator must not absorb too many of the neutrons itself.

As is described in Section 5.5.2, neutrons are slowed down mainly by successive scattering collisions with nuclei. This is done most effectively if the scattering nuclei are light, and there are three suitable materials which are widely used for this purpose: water, heavy water and graphite. Relevant properties of these are listed in Table 10.3.

A good moderator should be effective for slowing down neutrons, which implies a high scattering cross section and a high logarithmic energy decrement ξ. Both these properties are taken into account in the *slowing-down power*, which is defined as SDP $= \xi \Sigma_s$, where Σ_s is the macroscopic scattering cross section. For a moderator to be useful, it should also be weakly absorbing. Boron ($A \approx 11$), for example, would be more effective than carbon for slowing neutrons, but it has a high absorption cross section, which makes it useless as a moderator. Indeed, a common control system for thermal reactors consists of rods containing a neutron-absorbing material such as boron.

Each of the three listed moderator materials has certain advantages and disadvantages. Light water is cheap and has the highest values of σ_s and ξ. However, its absorption cross section is high and enriched uranium is needed in a reactor moderated with light water, which adds to the capital cost. Technically, heavy water (D_2O) is a very good moderator but its vastly greater cost, compared with either graphite or light water, has restricted its use to only a few commercial reactors, notably the Canadian CANDU (Canadian deuterium–uranium) series. Graphite is inexpensive but its mass number is rather high and, hence the logarithmic energy decrement ξ is comparatively small for graphite to be regarded as the ideal moderator. However, as we shall see below, it is possible to construct a reactor using only natural uranium and graphite.

Table 10.3   Properties of materials used as moderators.

Material	M.Wt	density ($g\,cm^{-3}$)	σ_s (b)	σ_a (b)	ξ	SDP (m^{-1})
H_2O	18.01	1.0	49.2	0.66	0.920	151
D_2O	20.02	1.1	10.6	0.001	0.509	18
Graphite	12.01	1.6	4.7	0.0045	0.158	6.0

10.3.4   Optimizing the design

We see from the above discussion that f is a decreasing function and p an increasing function of moderator-to-fuel ratio N_M/N_F. Thus, for a given enrichment of ^{235}U, there is a value of the ratio N_M/N_F for which k_∞ is maximum. In Table 10.4, values of η, f, p and k_∞ for a homogeneous assembly of graphite and uranium are given for various values of N_M/N_F and for two enrichments of ^{235}U: natural (0.72%) and 1.6%. From the table, it is evident that, for natural uranium, there exists no concentration at which the reactor would be self-sustaining. This is caused by the low values of η and p. The low values of p are because the slowing-down power of graphite is small (see Table 10.3) and thus, there is a significant probability that neutrons will be captured by ^{238}U in the resonance region. On the other hand, even with a small increase in enrichment, k_∞ can exceed unity. This is not due to an increase in p but to an increase in η. Note that the ^{235}U thermal fission cross section of 579 b is much higher than the ^{238}U thermal-neutron capture cross section of 2.72 b and, hence, η increases rapidly with enrichment.

A way of increasing p is to *clump* the fuel in the form of rods. With the rods well separated, the intervening graphite moderator thermalizes more of the neutrons leaving one fuel rod before they reach the next, thus reducing resonance absorption. Even more important is the fact that any neutrons with energies in the resonance region travel a very short distance in the fuel and, therefore, the uranium inside the surface layer of a fuel rod contributes relatively little to resonance absorption. This reduces the effective value of N_{238} in Equation (10.7), which increases p. In fact, a heterogeneous, graphite–natural uranium reactor can be made to be self-sustaining. The very first reactor, built in 1942 by a team working under Enrico Fermi in Chicago, was of this type. Of course, the fuel rods must not be too thick or thermal neutrons will not reach the interior and f will be reduced unnecessarily because the effective number of fuel atoms is reduced.

Any practical reactor is finite, and k_∞ must be greater than unity because neutron flux is lost by leakage at the surface. The size of a reactor depends on two principal quantities: (i) the amount by which k_∞ exceeds unity and (ii) the average distance neutrons are able to diffuse from their points of origin before being absorbed. Since k

Table 10.4   Uranium – graphite assemblies.

N_M/N_F	^{235}U: 0.72%; $\eta = 1.328$			^{235}U: 1.600%; $\eta = 1.654$		
	f	p	k_∞	f	p	k_∞
100.0	0.944	0.480	0.602	0.968	0.482	0.771
200.0	0.894	0.599	0.712	0.938	0.601	0.931
300.0	0.849	0.660	0.744	0.909	0.661	0.995
400.0	0.808	0.699	0.751	0.883	0.700	1.022
500.0	0.771	0.727	0.745	0.858	0.728	1.032
600.0	0.738	0.748	0.733	0.834	0.749	1.033
700.0	0.707	0.765	0.718	0.811	0.766	1.027
800.0	0.678	0.778	0.701	0.790	0.779	1.018
900.0	0.652	0.790	0.684	0.770	0.791	1.007
1000.0	0.628	0.800	0.667	0.751	0.801	0.994

must be unity for steady operation, $k_\infty - 1$ is a measure of the fraction of neutrons lost by leakage [see Equation (10.10)]. The smaller we build the reactor, the greater will be the leakage and the greater k_∞ has to be for steady operation. Furthermore, since only neutrons that reach the surface can escape, the more easily neutrons diffuse, the larger the reactor must be to maintain the leakage fraction at a given level. The next section describes in more detail the basic principles determining the size of a reactor.

10.4   THE FINITE REACTOR

Real reactors are finite in size. They are also mostly heterogeneous, as we have described earlier, with fuel rods arranged in a regular lattice surrounded by a moderator. Analysing such a reactor is very complex and we shall restrict ourselves to examining a homogeneous system that is simple enough for us to calculate. We shall derive the basic expressions which show how the design parameters, as outlined in the previous section, determine how large such a reactor has to be in order to go critical. The new feature we need to address is the diffusion of a neutron from its point of origin to its eventual fate, either by absorption within the reactor or by leakage out of it.

10.4.1   Diffusion

The classical theory of diffusion is based on the concept that moving particles, diffusing as a result of random collisions in a medium, will tend to become uniformly distributed within a given volume. If we consider a plane in the medium, e.g. $z =$ constant, there will be a flow of particles across it in either direction and we can define a current density J_z as the net number of particles crossing a unit area of the plane in the $+z$ direction per unit time. If the particle density n is uniform, $J_z = 0$, but if n is not uniform, e.g. if $n = n(z)$, there will be a net flow across the plane in an attempt to restore equilibrium at a rate which is proportional to the concentration gradient. Thus,

$$J_z = -D\frac{\partial n}{\partial z} \tag{10.13}$$

where D is a constant of proportionality called the *diffusion coefficient*. If $\partial n/\partial z > 0$, flow is in the negative direction, i.e. from high to low concentration. Similar equations determine current densities J_x and J_y due to concentration gradients in the x and y directions, respectively. The total current density is the vector sum:

$$\mathbf{J} = \mathbf{i}J_x + \mathbf{j}J_y + \mathbf{k}J_z = -D\nabla(n) \tag{10.14}$$

where $\nabla = \mathbf{i}\partial/\partial x + \mathbf{j}\partial/\partial y + \mathbf{k}\partial/\partial z$ and $\mathbf{i}$, $\mathbf{j}$ and $\mathbf{k}$ are unit vectors in the x, y and z directions.

We can use a simplified mean-free-path argument to obtain an estimate for D. First, we assume that all particles move at the same speed equal to their mean speed v. If there are n particles per unit volume, then on average one-third of them move predominantly along a given direction, e.g. the z direction. Half of these, or $n/6$ particles, have a velocity v in the $+z$ direction and $n/6$ have a velocity v in the $-z$ direction.[2] In a uniform situation, $nv/6$ particles cross the plane from below per unit time, and $nv/6$ cross from above and the net flow is zero. From the definition of the scattering mean free path λ_s, the particles crossing from below had their preceding scattering on average at a point one mean free path below z, i.e. at $(z - \lambda_s)$ and those crossing from above were last scattered at $(z + \lambda_s)$. If there is a concentration gradient, we can obtain from this the net current density crossing the plane in the $+z$ direction as

$$J_z = \frac{v}{6}n(z - \lambda_s) - \frac{v}{6}n(z + \lambda_s) = \frac{v}{6}\left[-2\lambda_s\frac{\partial n}{\partial z}\right] = -\frac{v\lambda_s}{3}\frac{\partial n}{\partial z} = -D\frac{\partial n}{\partial z} \qquad (10.15)$$

which gives a value for the diffusion coefficient D in terms of microscopic quantities.

10.4.2 The continuity equation

Equation (10.14) enables us to obtain a differential equation for the quantity n due to diffusion. Consider a volume element dV of dimensions dx, dy and dz as shown in Figure 10.7. The number of particles per unit time (current) entering the volume element through the bottom face is $J_z dx dy$ and the current leaving through the top face is $J_{z+dz}dx dy = (J_z + dJ_z)dx dy$. Thus, the net flow into the volume element due to current along the z axis is given by

$$J_z dx dy - (J_z + dJ_z)dx dy = -\frac{\partial J_z}{\partial z}dz dx dy = -\frac{\partial J_z}{\partial z}dV. \qquad (10.16)$$

Similar expressions can be derived for current flow along the x and y directions which, when combined, give the total net flow of particles per unit time into the volume. Dividing this by the volume element dV, we obtain the rate of increase of the particle density as

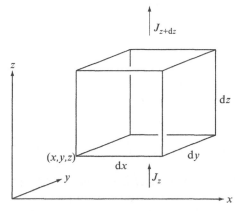

Figure 10.7 Volume element in Cartesian space showing net current density flowing in the z direction.

---

[2] We are taking the positive z-axis as pointing vertically upwards.

$$\frac{\partial n}{\partial t} = -\left[\frac{\partial J_x}{\partial x} + \frac{\partial J_y}{\partial y} + \frac{\partial J_z}{\partial z}\right] = -\nabla \cdot \mathbf{J}. \tag{10.17}$$

Substituting for $\mathbf{J}$ from Equation (10.14), we obtain

$$\frac{\partial n}{\partial t} = D\nabla^2 n \tag{10.18}$$

where

$$\nabla^2 = \frac{\partial^2}{\partial x^2} + \frac{\partial^2}{\partial y^2} + \frac{\partial^2}{\partial z^2}.$$

Equation (10.18) is the continuity equation for particles diffusing in a medium in which particles are being neither absorbed nor created. In order to describe neutrons in a reactor, we must add extra terms on the right-hand side of Equation (10.18) to account for neutron production and absorption.

From the definition of cross section given in Section 1.6.3, the reaction rate per absorbing nucleus is equal to the absorption cross section σ_a multiplied by the incident flux Φ. Thus, if we have N nuclei per unit volume, the rate of neutron loss per unit volume is given by $\Sigma_a \Phi$, where $\Sigma_a = N\sigma_a$ is the total macroscopic neutron absorption cross section. We denote the rate of neutron production per unit volume at $\mathbf{r}$ and time t as $S(\mathbf{r}, t)$. The continuity equation then becomes

$$\frac{\partial n}{\partial t} = D\nabla^2 n(\mathbf{r}, t) - \Sigma_a \Phi(\mathbf{r}, t) + S(\mathbf{r}, t). \tag{10.19}$$

Equation (10.19) is the time-dependent diffusion equation for neutrons in a medium. If the source term is time-independent, the neutron density distribution will eventually reach a steady-state situation with $\partial n / \partial t = 0$, giving the steady-state diffusion equation:

$$\frac{D}{v}\nabla^2\Phi(\mathbf{r}) - \Sigma_a \Phi(\mathbf{r}) + S(\mathbf{r}) = 0 \tag{10.20}$$

where we have used the fact that the neutron flux $\Phi(\mathbf{r}) = vn(\mathbf{r})$ (see Section 1.6.3) and assumed that the mean speed v does not vary with position.

10.4.3   Diffusion length

An important property of a moderator is the *diffusion length L*, which is related to the root-mean-square distance a neutron will diffuse in the medium before being absorbed. We can establish this relationship by considering how the neutron flux varies from a point source located at the origin in an infinitely extended medium. At points away from $\mathbf{r} = 0$ there are no source terms and Equation (10.20) becomes

$$\nabla^2\Phi(\mathbf{r}) = \frac{v\Sigma_a(M)}{D}\Phi(\mathbf{r}) = \frac{\Phi(\mathbf{r})}{L^2} \tag{10.21}$$

where $\Sigma_a(M)$ is the macroscopic absorption cross section for the moderator and we have introduced the quantity:

$$L^2 = D/v\Sigma_a(M) \tag{10.22}$$

where L has the dimension of length and is called the diffusion length.

For a point source at the origin, the system has spherical symmetry and we can look for solutions which are functions of r only: $\Phi = \Phi(r)$. Expressing $\nabla^2\Phi$ in spherical co-ordinates, Equation (10.21) becomes

$$\nabla^2\Phi = \frac{d^2\Phi}{dr^2} + \frac{2}{r}\frac{d\Phi}{dr} = \frac{\Phi}{L^2}. \tag{10.23}$$

Defining $u(r) = r\Phi(r)$ and substituting, we obtain

$$\frac{d^2u(r)}{dr^2} - \frac{u(r)}{L^2} = 0. \tag{10.24}$$

The general solution to this equation is

$$u(r) = ae^{-r/L} + be^{+r/L} \tag{10.25}$$

or

$$\Phi(r) = a\frac{e^{-r/L}}{r} + b\frac{e^{+r/L}}{r} \tag{10.26}$$

where a and b are constants. The particle flux vanishes infinitely far from the source, i.e. we have the boundary condition $\Phi(r) \to 0$ as $r \to \infty$, therefore, $b = 0$.

The probability of finding a neutron in the range r to $r + dr$ from the source is proportional to $\Phi(r)dV$, where dV is the volume element $4\pi r^2 dr$. Hence, the mean-square distance $\langle r^2 \rangle$ travelled by the neutron between its birth (at $r = 0$) and its absorption in the medium is given by

$$\langle r^2 \rangle = \frac{\int_0^\infty r^2\Phi(r)dV}{\int_0^\infty \Phi(r)dV} = \frac{\int_0^\infty r^3 e^{-r/L}dr}{\int_0^\infty r e^{-r/L}dr} = 6L^2. \tag{10.27}$$

10.4.4 Reactor equation

The steady-state neutron flux in a homogeneous reactor medium satisfies an equation of the form of Equation (10.20):

$$\frac{D}{v}\nabla^2\Phi(\mathbf{r}) \; - \; \Sigma_a(C)\Phi(\mathbf{r}) \; + \; S(\mathbf{r}) = 0 \tag{10.28}$$

(leakage)   [(absorption)]   [(production)]

where $\Sigma_a(C)$ is the macroscopic absorption cross section for the reactor core. For dilute reactors, the value of D is assumed to be that of the pure moderator.

The infinite neutron multiplication factor may be expressed as

$$k_\infty = \frac{\text{(total rate of neutron production in the core)}}{\text{(total rate of neutron absorption in the core)}} = \frac{\int S(\mathbf{r})dV}{\int \Sigma_a(C)\Phi(\mathbf{r})dV}.$$

The composition of a homogeneous reactor is uniform within its boundaries. Therefore, macroscopic cross sections are independent of position, and the rate of neutron production is proportional to the flux, as is the rate of neutron absorption. Hence, the *ratio* of production rate to absorption rate (k_∞) is also independent of position and we can write

$$k_\infty = \frac{S(\mathbf{r})}{\Sigma_a(C)\Phi(\mathbf{r})} \quad \text{for all } \mathbf{r} \tag{10.29}$$

whence

$$\frac{D}{v}\nabla^2\Phi(\mathbf{r}) + (k_\infty - 1)\Sigma_a(C)\Phi(\mathbf{r}) = 0.$$

This is usually written:

$$\nabla^2\Phi + \frac{(k_\infty - 1)\Phi}{L_c^2} = \nabla^2\Phi + B^2\Phi = 0 \tag{10.30}$$

where

$$L_c^2 = D/v\Sigma_a(C) \tag{10.31}$$

and

$$B^2 = (k_\infty - 1)/L_c^2. \tag{10.32}$$

Equation (10.30) is known as the *one-group* reactor equation, since it was derived by assuming that all neutrons form a single group moving at an average thermal speed. The quantity L_c is the thermal-neutron diffusion length in the core.

If the core can be regarded as fuel and moderator only, we have $\Sigma_a(C) = \Sigma_a(F) + \Sigma_a(M)$ and we can express L_c in terms of the diffusion length for the moderator L and the thermal utilization factor f. From Equation (10.9), we can write

$$(1 - f) = \frac{\Sigma_a(C) - \Sigma_a(F)}{\Sigma_a(C)} = \frac{\Sigma_a(M)}{\Sigma_a(C)} \tag{10.33}$$

whence, from Equations (10.22) and (10.31) we have our result:

$$L_c^2 = (1 - f)L^2 \tag{10.34}$$

where L_c refers to the diffusion length of neutrons in the reactor after they have been thermalized and neglects the distance they travel during moderation. We can take into account the distance travelled by the neutrons as they thermalize by introducing the *slowing-down* length L_s and replacing L_c^2 in Equation (10.30) by $L_c^2 + L_s^2$.
   Values of D, L^2 and L_s^2 for three common moderators are given in Table 10.5.

Table 10.5   Diffusion and slowing-down constants for moderators.

Moderator	D (cm)	L^2 (cm^2)	L_s^2 (cm^2)
H_2O	0.16	8.1	27
D_2O	0.87	30 000	131
Graphite	0.84	2650	368

10.4.5   Solving the reactor equation

The solution to the reactor equation links the basic reactor design parameters (k_∞, L_c) to the size of the reactor for criticality, i.e. steady-state operation. The critical size is determined by solving Equation (10.30) with the appropriate set of boundary conditions. This is straightforward to do for two particular geometries:

Rectangular geometry

Equation (10.30) becomes

$$\nabla^2 \Phi = \frac{\partial^2 \Phi}{\partial x^2} + \frac{\partial^2 \Phi}{\partial y^2} + \frac{\partial^2 \Phi}{\partial z^2} = -B^2 \Phi \tag{10.35}$$

which is solved by following a procedure similar to that outlined in Appendix B. The variables are separable, so we can write: $\Phi = X(x)Y(y)Z(z)$ which, after substitution, gives three independent equations:

$$\frac{1}{X}\frac{d^2 X}{dx^2} = \text{const.} = -\alpha^2 \text{(say)}, \quad \frac{1}{Y}\frac{d^2 Y}{dy^2} = -\beta^2 \quad \text{and} \quad \frac{1}{Z}\frac{d^2 Z}{dz^2} = -\gamma^2$$

where $B^2 = \alpha^2 + \beta^2 + \gamma^2$.
   The solutions are $X(x) = a\sin(\alpha x) + b\cos(\alpha x)$ (where a and b are constants of integration) and similarly for Y and Z. We now apply a boundary condition that the

flux vanishes only at the surfaces of the reactor, which, for a cubic reactor of length L, means that $\Phi \to 0$ at $x = 0$ and L. Thus, $b = 0$ and $\alpha L = \pi$. Similarly, $\beta L = \gamma L = \pi$, whence $B^2 = 3\pi^2/L^2$ and the volume of the reactor is

$$L^3 = 161/B^3.$$

Spherical geometry

In this case, we proceed, as we did in deriving Equation (10.24), by changing to spherical co-ordinates and seeking a solution with spherical symmetry, i.e. $\Phi = \Phi(r)$. Then, by substituting $u(r) = r\Phi(r)$ as before, we obtain

$$\frac{d^2u}{dr^2} + B^2u = 0. \tag{10.36}$$

The general solution to this equation leads to

$$\Phi(r) = u(r)/r = a\sin(Br)/r + b\cos(Br)/r$$

where a and b are the constants of integration. Since the flux density at the origin is finite, we must take $b = 0$. Assuming (as above) that the flux vanishes at the boundary of the reactor, we then obtain the condition $\Phi(R) = 0$, whence $BR = \pi$. Finally, we obtain for the volume V of the reactor:

$$V = 4\pi R^3/3 = 130/B^3.$$

In a real reactor, the flux does not fall exactly to zero at the physical boundary because there are neutrons leaking out. Also, if there is a reflector, some of the escaping neutrons will be scattered back into the core. The value of R we obtain using the above boundary condition is, therefore, larger than the actual physical radius of the core. However, the difference is typically only a few centimetres which, usually, is small compared with R and we neglect it here.

As an example, we shall calculate the radius of a spherical, uranium–graphite reactor, with a moderator-to-fuel ratio of 600 and a ^{235}U enrichment of 1.6%. From Tables 10.4 and 10.5, we take $k_\infty = 1.033$, $f = 0.834$, $L^2 = 2650\,\text{cm}^2$ and $L_s^2 = 368\,\text{cm}^2$, whence, we find that

$$B^2 = \frac{(k_\infty - 1)}{(1 - f)L^2 + L_s^2} = \frac{0.033}{0.166 \times 2650 + 368} = 4.08 \times 10^{-5}\,\text{cm}^2$$

and the radius of the reactor $R = \pi/B = 492\,\text{cm}$ or about 5 m.

Thus, we have found the size of a reactor based on our design of the previous section. Knowing its size and composition determines the total mass of the reactor and, hence, the mass of fuel we will need to put in it. The power of our reactor, however, depends also on the flux Φ and this is set by the reactor operating conditions.

10.5 REACTOR OPERATION

We now consider some aspects which affect the way a reactor operates. We shall begin with the all-important calculation of the power and fuel consumption. We shall then go on to consider the time variation of the neutron density in a reactor (which affects the ability to operate it in a steady state) and, finally, some problems caused by reactor poisoning, i.e. the production in the reactor of fission fragments with large neutron-capture cross sections.

10.5.1 Reactor power and fuel consumption

When a reactor is operating steadily, the neutron multiplication factor is held constant at unity. At startup, k is allowed to exceed this to allow the neutron flux to grow with time. Once the desired level is reached, k is then brought down and levelled off at unity to maintain a steady power output. Important parameters of an operating reactor are its power output, its rating (power output per unit mass) and its fuel consumption. These quantities depend on the total mass of fissile fuel and the cross sections for neutrons interacting with the fuel. They also depend on the neutron flux Φ which, following the definition given in Section 1.6.3, is the number of neutrons crossing unit area in unit time.

The reaction probability per unit time for N nuclei [from Equation (1.22)] is

$$R = N\Phi\sigma = MN_A\Phi\sigma/A \tag{10.37}$$

where σ is the microscopic cross section, M is the mass of fissile material, A is the mass number of the reacting nuclei, and N_A is Avogadro's number. By substituting the fission cross section σ_f for σ in this equation, we obtain the fission reaction rate R_f and if each fission liberates an amount E of recoverable energy, we obtain the power output as

$$P = R_f E. \tag{10.38}$$

For example, we shall calculate the power output, rating and fuel consumption for a thermal reactor containing 150 tonnes of natural uranium operating with a neutron flux of 10^{13} cm^{-2} s^{-1}. The basic data we need are: $\sigma_f(^{235}\text{U}) = 579$ b, $\sigma_c(^{235}\text{U}) = 101$ b (from Table 10.2) and recoverable energy per fission $E = 200$ MeV (see Section 10.2.2).

The fraction of ^{235}U atoms in natural uranium (mass number 238.03) is 0.72%. Therefore, the number of ^{235}U atoms in the reactor is

$$N(^{235}\text{U}) = \frac{1.5 \times 10^5 \times 0.0072 \times 6.022 \times 10^{26}}{238.03} = 2.73 \times 10^{27} \text{ nuclei.}$$

The reactor power is obtained immediately from Equations (10.37) and (10.38) as $P = N(^{235}\text{U})\Phi\sigma_f E = 507$ MW. The rating, or power output is about 3.4 MW per tonne of uranium.

Fuel consumption, leading to a loss of ^{235}U, depends on the total ^{235}U absorption cross section $\sigma_a = 680$ b. Therefore, the number of ^{235}U nuclei used up per second is equal to

$$N(^{235}\mathrm{U})\sigma_a\Phi = 2.73 \times 10^{27} \times 680 \times 10^{-24} \times 10^{13} = 1.86 \times 10^{19}\ \mathrm{s}^{-1}.$$

In one year ($\sim 3 \times 10^7$ s), this becomes 5.9×10^{26} nuclei or approximately one-fifth of the initial amount of ^{235}U.

10.5.2   Reactor kinetics

It is essential to be able to operate a reactor in a steady state with $k = 1$. In practice, the neutron concentration varies with time if k deviates from unity and these variations must be slow if we are to control them. This is what we shall examine now.

In a reactor operating with $q = k - 1 > 1$, the number of neutrons n will increase by an amount nq in a time known as the prompt neutron lifetime $t_p = t_s + t_d$, where t_s is the time for fast neutrons to become thermalized and t_d is the time they diffuse before being captured. Thus, $dn/dt = nq/t_p$ and the the number of neutrons will grow exponentially with time as

$$n(t) = n_o \exp(qt/t_p). \tag{10.39}$$

Generally, for most reactors, $t_s \ll t_d$ so that $t_p \approx t_d = \lambda_a/v$, where λ_a is the absorption mean free path of a thermal neutron in the core and v is its average speed. The absorption mean free path is equal to $1/\Sigma_a(\mathrm{C})$ (see Section 5.5.1), where $\Sigma_a(\mathrm{C})$ is the macroscopic absorption cross section for a neutron in the reactor core. If we need to take into account fuel and moderator only, $\Sigma_a(\mathrm{C}) = \Sigma_a(\mathrm{F}) + \Sigma_a(\mathrm{M})$ and we can use Equation (10.33) to write:

$$t_p \approx 1/v\Sigma_a(\mathrm{C}) = (1 - f)/v\Sigma_a(\mathrm{M}). \tag{10.40}$$

Consider, for example, a uranium–graphite reactor with $f = 0.9$, the macroscopic absorption cross section of the moderator (density $\rho = 1600\ \mathrm{kg\ m}^{-3}$) is

$$\Sigma_a(\mathrm{M}) = \rho\sigma_a/A\,m_u = (1600 \times 0.0045 \times 10^{-28})/12\,m_u = 0.036\ \mathrm{m}^{-1}$$

and substituting $\Sigma_a(\mathrm{M})$, f and $v = 2200\ \mathrm{m\,s}^{-1}$ into Equation (10.40), gives $t_p \approx 10^{-3}$ s. Then, if $k = 1.001$, $n(t) = n_o e^t$ and, in only 10 s, the reactor flux increases by $e^{10} \approx 22\,000$. This is a very rapid growth for even a small increase in k and under such circumstances it is not possible to control the reactor safely.

Fortunately, a fraction d of about 0.0065 of the neutrons from fission are delayed by β emission (see Section 10.2). In this case, Equation (10.39) is modified to have two terms, one depending on the prompt lifetime and one depending on the mean time τ of the delayed neutrons. An approximate expression is given by

$$n(t) = n_o \left[\frac{d}{d-q} \exp\left(\frac{qt}{(d-q)\tau}\right) - \frac{q}{d-q} \exp\left(\frac{-(d-q)t}{t_p}\right) \right]. \tag{10.41}$$

This equation reverts to Equation (10.39), if $d = 0$.

Provided that q is always less than d, the rate of change of $n(t)$ will be governed by the mean delay time, $\tau \approx 12.5$ s, which is much longer than t_p. If $q \ll d$, Equation (10.41) is dominated by the first term and

$$n(t) \sim n_o \exp(qt/(d-q)\tau) \tag{10.42}$$

which is a slow rate of growth or decay, depending on the sign of q. Clearly, q must be kept less than d, otherwise the second term in Equation (10.41) becomes a positive exponential growing rapidly according to the prompt lifetime.

For the uranium–graphite reactor considered above, we have $q = 0.001$ and Equation (10.42) gives $n(t)/n_o \approx \exp(t/70)$, which only increases by a factor e (≈ 2.7) in about 70 s.

10.5.3 Reactor poisoning

In an operating reactor, certain fission fragments (notably xenon and samarium) are produced, which have high neutron-capture cross sections. As these *reactor poisons* build up with time, they absorb neutrons and cause the neutron multiplication factor for the core to decrease. This decrease has to be compensated for by withdrawing neutron-absorbing control rods to restore equilibrium at $k = 1$. Xenon-135, with a thermal-neutron capture cross section of 2.75×10^6 b, is an important example of such a poison. Figure 10.8 shows schematically how it is produced and eliminated. It is produced directly as a fission fragment at the rate of $\gamma_X = 0.003$ atoms per fission. It is also produced as the daughter of the fission product ^{135}I, which itself is produced at the rate $\gamma_I = 0.061$ atoms per fission. During operation, ^{135}Xe builds up and, eventually, reaches a stable value when the rate of ^{135}Xe production is balanced by its rate of loss by neutron capture and decay (decay constant λ_X). In equilibrium, the rate of production of ^{135}I is also equal to its rate of loss by decay into ^{135}Xe (with decay constant λ_I).

The total fission rate in the reactor is $\Sigma_f \Phi$, where Σ_f is the macroscopic fission cross section and Φ is the neutron flux. We can use this to obtain expressions giving the equilibrium amounts of ^{135}I (N_I) and ^{135}Xe (N_X) as follows:

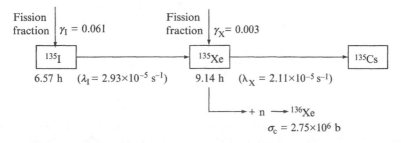

Figure 10.8 Formation and decay data for the reactor poison ^{135}Xe.

For ^{135}I, equating the rates of production and loss gives

$$\gamma_I \Sigma_f \Phi = \lambda_I N_I.$$

For ^{135}Xe, there are extra terms for both production and loss:

$$\lambda_I N_I + \gamma_X \Sigma_f \Phi = \lambda_X N_X + N_X \sigma_c \Phi.$$

Combining these two expressions gives

$$N_X = \frac{(\gamma_I + \gamma_X)\Sigma_f \Phi}{(\lambda_X + \sigma_c \Phi)}. \tag{10.43}$$

A change in neutron multiplication factor (from k to k') arises because there is a change in the thermal utilization factor (from f to f') as the xenon builds up. We define a fractional change in k, due to this effect, as

$$q_X = \frac{k' - k}{k'} = \frac{f' - f}{f'} \tag{10.44}$$

where $f = \Sigma_a(F)/\Sigma_a(C)$ and $f' = \Sigma_a(F)/(\Sigma_a(C) + N_X \sigma_c)$. The term $N_X \sigma_c$ is the additional contribution to the macroscopic absorption cross section in the reactor core due to the poison.

Using the expression for N_X given in Equation (10.43), we can write:

$$\frac{q_X}{f} = \frac{1}{f} - \frac{1}{f'} = -\frac{N_X \sigma_c}{\Sigma_a(F)} = -\frac{(\gamma_I + \gamma_X)\Sigma_f \Phi \sigma_c}{(\lambda_X + \sigma_c \Phi)\Sigma_a(F)}. \tag{10.45}$$

Now $\Sigma_f/\Sigma_a(F)$ is the fraction of neutrons absorbed in the fuel which lead to fission and, from the definition of η (Section 10.3.2), this is equal to η/v. Substituting this into Equation (10.45) gives, finally

$$\frac{q_X}{f} = -\frac{(\gamma_I + \gamma_X)\eta}{(1 + \lambda_X/\Phi \sigma_c)v}. \tag{10.46}$$

For example, taking values from Figure 10.8 and $\eta = 1.33$, $v = 2.42$, $\Phi = 10^{14}\,\mathrm{cm}^{-2}\mathrm{s}^{-1}$ and $f = 0.9$ as typical values, we find $\lambda_X/\Phi \sigma_c = 0.077$ and obtain the result:

$$\frac{q_X}{f} = -\frac{0.064 \times 1.33}{1.077 \times 2.42} = -0.033 \text{ or } q_X \approx -0.03$$

i.e. the poison decreases k by about 3%.

Figure 10.9 shows the gradual fractional decrease in k with time to a steady-state value as the poison builds up in the reactor in the period 'before shutdown'. During operation, the effective value of k must be maintained at unity and this is achieved by

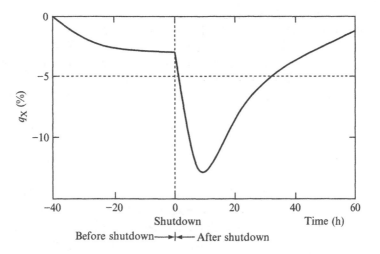

Figure 10.9   Effect of the buildup of ^{135}Xe concentration on the fractional change q_X [defined in Equation (10.44)] in the multiplication factor k during operation before shutdown, at a neutron flux level of 10^{14} cm^{-2} s^{-1}, and after shutdown. The horizontal dashed line indicates a maximum 5% decrease in q_X that can be compensated for by using control rods.

gradually withdrawing neutron-absorbing control rods to compensate for the effect of the poison. After the reactor is shutdown, the amount of the poison continues to increase for a time because a major part of the ^{135}Xe production comes from the β decay of ^{135}I. In this example, the decrease due to ^{135}Xe buildup reaches a maximum of about 13%. If the reactor control rods can compensate for a change no greater than 5% (indicated by the dashed line in Figure 10.9), it would not be possible to restart the reactor for over a day after a shutdown until most of the poison had decayed away.

10.6   COMMERCIAL THERMAL REACTORS

After World War II (WW II), although many reactors were built primarily to produce plutonium for nuclear weapons and for scientific research, they continued to be developed as power generators. They became economically viable in the 1960s and today fission reactors are a major source of energy in many countries. In this section, we present brief descriptions of some of the more common types of commercial, thermal-fission reactor. So-called fast reactors have been designed and built, which use no moderator at all and are intended for breeding new fuel. They do not yet provide significant amounts of commercial power, but may do so in the future. A brief discussion of fast reactors is given in the final section of this chapter.

10.6.1   Early gas-cooled reactors

In Britain, shortly after WW II, neither heavy water nor enriched uranium was available and the only reactors which could be built used natural uranium moderated with graphite. Uranium fuel was encased in tubes made of a magnesium alloy called

Magnox and these reactors, of which 26 were built in 11 power stations, are known as the Magnox reactors. They are cooled by circulating CO_2 gas through the core. The original design was for coolant operating temperatures up to 450°C, but this had to be reduced to limit corrosion of certain components. An example of one of the later Magnox reactors is at the Wylfa power station in Anglesey, Wales. Its general layout is similar to that shown in Figure 10.6. It has a large core, 10.3 m high and 15.9 m in diameter. This is supported inside a spherical, prestressed concrete pressure vessel, about 30 m in diameter, which also acts as a biological shield. The CO_2 gas is circulated through the core with an outlet temperature of 365°C. There are two reactors at the power station. Each has a thermal output of about 900 MW, giving a total thermal input, from both reactors together, of 1800 MW to the steam turbines. These deliver a total electric power output of 500 MW, i.e. at an efficiency of 500/1800 or about 28%.

A number of Magnox reactors have been closed down now and the others are approaching the end of their lives. Overall, their reliability has been excellent but, because of improvements in reactor design and the availability of enriched fuel, no more of this particular type will be built again.

10.6.2   Advanced gas-cooled reactor (AGR)

The need to improve the efficiency of the gas-cooled reactor by increasing the coolant temperature led to the use of more corrosion-resistant materials for structural components and a change in fuel cladding to stainless steel or a zirconium alloy (Zircalloy), both of which can withstand much higher temperatures than Magnox. Also, the fuel was changed from uranium metal to the oxide or carbide forms, which have extremely high melting points. Moderator and coolant are graphite and CO_2, as in the Magnox reactor. A simplified layout of an AGR is shown in Figure 10.10. Heat exchangers are located in the annular space between the core and the inside of a prestressed concrete pressure vessel. The coolant temperature can reach a maximum of about 650°C. The thermal output of a typical AGR is about 1500 MW and it operates with a thermal efficiency of about 42%, which is considerably higher than that of a Magnox reactor.

Unfortunately, there were many delays that beset the development and construction of AGRs. Also, their capital cost compared unfavourably with that of water-moderated reactors (see below) and it is mainly for this reason that no more AGRs are likely to be built in the future.

10.6.3   Pressurized-water reactor

Most modern reactor designs capitalize on the good qualities of light water (H_2O) as a moderator (see Tables 10.3 and 10.5). Its excellent slowing-down properties and relatively low diffusion and slowing-down lengths (compared with carbon) enables the size of the core to be reduced considerably to one which is, typically, about 3 m in height and diameter. However, enriched fuel must be used to offset the reduction in thermal utilization factor f, due to the greater tendency of water to absorb neutrons.

In order to maintain its effectiveness as a coolant, water must be prevented from boiling as far as reasonably possible and it is subjected to a very high pressure (up to 160 bar) to raise its boiling point. It is circulated through external heat exchangers,

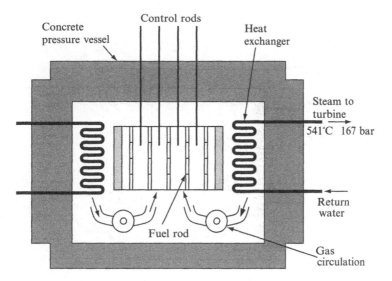

Figure 10.10   Layout of the core, heat exchangers and pressure containment of an advanced gas-cooled reactor.

where steam for turbines is generated. Such a reactor, using water as both moderator and coolant, is called a pressurized-water reactor (PWR).

The PWR was developed in the USA, where enriched uranium was more readily available after WW II. Subsequently, many PWRs have been built throughout the world and most new reactors are of this type. An example in the UK is at Sizewell B power station. Its layout is shown schematically in Figure 10.11(a). The fuel consists of 4.5% enriched uranium in the form of UO_2 pellets encased in tubes, which are arranged in a core of diameter 3.4 m and height 3.7 m. The pressure vessel (height 13.5 m and inside diameter 4.4 m) has a wall thickness of 21.5 cm and the operating pressure is 156 bar. A net electrical power of just over 1 GW is produced at an efficiency of about 32%. This is below that of the AGR, but the PWR is cheaper to build and since there is no shortage of fuel at present, there is no great incentive for achieving optimum efficiency.

As we have noted, the core of a PWR is much more compact than one consisting of uranium and graphite, and the power density is much higher. This reduction in size is why the PWR is cheaper to build than a uranium–graphite reactor.

The ability to construct a small, water-moderated reactor led to its early use in the USA for powering naval vessels. The strategic advantage of a submarine, for example, being able to remain at sea for long periods without the need for refuelling is considerable. Today, many modern submarines are nuclear powered; the first one, the USS Nautilus, was launched in 1955.

10.6.4   Boiling-water reactor

In the boiling-water reactor (BWR), as in the PWR, water serves as both moderator and coolant. However, in the BWR, water is allowed to boil, and steam, after first

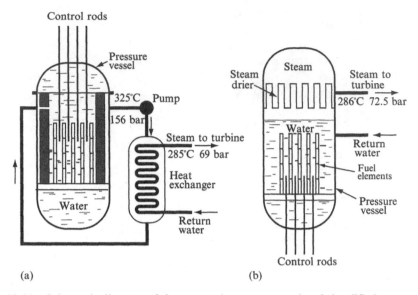

Figure 10.11   Schematic diagram of the core and pressure vessel and simplified steam cycle of (a) a pressurized-water reactor and (b) a direct-cycle, boiling-water reactor.

being dried, is passed directly to the turbines. This eliminates the need for heat exchangers with their cost and inevitable reduction of thermodynamic efficiency. Part of a schematic layout of a BWR is shown in Figure 10.11(b).

Neither the stability nor safety of a BWR is seriously affected when water boils in the core because the reactor is designed with what is called a negative void coefficient of reactivity. This means that if, for some reason, an increase in power output causes more boiling and, therefore, more steam replaces water as moderator, the value of the neutron multiplication factor k for the core *decreases*, leading to a drop in power. This is a self-stabilizing behaviour, and a BWR can be operated to follow a load demand. If less power is needed, coolant flow is reduced. This increases the temperature in the core, which leads to an increase in steam voidage and a consequent decrease in reactor power.

The possibility of the production of radioactive contamination in the coolant, which could reach the turbines, is a concern. This is largely overcome by using water with a very high level of purity.

A typical BWR contains 140 tonnes of fuel, enriched to between 1.7 and 2.5%, in a core approximately 4.7 m diameter and 3.7 m high – slightly larger than that of a PWR. Operating pressure is about 70 bar, and the BWR provides power with an efficiency of about 35%.

10.6.5   Heavy-water reactors

As can be seen in Table 10.3, high values of both scattering cross section σ_s and logarithmic energy decrement ξ make heavy water (D_2O) a better moderator than graphite and it can also serve as the coolant. Furthermore, its very low neutron-capture cross section means that a heavy-water moderated reactor can be fuelled with

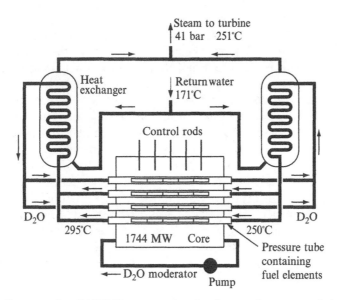

Figure 10.12   Layout of a CANDU reactor showing heat exchangers and simplified steam cycle.

natural uranium, leading to low fuel costs. However, heavy water is expensive and, as in light-water reactors, it is necessary to operate at high pressures. Thermodynamically, heavy- and light-water reactors are very similar.

At the end of WW II, only Canada had the capability of producing heavy water in sufficient quantities to develop a reactor moderated with D_2O. These are the CANDU reactors. Figure 10.12 illustrates the basic design, which uses D_2O as both moderator and coolant. An alternative version uses light water as coolant, which reduces the capital cost and eliminates the operating cost of D_2O leakage from the coolant circuit. Also, the light water can be allowed to boil in the pressure tubes, and a direct system for sending (dried) steam to the turbines can be used, as in the BWR.

10.7   FUTURE OF NUCLEAR FISSION POWER

Nuclear power has been in use commercially for nearly half a century, but not without problems. There have been accidents, of which the one at Chernobyl in the Ukraine has been the worst to date, and there is genuine concern about the proliferation of reactors all over the world. Yet, despite this concern, the demand for energy will continue and it is likely that nuclear power will be used to satisfy part of that demand. It is important, therefore, that it be used safely and there are many aspects of safety, regulation and control of nuclear installations and nuclear material which have been and must continue to be addressed. It is not possible to deal with all these issues here, but we shall consider briefly two particular ones, which become more important if nuclear power is envisaged as a long-term energy option. One is how to deal with the radioactive waste and the other is how to ensure there is enough fuel.

When a reactor is refuelled, and especially when it is decommissioned, there is a great deal of radioactive waste, which must be extracted and stored safely until its activity has decayed to a safe level. The majority of this waste consists of radioactive fission fragments with intermediate half-lives up to a few tens of years. These will require safe storage for several hundred years. Unfortunately, however, a significant fraction of the waste products are radioactive nuclides with very long lifetimes, which would need to be stored for hundreds of thousands of years. The most important of these long-lived activities are not fission fragments but nuclei heavier than uranium formed by successive neutron capture not followed by fission. Examples are the actinide nuclei, ^{239}Pu, ^{242}Pu and ^{243}Am, which have half-lives of 24 000, 373 000 and 7 400 years, respectively. However, if these nuclei remain in a reactor long enough, further neutron captures will take place. Eventually, they will be transformed into fissile nuclei, undergo fission and be converted into fission fragments which generally have much shorter half-lives. This is one of the aims of the proposed accelerator-driven system (ADS) or hybrid reactor, which is specially designed to produce useful energy while consuming the bulk of the most long-lived radio-toxic waste from fission energy production.

In principle, nuclear fission represents a vast potential source of energy for the future. However, the only naturally occurring, fissile material is ^{235}U, which constitutes less than one percent of natural uranium. Therefore, in order to realize the potential, more fissile fuel will need to be created. This is the function of the breeder reactor, which uses neutrons to convert ^{238}U and ^{232}Th into new fissile material: ^{239}Pu and ^{233}U (see Section 10.2.5). Various forms of fuel breeder have been built, based on a fast-reactor design, and operated successfully. As we shall see below, they are technically more challenging and more expensive than a conventional power reactor. Work on their development has been cut back because they cannot produce fuel at competitive prices at the present time. In principle, an ADS could operate as a fuel breeder as well as a disposer of reactor waste. However, the ADS is a much more speculative concept, which will need considerable effort and resources to prove and then to develop. The basic principles of the breeder reactor and ADS are summarized in the following subsections.

10.7.1   The breeder reactor

All reactors produce new fissile material. However, most of them do this at the cost of consuming an even greater quantity of fissile fuel. A true breeder reactor produces more new fuel than it consumes. Its effectiveness is measured by the *breeding ratio B*, which is defined as the number of new fissile atoms formed per atom of existing fuel consumed. If $B = 1$, fuel is replaced, if $B > 1$, the amount of fuel is increased and, if $B < 1$, there is a net decrease.

Recall that η is defined as the number of neutrons produced per neutron absorbed in fuel. One of these neutrons is required to maintain reactor operation at $k = 1$, while others are lost due to capture (C) in non-fuel components and leakage (L). The remainder are available for breeding. Thus, $\eta = 1 + B + C + L$. Values of η for pure fuel are listed in Table 10.6 for both thermal and fast neutrons. A typical value for $C + L$ is 0.2. Hence, only ^{233}U offers a realistic possibility of B being greater than

Table 10.6   Values of η for fissile nuclei.

Nucleus	Thermal neutrons (0.025 eV)	High-energy neutrons (0.5 MeV)
^{235}U	2.06	2.35
^{239}Pu	2.16	2.90
^{233}U	2.29	2.40

unity in a thermal breeder reactor. Indeed, these figures indicate a distinct advantage for using fast neutrons and, so far, only the fast reactor has been considered for breeding.

The core of a fast-breeder reactor consists of an inner region containing fissile fuel where energy and neutrons are generated. This is surrounded by a blanket of ^{238}U or ^{232}Th, where the bulk of new fissile material is produced. There is no neutron loss in a moderator but, because there is no moderator, the inner core must contain highly enriched fuel, which is expensive. Compared with a moderated reactor, a fast reactor has a very small core and, therefore, operates at a very high power density, which may be 100 times that of a conventional power reactor. This creates technical problems for heat transfer. The coolant must have excellent heat transfer properties and be non-moderating, which rules out water and heavy water. Most designs use liquid metals: e.g. sodium or a sodium/potassium eutectic alloy, which is liquid at room temperature. These are highly reactive in air and water but, because they have high boiling points, the reactor need not be pressurized.

With the high initial cost of a breeder reactor and the need for reprocessing, it is likely that international co-operation will be required for their development. Also, because there is an excess of ^{239}Pu at present, there is little demand for breeder reactors and it may not be until the middle of the 21st century that fuel breeding will be required. However, if an expanding programme is desired in the future, it is important to realize that it takes a considerable time to breed fuel. The *doubling time* is the time needed for a reactor to produce an excess of fuel equal to its initial charge and, therefore, to provide fuel for another reactor. It is calculated as follows.

Consider a reactor, with breeding ratio B, containing N_F fissile atoms per unit volume. The rate of fuel consumption is $N_F \sigma_a \Phi$, where σ_a is the absorption cross section in the fissile material. From the definition of B, the rate of fuel production is $B N_F \sigma_a \Phi$. Therefore, the net fuel production rate is $dN_F/dt = (B-1)N_F \sigma_a \Phi$. The doubling time T_d is given by $T_d \times dN_F/dt = N_F$, or

$$T_d = \frac{1}{(B-1)\sigma_a \Phi}. \tag{10.47}$$

Substituting values for a reasonably high-rated reactor: $B = 1.2$, $\sigma_a = 1.2\,\mathrm{b}$ and $\Phi = 3 \times 10^{15}\,\mathrm{cm^{-2}s^{-1}}$ gives $T_d = 1.4 \times 10^9$ s (about 44 years). This time is comparable with some estimates of the lifetime of existing oil and gas reserves. Clearly, a breeder programme cannot be accelerated quickly and so some key decisions on the future use of nuclear fission power will need to be made in the not too distant future.

10.7.2   Accelerator-driven systems

An accelerator-driven system (ADS) consists of a nuclear particle accelerator, injecting a high-intensity beam of protons or deuterons into a subcritical, fission reactor
unit. The particles are of very high energy and when they bombard a heavy target
such as lead or uranium they produce copious neutrons. Some of these neutrons
cause fission, which generates energy and amplifies the neutron flux, depending on
the value chosen for the neutron multiplication factor k. These neutrons react in the
core and may be used to convert waste, breed new fuel and generate energy.

There are a number of attractions to the scheme. A critical accident should be
impossible, since the main reactor assembly is operated at a subcritical level.
There should be no conditions in which the fission reactions would be self-sustaining;
if the accelerator stops, the reactions stop. If successful, some of the proposed
ADS designs would incinerate waste actinides and certain long-lived, fission
fragments to such an extent that the need for long-term (geologic) waste disposal
would be greatly reduced or even eliminated altogether. Also, if ADSs were located
near conventional reactors, waste could be processed and incinerated on site, thereby
reducing concerns about transporting dangerous radioactive material for off-site
waste disposal.

A number of different ADSs have been proposed. Some use a moderated-reactor
assembly, for which the technology is well developed and which, therefore, would be
more predictable. Other designs are more ambitious and emphasize fuel-breeding
capability and energy production as well as waste disposal. They anticipate the
development of fast-reactor technology which, as noted above, means operation at
high flux and high power density. Neutron loss is reduced if there is no moderator,
but a liquid metal would be needed for cooling. However, this offers the possibility of
effective heat exchange at a high temperature with consequent better thermodynamic
efficiency for energy conversion into electric power than is obtained with a conventional power reactor.

A schematic layout of a proposed fast-neutron ADS is shown in Figure 10.13. A
particle accelerator, shown as a cyclotron, delivers an intense beam of 1-GeV protons
into a subcritical, fast-fission unit ($k \approx 0.95$). Nuclear reactions induced by each
proton convert about half the incident energy into neutrons and protons (at about
10–12 MeV per nucleon) and half into larger fragments. The neutrons would induce
an estimated 13 fissions per proton in the first cycle; then, from the definition of k, we
obtain $13k$ neutrons in the second generation, $13k^2$ in the next and so on, giving a
total of $13 \times (1 + k + k^2 + \cdots) = 13/(1 - k)$ fissions per 1-GeV proton. At 200 MeV
per fission and $k = 0.95$, this example gives an energy gain of about 50. It has been
estimated that the heat produced might be converted with an efficiency of 45% into
electricity and only a relatively small part (10–20%) of this would be needed to power
the accelerator, leaving a substantial surplus.

For fuel breeding, the reactor would be loaded with ^{232}Th, which the excess
neutrons would convert into fissile ^{233}U. Thorium is even more plentiful than
uranium in the earth's crust and thus constitutes a huge potential source of energy
into the far distant future. Realistically, for waste incineration, the ADS must
operate at high flux. Even in a flux as high as 3×10^{15} cm^{-2} s^{-1}, the mean life of a
nucleus, with a typical transmutation cross section of 1 b, is about 10 years.

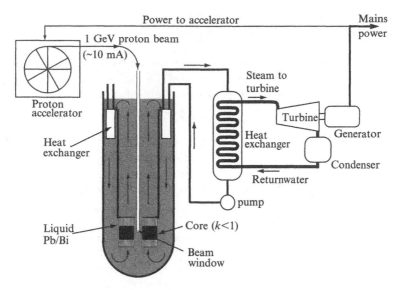

Figure 10.13 Schematic layout of a proposed accelerator-driven (hybrid) reactor. A proton beam from a cyclotron is injected down a long vertical beam pipe into a subcritical assembly containing thorium fuel. The core and the region where the beam emerges from the pipe through a specially designed window are cooled by convection in a liquid lead/bismuth mixture.

The ADS is an ambitious scheme and there are considerable technical challenges facing its proponents. Intense proton beams of tens of milliamps would be required to give a high flux in the reactor region. This beam intensity at 1 GeV is over an order of magnitude greater than the current world record, but is considered to be an achievable target. It will need to be produced reliably and transported with negligible beam loss into the reactor core through a special window capable of withstanding high pressure and high temperature under the intense bombardment of the ion beam. Also, the efficiency of converting electrical energy into beam energy in the accelerator, which has been quoted at 40–50%, is an estimate based on an extension of current technology. It has yet to be realized in practice.

Evidently, major technical advances will need to be made, requiring a considerable investment of R&D effort, before the full potential of an accelerator-driven (hybrid) reactor can be properly assessed. However, as we shall see in Chapter 11, compared with the problems facing attempts to develop fusion power for commercial use, none of these advances appears to be a great extrapolation from what has already been attained. The possibility of safer, nuclear power without the need for long-term waste disposal may be sufficiently attractive to warrant the effort.

PROBLEMS 10

10.1 Estimate the distance separating the centres of two fission fragments, when they break apart from each other in fission, if they acquire 168 MeV from the conversion of Coulomb energy into kinetic energy. Assume each fragment is spherical and has half the charge of ^{236}U.

10.2 Calculate the η factor and the microscopic thermal-neutron absorption cross section $\sigma_a(U)$ for uranium enriched to 6% in ^{235}U.

10.3 Calculate the resonance escape probability for a mixture of uranium (enriched to 1.6% in ^{235}U) and graphite in the ratio of 1 atom of uranium to 600 atoms of carbon. You may assume that the values of the logarithmic energy decrement for ^{235}U and ^{238}U are the same.

10.4 On average, a thermal neutron travels about 27 m in pure graphite before it is absorbed. By what factor will this change if the graphite is uniformly mixed with 2% enriched uranium in the ratio of 1 atom of uranium to 400 atoms of graphite. Assume the density of atoms in the mixture is the same as it is in the pure moderator.

10.5 Neutrons diffusing in an absorbing medium obey Equation (10.21). If the neutron concentration varies in the z direction only, show that the mean distance travelled by a neutron in the z direction is equal to the diffusion length L.

10.6 Calculate the neutron multiplication factor k_∞ for a dilute solution of (natural) uranyl sulphate in heavy water (1 molecule of UO_2SO_4 to 750 molecules of D_2O). Data: sulphur: $\sigma_c = 0.52\,b$, $\sigma_s = 1.1\,b$, $\xi = 0.061$; oxygen: $\sigma_c =\sim 0\,b$, $\sigma_s = 3.8\,b$, $\xi = 0.120$.

10.7 Using the results of Problem 10.6, determine the minimum diameter of a spherical reactor consisting of the specified homogeneous mixture of uranyl sulphate in heavy water.

10.8 If a natural uranium, thermal fission reactor is operating at a thermal power output level of 2 GW, calculate the total rate of consumption of ^{235}U (in $kg\,y^{-1}$). Take the energy release per fission to be $E = 200\,MeV$.

10.9 Calculate the rate of production (in $kg\,y^{-1}$) of ^{239}Pu due to thermal-neutron capture only in the reactor of Problem 10.8, assuming that all neutrons captured by ^{238}U lead to ^{239}Pu.

10.10 A reactor consisting of 1 atom of 1.5% enriched uranium to 500 atoms of graphite is operating at a rating of 4 MW tonne^{-1} of uranium. Calculate the neutron flux and the thermal utilization factor f. Energy release per fission $= 200\,MeV$.

10.11 Calculate the prompt lifetime t_p of a thermal neutron in the reactor described in Problem 10.10. Assume that the density of carbon atoms in the reactor is the same as that in graphite. Atomic weight of graphite $= 12.01$ and the speed of a thermal neutron $= 2200\,m\,s^{-1}$.

10.12 A 2-MeV neutron scatters on average about 115 times in graphite before becoming thermalized (see Table 5.1). Estimate the slowing-down time and compare it with the diffusion time for a thermal neutron in graphite.
    Hint: The fractional average energy change per collision, during moderation, is given by the logarithmic energy decrement ξ (see Section 5.5.2).

10.13 As described in Section 10.5.3, ^{135}I and the reactor poison ^{135}Xe are produced as fission fragments in a reactor. ^{135}Xe is lost by decay and by neutron capture.
(a) Obtain an expression for the ratio of the concentrations of ^{135}I (N_I) and ^{135}Xe (N_X) in equilibrium, in terms of the symbols used in Section 10.5.3, and determine its value if the neutron flux $\Phi = 7.5 \times 10^{12}$ cm^{-2} s^{-1}, using the data given in Figure 10.8.
(b) If $\eta = 1.4$, $v = 2.4$ and $f = 0.8$ for the reactor, show that there is a fractional decrease q_X in the neutron multiplication factor of about 1.5% due to the buildup of ^{135}Xe.

10.14 The reactor poison, ^{149}Sm, is formed as the final daughter in the decay chain of the fission product ^{149}Nd as follows:

$$^{149}\text{Nd}(t_{1/2} = 2\text{ h}) \rightarrow {}^{149}\text{Pm}(t_{1/2} = 53\text{ h}) \rightarrow {}^{149}\text{Sm(stable)}$$

It is removed by neutron capture for which the cross section $\sigma_c = 58$ kb. Derive an expression for N_{Sm}, the equilibrium concentration of ^{149}Sm nuclei, in terms of the macroscopic fission cross section in the reactor Σ_f, the ^{149}Nd fission fragment yield γ and σ_c.
    Hence, determine the fractional change in k (q_{Sm}) after the samarium has built up to equilibrium in a reactor operating with $f = 0.9$ and $\eta = 1.32$. Take the number of neutrons produced per fission to be $v = 2.42$ and $\gamma = 0.0113$.

10.15 Calculate the absorption mean free paths (see Section 5.5.1) of (a) 2-MeV neutrons and (b) thermal neutrons in natural uranium. What fraction of them will be captured by ^{238}U in each case? Take the absorption cross sections at 2 MeV to be $\sigma_a(^{235}\text{U}) = 2.6$ b and $\sigma_a(^{238}\text{U}) = 0.6$ b. Other relevant data are given in Table 10.2.

10.16 A breeder reactor operates at a rating of 500 MW per tonne of ^{233}U. Calculate the neutron flux and hence (or otherwise), estimate the doubling time if neutron losses can be kept down to 10%.
    Take the fission and absorption cross sections for ^{233}U to be $\sigma_f = 1.8$ b and $\sigma_a = 2.2$ b, respectively, and the energy release per fission $E = 200$ MeV.

10.17 In an accelerator-driven system (ADS), the spallation reaction generates an average of 25 neutrons per 800 MeV proton incident on a heavy target contained in a subcritical ($k = 0.9$) assembly. Determine the fraction F of these neutrons that will go on to induce fission and, hence, estimate the energy gain of the reactor. Take the number of neutrons produced per fission to be $v = 2.5$.

10.18 A proton beam ($I = 10$ mA) is injected into a subcritical assembly ($k < 1$), which contains fissile fuel and a ^{232}Th blanket for breeding. Each proton generates 30 neutrons of which a fraction $F = 0.38$ induce fission in the fuel. If 10% of the neutrons not absorbed by fissile fuel are lost and the rest lead to fuel breeding by capture in ^{232}Th, calculate how much new fuel (^{233}U) is produced per year. Does the breeding ratio exceed unity? For the fissile fuel, take $\sigma_f = 1.9$ b and $\sigma_a = 2.2$ b and $v = 2.5$.

11

Thermonuclear Fusion

11.1 INTRODUCTION

Energy is liberated when light nuclei combine together in a fusion reaction, provided that the product nucleus has a mass number below about $A = 56$ (see Figure 1.1). The energy released in fusion per unit mass of material is comparable with that released in fission ($\approx 1\,\text{MeV}\,\text{u}^{-1}$) and, in some cases, exceeds it by a considerable factor. Indeed, as a controlled source of power on earth, fusion has even greater potential than fission and the rewards for harnessing it are great. Light nuclei are more plentiful than fissile nuclei, and there would be much less radioactive waste from a fusion reactor than from a fission reactor. Furthermore, any radioactivity which might be produced would decay away relatively rapidly and there would be no need to store the waste for geological periods of time.

However, there is a major technical difficulty. All nuclei are charged and an initial amount of kinetic energy is needed to increase their probability of penetrating the Coulomb barrier, which, normally, keeps them apart and prevents fusion from taking place. In a fusion reactor, it is intended to generate this energy by heating the reactants. When fusion is driven by heat energy, the process is called thermonuclear fusion. It is the principle behind the thermonuclear bomb in which detonation of a fission bomb raises the temperature enough to trigger fusion in a core made of light nuclear material. Thermonuclear fusion has been achieved in the laboratory, but it requires a temperature of about a 100 million degrees and it is proving exceedingly difficult to create and maintain the required conditions stably and efficiently.

Elsewhere in the Universe, suitable conditions for thermonuclear reactions exist deep within stars. Nuclear fusion generates the energy a star radiates into space and determines how it evolves with time. Fusion reactions also took place during a critical period of the early Universe a few minutes after the Big Bang.

In this chapter, we review some of the possible, exothermic fusion reactions being considered for thermonuclear power generation and outline the main problems facing current attempts to develop a practical fusion reactor. We then describe the part played by fusion reactions, first, in the early Universe and, later, in stars leading to the creation of all the naturally occurring elements necessary for life on earth to have evolved.

11.2   THERMONUCLEAR REACTIONS AND ENERGY PRODUCTION

11.2.1   Basic reactions and Q values

The following is a list of fusion reactions which might be considered for nuclear power production. All are exothermic and the energy released (Q value) in MeV is indicated in each case.

	Reaction			Q value (MeV)
(1)	$p + d$	$\rightarrow$	$^3He + \gamma$	5.49
(2)	$d + d$	$\rightarrow$	$^4He + \gamma$	23.85
(3)	$d + d$	$\rightarrow$	$^3He + n$	3.27
(4)	$d + d$	$\rightarrow$	$t + p$	4.03
(5)	$d + t$	$\rightarrow$	$^4He + n$	17.59
(6)	$d + {}^3He$	$\rightarrow$	$^4He + p$	18.35

Both interacting nuclei, with one exception, are isotopes of hydrogen ($Z = 1$) because this minimizes the Coulomb repulsive force, which hinders fusion. The fusion of two protons is not listed because the diproton is unbound. However, as we shall see in Section 11.6, proton–proton fusion is the primary astrophysical reaction, but it occurs at much too slow a rate for it to be considered as a source of thermonuclear power on earth.

Reaction (1) has a small cross section. Also, it is not ideal because essentially all the energy output would be carried by γ rays (see Problem 11.1), which are penetrating. They would escape, taking their energy from the reaction zone, and an external source of power would have to be used to maintain the temperature of the reacting material. The same is true of the second reaction. However, reactions (3) and (4) are also possible when deuterons interact with each other. These are known as the D–D reactions and are much more likely than either (1) or (2). Also, they are much more suitable candidates for fusion power because part of the output energy is carried by charged particles, which can be retained within the reactor to compensate for energy losses and maintain the temperature. Even more promising is the deuteron–triton (D–T) reaction. This has a Coulomb barrier similar to that of D–D, but its cross section is larger. Also, there is a much greater release of energy because one of the final products is the very tightly bound α particle. The energy efficiency of the reaction is $17.6/5 = 3.5\,\mathrm{MeV\,u^{-1}}$, which is about four times that of uranium fission. Unfortunately, it requires tritium as a fuel component, which is radioactive and would have to be produced in a fusion reactor since it does not occur naturally. Reaction (6) is attractive because it also has a high Q value and both final products are charged, which makes it relatively easy to contain more of their energy in the reactor if required. Also, the fuel is not radioactive and no neutrons are produced which, in a D–T reactor, will inevitably produce some radioactivity in the reactor structure. The disadvantage is the higher Coulomb barrier and, as we shall see, the reactor temperature to achieve a given reaction rate, using a deuterium-^{3}He mixture, would need to be about six times higher than in a D–T reactor.

The total kinetic energy of the reaction products is equal to the sum of the Q value and the initial kinetic energy of the fusing particles. In reactions (3)–(6), which do not produce a γ ray, this energy is shared between the two product particles according to their masses and outgoing angles. However, if the initial kinetic energy is small compared with the Q value, as it usually is at thermonuclear temperatures, the final momenta are approximately equal and opposite and the ratio of the outgoing kinetic energies is the inverse of the ratio of the masses. In the D–T reaction, for example, $E_n/E_\alpha \sim m_\alpha/m_n = 4$. Thus, the output energy of 17.6 MeV appears as a 14.1-MeV neutron and a 3.5-MeV α particle. In the D–D reactions, about 75% of the energy is taken by the proton or neutron.

11.2.2   Cross sections

The Coulomb barrier between two hydrogen nuclei is about 200 keV and, classically, the cross section should fall to zero when the bombarding energy is at or below this value. However, fusion does occur at lower energies and the reason it does so is because of quantum-mechanical barrier penetration.

In quantum mechanics, the wave nature of matter allows the wave function ψ to penetrate into the classically forbidden region under a finite energy barrier. Inside the barrier, i.e. in the region where E is less than the barrier height, the wave function ψ is decaying exponentially, as is shown schematically in Figure 11.1. Here, the particles are approaching a simple, rectangular barrier, height B, from the left. In the outside regions, $x < x_1$ and $x > x_2$, ψ is oscillatory, with the amplitude on the far side of the barrier $(x > x_2)$ decreased; the decrease depending on the energy difference $(B - E)$ and the thickness of the barrier.

Barrier penetration means that there will be some probability that two hydrogen nuclei can come into contact and fuse even when, classically, there is not enough energy for them to do so. For students who are interested, the mathematics of barrier penetration is discussed in Section 3.4. Here, we need only the general result that the

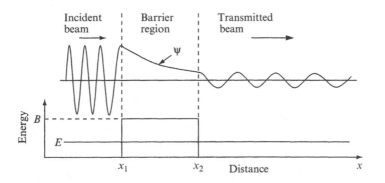

Figure 11.1   Penetration through a rectangular energy barrier (height B) of a particle beam, of kinetic energy E ($< B$), incident from the left. The form of the wave function ψ is sketched in the upper part of the figure. Inside the barrier, ψ is an exponentially decaying function of x.

probability of tunnelling through the barrier decreases strongly with decreasing energy but that, even at energies well below the barrier, the cross section remains finite, albeit small.

11.3   FUSION IN A HOT MEDIUM

11.3.1   Reaction rate

Atoms and molecules in a gas at a finite temperature are in thermal motion and constantly colliding with each other. Their velocity spectrum varies according to the Maxwell–Boltzmann distribution: $p(v) \propto v^2 \exp(-mv^2/2kT)$, where $p(v)dv$ is the probability that the speed lies between v and $v + dv$, k is the Boltzmann constant, and T is the absolute temperature. The kinetic energy corresponding to the most probable speed is kT. At room temperature, kT is about 0.025 eV and, even in a hot furnace, it is only a few tenths of an electron volt. This is enough to allow a repulsive barrier of a few electron volts between two molecules to be overcome and enable chemical reactions to proceed, but (fortunately) it is totally inadequate for thermonuclear fusion which, as we have already noted, requires a temperature of at least 10^8 K ($kT \approx 10$ keV). It is the dream of fusion reactor scientists and engineers to be able to raise the temperature of a confined quantity of gas to this level so that thermonuclear fusion occurs at a rate which would not only be able to sustain the required reactor conditions of temperature and density but also provide an excess of energy for commercial use. In order to determine the required conditions, we need to examine in more detail the factors that govern the reaction rate.

Consider a mixture of two gases consisting, respectively, of n_1 and n_2 particles per unit volume. From the definition of cross section σ, the probability for a particle in the first gas to react with one in the second, per unit distance travelled, is $n_2\sigma$. The distance travelled per unit time is the speed v of the particle. Therefore, the reaction probability per unit time is $n_2\sigma v$ and since n_1 is the density of the particles of gas 1, the total reaction rate per unit volume is $R = n_1 n_2 \sigma v$.

In arriving at this expression, we have naïvely assumed that all n_1 particles have the same speed and that the n_2 particles of the second gas are stationary. In reality, particles in both gases will have a range of speeds of the form given by the Maxwell–Boltzmann distribution and, to obtain the actual reaction rate, we must evaluate the average value of $v\sigma$ over all relative speeds. Thus,

$$\langle v\sigma \rangle = \int p(v)\sigma(v)v\,dv \tag{11.1}$$

and the reaction rate becomes

$$R = n_1 n_2 \langle v\sigma \rangle. \tag{11.2}$$

The integrand in Equation (11.1) is finite in a region where both functions $p(v)$ and $\sigma(v)v$ are finite, as shown in Figure 11.2. It has a maximum at a particular speed v_m, which corresponds to an energy known as the effective thermal energy E_m, where the reaction cross section is rising sharply with speed (or energy) and the tail of the

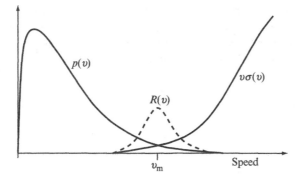

Figure 11.2 Qualitative plots showing the variation with speed of the Maxwell–Boltzmann probability distribution $p(v)$ and the fusion reaction rate $v\sigma(v)$. Their product $R(v)$ (shown dashed) is the integrand of Equation (11.1), which has a maximum at v_m corresponding to an effective thermal energy E_m.

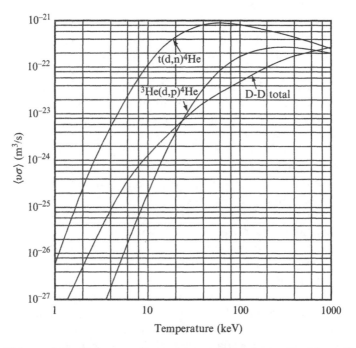

Figure 11.3 Values of the fusion reaction rate $v\sigma$, averaged over the Maxwell–Boltzmann distribution, for several fusion reactions. From Keefe (1982) p. 395.

Maxwell–Boltzmann distribution is falling. If the temperature is increased, the exponential term in $p(v)$ falls off less rapidly with energy and there is a greater probability of having collisions at higher relative speeds. This increases E_m and, if the temperature is not too high, it will also increase $\langle v\sigma \rangle$. Graphs of the variation of the rate factor $\langle v\sigma \rangle$ with temperature (in keV) for several fusion reactions are plotted in Figure 11.3. Different reactions peak at different temperatures. The D–T reaction peaks at about 60 keV and becomes less favourable compared with some other reactions at very high temperatures. However, kT in a practical thermonuclear reactor is likely to be between 10 and 30 keV and, in this range, the D–T reaction is superior to all the others by well over an order of magnitude.

11.3.2   Performance criteria

There are a number of intermediate stages which designers are trying to achieve on their way to developing fusion power. These all relate to the amount of fusion energy generated compared with that required as input to heat the reacting gas and replace losses. At temperatures required for fusion, atoms are ionized and the fuel will be in the form of a plasma, which is an electrically neutral cloud of positive ions and negative electrons. The plasma will radiate energy to its surroundings at a rate that depends on its temperature T. The primary mechanism for this power loss is bremsstrahlung.

Bremsstrahlung is emitted when charged particles interact with each other and undergo acceleration (see Section 5.3). It can be shown that the bremsstrahlung power loss per unit volume in a plasma is proportional to $\sqrt{T}$ and to Z^2 where Z is the atomic number of the ionized atoms. Bremsstrahlung exceeds fusion power output below a certain temperature, and the temperature in a fusion reactor must be greater than this in order that there is excess of energy produced which could be used to maintain the plasma. This temperature depends on the ion density in the plasma and also on the nature of the plasma constituents. For example, in the case of a D–T plasma at an ion density of 10^{21} m^{-3}, kT must be greater than about 4 keV, but it must exceed 40 keV for a D–D plasma, which is further indication of the superiority of a deuterium/tritium mixture as fuel. The dependence on Z^2 means that the use of nuclei with $Z > 1$, such as ^{3}He, as fuel is less favoured because there will be more bremsstrahlung loss as well as a higher Coulomb barrier to be overcome, both of which drive up the required temperature.

Power from a thermonuclear reactor comes from the energy of the nuclear reaction products. The so-called break-even point is when the fusion power generated is equal to the power needed to maintain the plasma conditions. In a D–T plasma, neutrons and α particles are the reaction products. Most of the neutrons escape with their energy and so, even when the break-even point is reached, external energy still has to be supplied to maintain the plasma temperature. A more advanced stage is the ignition point. This is when self-heating from the energy deposited by the α particles, which must be retained in the plasma, is sufficient to compensate for all the energy losses. External sources of plasma heating are then no longer necessary in principle and the reactor starts to become self-sustaining.

A preliminary stage on the way to either the break-even or ignition points is to be able to confine a hot, reacting plasma long enough that the nuclear energy produced exceeds the energy required to create the plasma. This leads to a requirement, known as the Lawson criterion, for the product of plasma density (n) and confinement time (τ). It can be estimated as follows:

The fusion energy output is

$$E_f = n_1 n_2 \langle v\sigma \rangle Q\tau \qquad (11.3)$$

which is the reaction rate [Equation (11.2)] multiplied by the energy per reaction (Q value) times the confinement time.

If particles 1 and 2 are different, as they are in a D–T plasma, and assuming equal numbers of each, we have $n_1 = n_2 = n/2$ and Equation (11.3) becomes

$$E_f = \frac{n^2}{4} \langle v\sigma \rangle Q\tau. \tag{11.4}$$

For a D–D plasma, there is an additional factor of 2 multiplying the right-hand side of this expression (assuming $n \gg 1$) because each deuteron can interact with all the remaining gas atoms and not just half of them.

There are n ions and n electrons in the plasma and, in equilibrium, each has to be given the same initial, average kinetic energy $\frac{3}{2}kT$. So, the energy required to create the plasma is

$$E_p = 3nkT. \tag{11.5}$$

The Lawson criterion requires that $E_f > E_p$. From Equations (11.4) and (11.5), this criterion can be written

$$n\tau > \frac{12kT}{\langle v\sigma \rangle Q}. \tag{11.6}$$

If a D–T plasma is operated at $kT = 20\,\text{keV}$, we obtain, from Figure 11.3, $\langle v\sigma \rangle = 4.5 \times 10^{-22}\,\text{m}^3\,\text{s}^{-1}$, which gives $n\tau > 3 \times 10^{19}\,\text{s}\,\text{m}^{-3}$. This means that if, for example, $n = 10^{20}\,\text{m}^{-3}$, the confinement time must exceed about 0.3 s.

A D–D plasma would need to be heated to a higher temperature because of bremsstrahlung losses. Operating at 100 keV leads to a value for the product $n\tau > 3 \times 10^{21}\,\text{s}\,\text{m}^{-3}$ to meet the Lawson criterion (see Problem 11.7), which is about 100 times larger than that for the D–T example.

Temperature, plasma density and confinement time all have to be attained simultaneously in an operating reactor, and a quantity called the triple product $n\tau T$ is often used by designers as a measure of the difficulty of meeting a particular target criterion. In our examples, the triple products to meet the Lawson criterion are $6 \times 10^{20}\,\text{s}\,\text{keV}\,\text{m}^{-3}$ for D–T at 20 keV and $3 \times 10^{23}\,\text{s}\,\text{keV}\,\text{m}^{-3}$ for D–D at 100 keV.

11.4 PROGRESS TOWARDS FUSION POWER

Two very different approaches are being used by scientists and engineers trying to develop a practical fusion reactor. One, which has received the most effort, is magnetic confinement fusion (MCF). It uses the fact that a plasma consists of charged particles and tries to confine them in a region thermally insulated from the surroundings by using a special configuration of magnetic fields. In the second method, called inertial confinement fusion (ICF), a small pellet of fusible material is caused to implode with such violence that the inner core becomes heated, undergoes a mini thermonuclear explosion and radiates usable energy. The following subsections briefly outline the main principles and problems of each of these techniques.

11.4.1   Magnetic confinement

A particle with charge q moving in a uniform magnetic field $\mathbf{B}$ experiences a Lorentz force $q\mathbf{v}\times\mathbf{B}$. If its velocity $\mathbf{v}$ is at right angles to $\mathbf{B}$, the particle moves in a circular orbit with a frequency known as the cyclotron frequency [Equation (6.5)]. This frequency depends on the mass of the particle but, at non-relativistic energies, it is independent of speed. Electrons orbit at one frequency and ions at another. If the particle has a velocity component $v_{\parallel}$ parallel to $\mathbf{B}$, there is no additional force due to this, and $v_{\parallel}$ remains constant. The particle moves in a helical path along the field direction with a pitch that depends on the ratio of its velocity components parallel to and perpendicular to $\mathbf{B}$. It is constrained in two dimensions. Two methods have been proposed to prevent particle loss along the field direction. One is to use magnetic mirrors to reflect the particles. The other is to use a closed-field geometry in which the particles could circulate indefinitely.

In a magnetic mirror, the field strength is arranged to be greater at the ends than it is in the middle. The situation at one end of such a device is sketched in Figure 11.4, which shows a positively charged particle moving in a helical orbit to the right approaching the region of increasing field. In this region, the field lines converge and the force on the particle, which is always at right angles to $\mathbf{B}$, has a component pointing away from the direction of increasing field, as shown. Under the influence of this force, the particle is reflected back towards the region of weaker field in the centre and thus is trapped. The earth's magnetic field, illustrated in Figure 11.5, is weaker above the equator than it is closer to the magnetic poles and fulfils the basic criterion of a magnetic mirror. Incoming cosmic-ray particles from the sun or outer space can reach the surface at high latitudes by travelling nearly parallel to the magnetic field. Particles approaching the equator are mostly deflected away. However, some particles have paths which cause them to become trapped high above the earth. They then orbit back and forth between northern and southern latitudes for

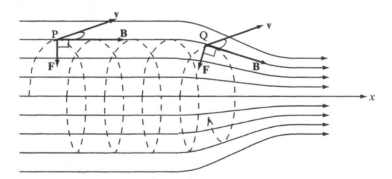

Figure 11.4   Motion of a positively charged particle moving from a region of uniform magnetic field on the left to one of increasing field on the right. The velocity of the particle $\mathbf{v}$ is tangential to the trajectory and lies in the horizontal plane at points P and Q. At P, the magnetic field $\mathbf{B}$ is uniform in the x direction and so the magnetic force $\mathbf{F}$ acts vertically downwards. However, at point Q, $\mathbf{B}$ has a vertical component, which results in the force having a component parallel to the x-axis directing the particle towards the region of lower field.

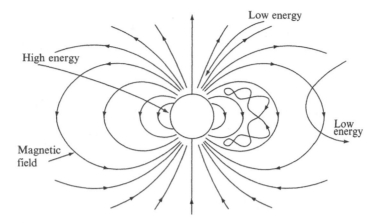

High energy

Low energy

Low
energy

Magnetic
field

Figure 11.5 Particle motion in the magnetic field of a planet like the earth. High-energy particles gain unrestricted access to the planet's surface. Low-energy particles are repelled near the magnetic equator, but can gain access in the polar regions. Those that become caught high above the planet execute trapped particle motion between northern and southern latitudes, as indicated by the closed trajectory on the right-hand side of the figure.

long periods in trajectories such as that shown in Figure 11.5 on the right-hand side of the planet. These trapped particles constitute the Van Allen belts of radiation. The particle flux within the belts is much higher than it is closer to the earth or in outer space and it is important that manned spacecraft spend as little time as possible passing through them in order not to expose the astronauts to dangerous doses of radiation.

Conceptually, the action of a magnetic mirror is straightforward, but it is proving difficult to create one suitable for a fusion reactor. This is because a hot, dense plasma in the low-field region of a simple, linear magnetic mirror is not properly confined against its tendency to expand outwards, and scientists are having to invent more complex field configurations in order to construct a magnetic bottle, based on the mirror principle, which is fully effective in all three dimensions.

The alternate method of magnetic confinement is to use a closed-field geometry. The simplest of these is the toroidal field, which is produced by passing a current through a solenoid wound into the form of a doughnut, as shown in Figure 11.6(a). In principle, plasma particles constrained in a uniform toroidal field could circulate endlessly, following helical paths along the direction of the field inside a vacuum vessel and be well insulated from its walls. However, in any practical arrangement, a simple, toroidal magnetic field is non-uniform and becomes weaker at larger radii (measured from the centre of the doughnut). A plasma in such a field is unstable to being lost at large radii and a second field, called a poloidal field, has to be included to prevent this occurring. The geometry of the poloidal field is sketched in Figure 11.6(b). It gives a twist to the toroidal field when added to it, and the resultant field lines trace out helices around the median axis of the torus as shown in Figure 11.6(c). For interested students, the main causes of the plasma instability in a simple toroidal field and how the addition of the poloidal field corrects for it and achieves effective plasma confinement are outlined in the inset.

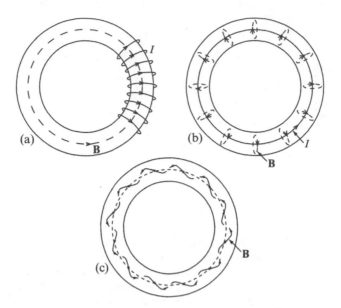

Figure 11.6 (a) Toroidal magnetic field **B** produced by passing a current *I* through a coil (part of which is shown) wound on a torus. (b) Poloidal field produced by passing a current *I* round the axis of the torus. (c) Result of combining the toroidal and poloidal fields is one in which field lines (solid curve) trace out helices around the axis of the toroid (dashed line).

Plasma confinement in a combined toroidal and poloidal field

Figure 11.7 shows two sectional views taken through a toroidal field in a torus (part of which is shown). The magnetic field **B** at the position of the section is directed into the page. Plasma particles moving in this field may have a velocity component parallel to **B**, which does not give rise to a force. We disregard this for a moment and consider the change of direction due to the velocity component in the plane of the page (perpendicular to **B**). A positively charged particle will move in an anticlockwise direction and a negatively charged particle in a clockwise direction. The curvature of the path in each case is proportional to **B**. Therefore, in a toroidal field, where **B** decreases with radius *r* (as noted above), the curvature also decreases with *r* and orbiting particles will shift sideways in a direction perpendicular to *r*, as shown in Figure 11.7(a). Ions and electrons orbit in opposite directions and, therefore, will tend to move apart. This would cause a separation of charge on opposite sides of a plasma and give rise to an electric field **E** across it.

Once an electric field becomes established, there will be an electric force acting on a moving charged particle as well as a magnetic force. If the particle is moving in the direction of increasing *r*, the two forces act in opposite directions and, therefore, tend to cancel each other. The situation is shown in Figure 11.7(b) for a positive ion, but the effect is independent of charge. For a negatively charged, moving particle, the forces would be reversed in direction, but would still act in opposition to each other. The result is that plasma

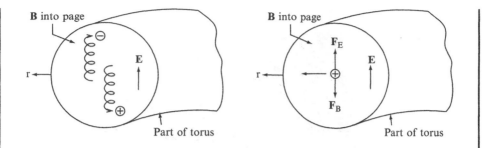

Figure 11.7 Cross sections through the toroidal magnetic field region of Figure 11.6(a). The magnetic field **B** (directed into the page) decreases with distance r from the centre of the torus. In (a), the motion of positive and negative particles is shown due to their velocity components perpendicular to **B**, i.e. the particle motion is in the plane of the paper. The curvature of their paths decreases with r and this causes the orbits of positive and negative charges to drift in opposite directions, as shown, giving rise to an electric field **E** across the region. In (b), magnetic (**F**$_B$) and electric (**F**$_E$) forces are shown acting on a positive ion moving in the direction of increasing r.

particles with a particular velocity are not constrained in the direction of increasing r (see Problem 11.10) and, over a period of time, the plasma will drift outwards until it reaches the outer wall of the containing vessel and loses its energy.

Adding a poloidal field produces a twisted toroidal field as shown in Figure 11.6(c). Plasma particles are constrained by the field and still trace out approximately helical paths. However, the axes of these helices follow the field lines, which wind around the axis of the torus. As a result, any particle moves constantly from one side of the torus axis to the other because of the twist in the field. This means that the sideways drift it experiences, due to the non-uniformity of the toroidal field, is alternately directed towards and away from the torus axis, and the net effect is zero. Thus, any tendency for electrons and ions to drift away from the axis and separate out is neutralized and the electric field is effectively 'shorted out'.

In experimental assemblies, the poloidal field is generated by passing a current either through external coil windings or along the axis of the toroid through the plasma itself. The latter method is the one used in the Tokamak design which was conceived in Russia. The name is a Russian acronym for 'toroidal magnetic chamber'. It is an effective arrangement, and most progress in the development of MCF has used the tokamak configuration.

The largest tokamaks currently in operation are JET (Europe), TFTR (USA) and JT-60 (Japan). Figure 11.8 shows a sketch of the main field components of the Joint European Torus (JET), which is a major project set up by several European countries to study the conditions and technical requirements for MCF. The strong, main toroidal field is generated with external coils as shown. The current generating the poloidal field is induced by transformer action on the plasma. The primary windings of the transformer are shown in the figure; the plasma itself forms a single secondary turn. A current pulse in the primary induces a large current of up to 7 MA in the

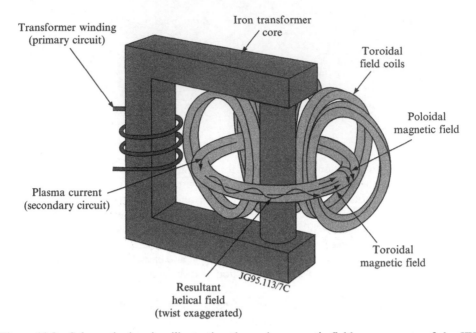

Transformer winding
(primary circuit)

Iron transformer
core

Toroidal
field coils

Poloidal
magnetic field

Plasma current
(secondary circuit)

Toroidal
magnetic field

Resultant
helical field
(twist exaggerated)

JG95.113/7C

Figure 11.8   Schematic drawing illustrating the main magnetic field components of the JET tokamak. By courtesy of EFDA-JET.

plasma. This current not only generates the required poloidal field, but also provides several megawatts of resistive heating to the plasma. Unfortunately, this form of ohmic plasma heating is not sufficient to raise the temperature high enough for fusion to occur because plasma resistance R decreases with temperature and power input ($i^2 R$) becomes less efficient as the plasma heats up.

In most tokamaks, additional heating is provided by a combination of powerful radio-frequency (r.f.) sources and neutral-beam injection (NBI). In r.f. heating, high-power radio- or micro-waves are directed into the plasma, which will absorb energy resonantly if it is delivered at either the electron or ion cyclotron frequencies. In neutral-beam heating, hydrogen or deuterium ions are accelerated to energies up to 100 keV and then neutralized by charge-exchange reactions in a region of hydrogen or deuterium gas before being injected into the plasma. Since these fast atoms are neutral, they are unaffected by the magnetic field until they become ionized or undergo charge exchange during collisions inside the plasma. They then become trapped in the field and transfer their energy to the plasma by collisional energy loss. Neutral-beam injection has produced ion temperatures up to 30 keV in the JET and TFTR tokamaks. It can also be used to refuel the plasma as the deuterium and tritium become depleted.

In the past decade, considerable progress has been made in approaching the basic criteria for achieving ignition in a magnetically confined plasma. Each of the required values of n, τ and T for ignition have been reached, but not all at the same time. The triple product has been increased by several orders of magnitude within the past decade and now is close to meeting the Lawson criterion and about a factor of six from ignition. In December 1997, a D–T plasma in JET reached a peak fusion power output of 16 MW and a power of 10 MW was sustained for at least half a second. The ratio of output power to net input power to the plasma was 0.65 (more than double the previous record) and the burning of the D–T fuel was as expected. These are impressive achievements. However, a great deal of work still needs to be done even to reach ignition for the first time and then much more to demonstrate that efficiencies can be improved to the level required for commercial fusion power production.

11.4.2   Inertial confinement fusion

Difficulties with achieving stable plasma confinement in a magnetic field led to a radical, alternative proposal for obtaining controlled thermonuclear power. In ICF, a pulse of energy is directed from several directions at once on to a small pellet of fusible material – such as a frozen D–T mixture. The energy is delivered with such power that material is heated and violently ejected from the surface. When this happens, material inside the surface is driven inwards, compressing the core and raising its temperature to the point at which fusion occurs at a high rate. The whole process is then repeated in what amounts to a series of micro thermonuclear explosions.

The number of pellets used per unit time need not be high and each one is small. For reference, complete conversion of 1 mg of D–T (containing 2.4×10^{20} atoms) liberates about 350 MJ of energy, and an ICF reactor might be designed to operate at 10 micro explosions per second, each consuming 1 mg of material. This would generate about 3.5 GW of raw power, which would be absorbed in a surrounding thermal blanket and converted into electric power by conventional means.

In ICF, no attempt is made to contain the reacting atoms although the time during which fusion occurs can be increased to some extent by using tampers of higher density materials to slow down the expansion of the ignited core. The problem of containment is replaced with the major technical difficulty of generating and directing sufficient power to trigger fusion. Proposals include the use of high-power, pulsed lasers or intense beams of charged particles to bombard the pellets.

To appreciate some of the technical challenges facing developers of ICF we will make crude estimates of the main parameters. These are the confinement time and the compression factor. The product has to be large enough so that a significant fraction of the material undergoes fusion before the reacting nuclei fly apart. We will assume an equal mixture of deuterons and tritons reacting at $kT = 20$ keV – the most favourable case.

The confinement time is determined by the relative speed v of the heated ions and the radius r of the compressed fuel pellet. At 20 keV, v is about 2×10^6 m s^{-1} and the radius of a compressed fuel pellet will be no more than about 0.2 mm, giving an estimate for the confinement time τ of $r/v \sim 10^{-10}$ s (100 ps). The condition for the product $n\tau$ is roughly the same for ICF as for MCF described above. So, taking the value $n\tau = 3 \times 10^{19}$ s m^{-3}, obtained from Equation (11.6) for the Lawson criterion for D–T at 20 keV, we obtain $n \approx 3 \times 10^{29}$ m^{-3}, which is about 10 times the atom density of ordinary liquid or solid hydrogen. However, meeting the Lawson criterion in this example requires only enough energy from fusion to raise the average energy of an atom in the pellet to 20 keV. Very little of the material in the pellet is used up in achieving this.

A more useful performance indicator, leading, in general, to a very different value of $n\tau$, is given by the requirement that a certain fraction f of the fuel be consumed in the time τ. The rate of depletion of fuel atoms $dn/dt = -2R$, where R is the reaction rate. The factor of two appears because two nuclei are consumed every time a reaction takes place. Therefore, we have from Equation (11.2):

$$dn(t)/dt = -\langle v\sigma \rangle n^2(t)/2 \tag{11.7}$$

where $n(t)$ is the number of fuel atoms at time t and we have again taken the case that $n_1 = n_2 = n/2$.

After a time $t = \tau$, the number remaining $n(\tau)$ is given by the integral:

$$-\int_0^\tau \frac{dn(t)}{n^2(t)} = \frac{1}{n(\tau)} - \frac{1}{n} = \frac{1}{2}\langle v\sigma \rangle \tau \tag{11.8}$$

where n is the initial number of atoms. Substituting $f = 1 - n(\tau)/n$, we obtain

$$n\tau = \frac{2f}{(1-f)\langle v\sigma \rangle}. \tag{11.9}$$

For a significant burnup of $f \approx 30\%$, and using the value of $\langle v\sigma \rangle = 4.5$ $\times 10^{-22}$ m^3 s^{-1} from Figure 11.3 for D–T at 20 keV, Equation (11.9) gives $n\tau \approx 2 \times 10^{21}$ m^{-3}. If $\tau \approx 100$ ps, the compressed density then needs to be about 2×10^{31} m^{-3}, which is about three orders of magnitude greater than normal density. Thus, successful burning of about one-third of a pellet requires that it be compressed in radius by about a factor of 10.

An important side benefit of the high fuel–compression ratio is that it would improve the effectiveness of the energy of the α particles from D–T fusion to heat the reactants. The range of an α particle, in units of mass per unit area (which is the range in units of length multiplied by the density), is independent of density ρ (see Section 5.2.2) and, therefore, $\rho x = \rho' x'$, where the prime refers to the compressed state. This means that, as the material is compressed and its density increases, the range x', in units of distance travelled, decreases. Even at a density corresponding to the Lawson criterion (see above) the range would be a small fraction of the pill radius (Problem 11.12) and α particles would be stopped very close to their point of origin. Thus, it may only be necessary to initiate fusion in a small central region of the compressed pill and, thereafter, the thermonuclear reaction would feed itself, radiating outward as a spherical wave of reacting material.

The energy gain is the ratio of the energy produced divided by the input energy to trigger the reaction. Each pair of reacting D–T atoms, which releases 17.6 MeV, needs an initial energy of 40 keV, giving a raw energy gain of 17 600/40 = 440. If a burnup of 25% is achieved, the gain is reduced to about 100. This is still large, but it will certainly be reduced by many sources of inefficiency. For example, most of the input energy goes into ejecting material from the surface of the pill and perhaps only 10% is converted into raising the temperature of the compressed material. Also, it is proving very difficult to obtain a high efficiency for converting energy into a laser pulse and it may not be possible in a laser-driven system to achieve anything close to or exceeding 10%, which is what would be required.

Alternative proposals being considered for providing the trigger energy are to use beams of charged particles. The energy needs to be deposited mainly within the pill's surface in order to cause the inner material to implode. This constrains the range and, therefore, the particle energy. For example, if a thin (0.1 mm) outer layer is made of high-Z material, such as lead, the energy of a proton beam must be about 5 MeV. For an energy input of about 7.7 MJ (see Problem 11.13), which assumes a 10% efficiency for heating 1 mg of fuel to an average energy per atom of 20 keV, the number of incident protons would need to be

$$\frac{7.7 \times 10^6}{5 \times 1.6 \times 10^{-13}} \approx 10^{19}. \tag{11.10}$$

This has to be delivered in about 100 ps, which corresponds to a current in the pulse of over 10^{10} A!

Using a beam of electrons would require even greater currents because their energy would need to be much lower for the same range in the material. Heavy ions, which are much more easily stopped, could have energies of the order of 1 GeV. The current requirement would be reduced considerably in this case, but it would still be very large and well beyond what is currently attainable from a particle accelerator.

11.5 FUSION IN THE EARLY UNIVERSE

It is generally accepted that our Universe is expanding and current theory suggests that it began about 10^{10} years ago as a tiny speck at enormous temperature and pressure. Initially, conditions were such that even neutrons and protons did not exist in their present form but were broken down into their constituent quarks which normally cannot exist as free particles. As the expansion proceeded, temperature and pressure decreased quickly to the point where neutrons and protons condensed out of the hot quark soup. Shortly after that, nuclear fusion took place producing, first, deuterons, then, tritons and helium-3 nuclei and, finally, α particles. The following is a brief description of the sequence of events during this early period when the first complex nuclei were born. All the fusion reactions leading to the creation of the primordial elements are thought to have taken place within the first few minutes of time.

When the Universe was 1 μs old, current cosmological theories predict that it was less than half a mile across and its temperature was about 10^{13} K ($kT \approx 1$ GeV). At this temperature, there was enough thermal energy in the radiation field to create neutron–antineutron and proton–antiproton pairs and, for a time, an equilibrium between nucleon pair production and annihilation was established, which meant there were approximately equal numbers of protons, neutrons and their antiparticles. With further cooling, there were fewer high-energy photons and annihilation began to dominate, leaving a Universe consisting mostly of photons. Fortunately, a small amount of matter (neutrons, protons and electrons) was left over because the rules of nature governing creation and annihilation are not exactly symmetric between matter and antimatter with the balance being slightly in favour of matter as we know it.

During the next 0.01 s, the temperature decreased to 10^{11}K ($kT \approx 10\,\text{MeV}$) and an equilibrium existed briefly between leptons and antileptons (e^-e^+, $\nu\bar{\nu}$), photons and nucleons. The abundance of energetic leptons enabled protons and neutrons to be transformed into each other via the weak interaction processes:

$$e^+ + n \rightleftharpoons p + \bar{\nu}$$

$$e^- + p \rightleftharpoons n + \nu.$$

Gradually, however, the ratio of the number of neutrons to protons began to change because their masses are not the same. The proton, being lighter, is the more stable and the equilibrium began to shift in its favour with the n/p ratio given by the Boltzmann factor as

$$\frac{N_n}{N_p} = \exp(-\Delta E / kT) \tag{11.11}$$

where $\Delta E = (m_n - m_p)c^2 = 1.29\,\text{MeV}$ is the neutron–proton mass–energy difference.

At a certain point (≈ 1 s), the Universe became transparent to neutrinos and they ceased to play a role in transforming nucleons into each other. Shortly thereafter, e^+e^- pair production ceased and positrons quickly disappeared by annihilation, leaving behind an excess of electrons equal to the number of protons. The net result was that, when the Universe was about 3 s old, the N_n/N_p ratio became frozen at about 1/5. The temperature was still too high for fusion to occur, however, and it was not until about 4 min had elapsed that primordial nucleosynthesis could begin. During this extra time, some of the neutrons decayed into protons via the β-decay process $n \rightarrow p + e^- + \bar{\nu}$ ($t_{1/2} = 10.24\,\text{min}$) and the n/p ratio decreased further to about 1/7.

The first fusion reaction was

$$n + p \rightleftharpoons d + \gamma.$$

The Q value of $+\,2.22\,\text{MeV}$ is the energy required to dissociate the deuteron. This is well above kT, even at $t = 3$ s, but, as we have seen, the radiation spectrum has a high-energy tail, and a small fraction of photons will have energies great enough to cause deuteron breakup. The density of these energetic photons is very sensitive to temperature and it was only after several minutes, when the Universe had cooled below a critical value, that the deuterons survived long enough to allow them to participate in reactions leading to the mass-3 nuclei:

$$d + n \rightarrow t + \gamma$$

$$d + p \rightarrow {}^3\text{He} + \gamma.$$

Both the triton and helium-3 are more stable than the deuteron. They were not dissociated in the radiation field and further nucleon capture led to the production of helium-4:

$$p + t \rightarrow \alpha + \gamma$$

$$n + {}^3He \rightarrow \alpha + \gamma.$$

The α particle is the most stable of these light nuclei, and helium-4 was the main end product of these early fusion processes.

Further reactions were severely limited by the fact that mass-5 and mass-8 nuclei are unstable and decay very quickly. So, protons and neutrons cannot combine with α particles, and two α particles cannot fuse together. Some mass-7 nuclei were formed via the reactions: $t + \alpha \rightarrow {}^7Li + \gamma$ and ${}^3He + \alpha \rightarrow {}^7Be + \gamma$. However, the Coulomb barrier restricted these reactions considerably and, although small amounts of these heavier nuclei remained at the end, essentially all the neutrons at the beginning of the process were consumed to form helium-4. Since $N_n/N_p \approx 1/7$ initially, this would have led to a mass ratio of helium-4 to hydrogen in the early Universe of 1:3. This is in excellent agreement with the observed ratio of 0.32 ± 0.13, obtained from observations of various nebulae and stars.

After only a few minutes, the density and temperature of the expanding Universe had decreased to the point where fusion stopped altogether. Nuclear reactions did not begin again until stars condensed out of the primordial H/He mixture. What happened then is described in the next section.

11.6   STELLAR BURNING

11.6.1   Hydrogen burning

As a cloud of interstellar gas contracts to form a star, its temperature increases as gravitational potential energy is converted into kinetic energy. When the interior heats up to about 10^7 K, the first thermonuclear reactions begin to take place converting hydrogen into helium in a sequence of reactions known as the *proton–proton chain*. Reactions involving helium-4 do not occur at this stage because both mass-5 and mass-8 nuclei are unbound, as we have already noted. Note that simple fusion of two protons is impossible because the diproton is unstable.

The first hydrogen-burning reaction to proceed in a star is the conversion of two protons into the only stable dinucleon system, namely, the deuteron. The reaction is written

$$p + p \rightarrow d + e^+ + \nu.$$

It requires a β decay to occur during the time the protons are interacting with each other. The probability of a β decay during this short period is extremely low and so the reaction has a very small cross section. Even at the centre of our sun, where conditions for this fusion reaction are favourable, the average time a proton exists before being converted is about 10^{10} years. Fortunately for us, this is about twice the estimated age of the sun.

Once the deuteron is formed, it very rapidly undergoes the reaction:

$$d + p \rightarrow {}^3He + \gamma.$$

At this point, no D–D reactions will occur because it is extremely unlikely that a deuteron will collide with another deuteron. Given that the average lifetime of a proton before fusion is about 10^{18} s, we know that only one deuteron is produced per second for every 10^{18} protons. Inside the sun, it can be calculated that the mean lifetime of a deuteron is about 2 s and, therefore, it is approximately 10^{18} times more probable that that a deuteron will collide with a proton than with another deuteron and so the deuteron concentration never reaches a significant level. Indeed, the concentration of deuterium in hydrogen which we find on the earth is much too high for it to have come from a star and it must have been created just after the Big Bang.

Reactions between ${}^3He$ and a proton are not possible because ${}^4Li$ is unbound. The most probable fate of a ${}^3He$ nucleus is to react with another ${}^3He$ in the reaction:

$$^3He + {}^3He \rightarrow \alpha + 2p.$$

The overall effect of the p–p chain of reactions is that four protons are converted into an α particle with an energy release equal to the binding energy of the α particle less twice the neutron–proton mass difference, or about 26.7 MeV.

As we shall see in the next section, heavy elements are formed in stars and ejected into interstellar space when a star explodes in a supernova. Second-generation stars contain some of this material and, in them, hydrogen burning can proceed in a cycle of fusion reactions quite different to those in a first-generation star, which condensed out of primordial material (H and He only). One of these reaction cycles, is known as the *CNO cycle*, the dominant reactions of which are listed below. It is named after the intermediate nuclei which are produced.

$$^{12}C + p \rightarrow {}^{13}N + \gamma$$

$$^{13}N \rightarrow {}^{13}C + e^+ + \nu$$

$$^{13}C + p \rightarrow {}^{14}N + \gamma$$

$$^{14}N + p \rightarrow {}^{15}O + \gamma$$

$$^{15}O \rightarrow {}^{15}N + e^+ + \nu$$

$$^{15}N + p \rightarrow {}^{12}C + \alpha.$$

The reactions are also illustrated in Figure 11.9, which shows graphically how the sequence of fusion and β-decay processes results in the formation of an α particle from four protons, as in the p–p chain and with the same Q value. The ${}^{12}C$ nucleus, at the beginning, acts as a catalyst and remains at the end available to continue the process over and over again. Unlike the first reaction in the p–p chain, none of the reactions in the CNO cycle requires a β decay to occur at the same time as fusion is taking place and

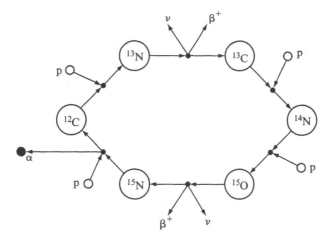

Figure 11.9   Representation of the reactions in the CNO cycle showing how four protons are brought together to produce an α particle. The ^{12}C nucleus at the beginning of the cycle remains at the end.

so, in principle, the CNO reactions can proceed faster. However, the Coulomb barrier also affects the reaction rates and it is much higher between a proton and a C, N or O nucleus than it is between two protons. As the stellar core heats up, therefore, the p–p reactions begin first. Later, when the core temperature has increased sufficiently, more protons have thermal energies able to penetrate the higher Coulomb barriers and the CNO reactions become more probable. Our sun's core temperature is about 10^7 K and the p–p cycle accounts for about 98% of the energy produced. However, if the core temperature were to increase to about 1.5×10^7 K, the contributions from the the p–p and CNO cycles would be approximately equal.

Of all the possible fusion reactions, hydrogen burning generates the most energy per unit mass of matter (≈ 7 MeV u^{-1}) and it is the source of power which maintains a star's stability for the greater part of its life. In the case of our sun, we can easily estimate its rate of fuel consumption. The flux of solar radiation reaching the earth (the solar constant) is 1.4×10^3 W m^{-2} from which we deduce that the total solar output is 4×10^{26} W, assuming it is radiated isotropically. As we have seen, any of the hydrogen-burning cycles of fusion reactions consumes four protons and liberates about 26 MeV of energy.[1] Thus, to generate its output, we estimate that the sun uses up

$$\frac{4 \times 10^{26} \times 4}{26 \times 1.6 \times 10^{-13}} \approx 4 \times 10^{38} \text{ protons per second.}$$

The sun contains about 7×10^{56} protons and, at its present age, has burned about 10% of its original hydrogen. It is expected to continue to shine, as it does now, with little apparent change for about another 6 billion years. At this point in its life, it will undergo a major change, swell up to become a red giant and begin to burn helium.

---

[1] This excludes the energy carried by neutrinos, which does not form part of the solar constant.

11.6.2   Helium burning

When the hydrogen in the stellar core becomes depleted, the energy produced by hydrogen burning is no longer sufficient to support the outer layers of material. Gravitational contraction begins again and the core temperature rises. Eventually, it reaches between 1 and 2×10^8 K when the Coulomb barrier between two α particles can more easily be overcome and helium burning begins. It proceeds in several stages.

As we have noted above, the reaction $\alpha + \alpha \rightarrow {}^8$Be does not lead to a stable nucleus. However, ^{8}Be is almost bound, requiring an energy of only 92 keV to form it. It decays back into two α particles in about 10^{-16} s. The existence of the ^{8}Be state acts as a resonance, increasing the interaction cross section between two α particles at a relative energy of 92 keV and causing them to stay together much longer than they would if they simply scattered off each other.

In the hot core, there will be an equilibrium between the two states:

$$\alpha + \alpha \rightleftharpoons {}^8\text{Be}$$

resulting in a small, but finite equilibrium concentration of ^{8}Be. This concentration is highly dependent on temperature, according to the Boltzmann factor $\exp(-\Delta E/kT)$, where $\Delta E = 92$ keV. This factor is about 5×10^{-3} at a temperature of 2×10^8 K ($kT \approx 17$ keV) and at 10^8 K, it is about 200 times smaller.

This small concentration of short-lived ^{8}Be nuclei is enough to allow the occasional capture of a third α particle to form ^{12}C via the reaction: ^{8}Be $+ \alpha \rightarrow {}^{12}$C. However, it was realized by Hoyle in 1954 that this process would not proceed fast enough unless there was a state in ^{12}C at a particular excitation energy which would act as a resonance to enhance the ^{8}Be $+ \alpha$ reaction cross section just as the transient ^{8}Be state enhances $\alpha + \alpha$ fusion. In fact, shortly after the suggestion was made, a new state, with all the right properties, was discovered in ^{12}C at an excitation energy of 7.654 MeV very close to the energy predicted by Hoyle.

The thresholds and energy levels involved in the burning of helium to carbon are shown in Figure 11.10. The Q value for the reaction $3\alpha \rightarrow {}^{12}$C is 7.275 MeV, which

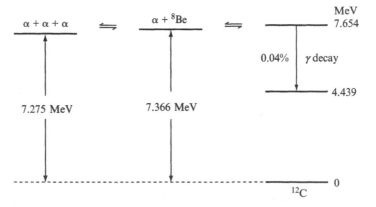

Figure 11.10   Energy-level diagram showing the ground and first two excited states of ^{12}C and the threshold energies required to break up ^{12}C in its ground state into $\alpha + {}^8$Be or into three α particles.

means that the ground state of ^{12}C lies at this energy below the state of three α particles at rest. The ^{8}Be $+ \alpha$ state is 92 keV above the 3α state and the critical ^{12}C excited state is at an energy of just 288 keV above that. Thus, an energy of 380 keV is needed from the thermal environment to allow three α particles to be able to combine to form ^{12}C in the 7.654-MeV state. This is well within the range of thermal energies inside the hot helium core at 2×10^{8} K.

Due to the relevant Boltzmann factors, there will be an equilibrium concentration of excited ^{12}C nuclei in the stellar core. The vast majority of these nuclei dissociate back into three α particles, but there will be a fraction which γ decay mainly to the first-excited state at 4.439 MeV (see Figure 11.10). This state then decays to the ground state of ^{12}C, which is stable. The γ-decay branch of the 7.654 MeV state is only 0.04%, but it is enough to account for the observed abundance of ^{12}C in the Universe.

Once ^{12}C is formed, oxygen is created by the α-capture reaction:

$$\alpha + {}^{12}\text{C} \rightarrow {}^{16}\text{O} + \gamma.$$

Further α captures leading to ^{20}Ne and ^{24}Mg also occur, but to a much lesser extent because the cross sections are inhibited by the higher Coulomb barriers. The end result of helium burning is a core consisting mainly of ^{12}C and ^{16}O with smaller amounts of ^{20}Ne and ^{24}Mg.

11.6.3   Beyond helium burning

Many stars come to the end of their lives when helium burning is complete and form white dwarf stars, but, in massive stars, there can be several more burning stages each preceded by gravitational contraction and an increase in the core temperature. The next stage after helium burning is reached when the carbon / oxygen core has shrunk until the temperature at its centre has reached about 5×10^{8} K and there is enough relative kinetic energy available to allow carbon nuclei to interact with each other at a significant rate producing ^{20}Ne, ^{23}Na and ^{23}Mg:

$$^{12}\text{C} + {}^{12}\text{C} \rightarrow {}^{20}\text{Ne} + \alpha$$

$$^{12}\text{C} + {}^{12}\text{C} \rightarrow {}^{23}\text{Na} + \text{p}$$

$$^{12}\text{C} + {}^{12}\text{C} \rightarrow {}^{23}\text{Mg} + \text{n}.$$

Note that two of these reactions produce neutrons and protons which, as we shall see in the next section, are used to synthesize heavy elements.

Oxygen burning occurs when the temperature reaches 2×10^{9} K, the most import-ant reaction being the one producing ^{28}Si:

$$^{16}\text{O} + {}^{16}\text{O} \rightarrow {}^{28}\text{Si} + \alpha.$$

The final stage is reached at $T \approx 3 \times 10^{9}$ K $(kT \approx 0.25$ MeV$)$, when silicon burning begins. However, even at this temperature, the Coulomb barrier is too high to allow a silicon nucleus to interact directly with another silicon nucleus. Instead, a series of reactions takes place beginning with thermal photodisintegration of silicon:

$$\gamma + {}^{28}\text{Si} \rightarrow {}^{24}\text{Mg} + \alpha$$

which requires a γ ray of at least 9.98 MeV. Photons of this energy lie in the extreme high-energy tail of the thermal radiation spectrum where the photon flux is very weak and so the reaction proceeds slowly. The α particles from this reaction then generate energy by building up heavier nuclei in successive (α, γ) capture reactions forming ${}^{32}\text{S}, {}^{36}\text{Ar}, {}^{40}\text{Ca}$, etc. Many other reactions occur during the silicon-burning phase, some of which produce protons and neutrons in addition to α particles. The result of this series of photodisintegration and radiative captures is the steady build-up of heavier elements until the mass number reaches $A \approx 56$. Up to this point, the energy required to photodisintegrate ${}^{28}\text{Si}$ is more than offset by the energy released by the subsequent capture reactions because the resulting heavier nuclei are increasingly more stable. The net energy produced is enough to stabilize the star, but not for long.

11.7   NUCLEOSYNTHESIS BEYOND $A \approx 60$

The sequence of stellar burning terminates when the core of the star is largely composed of nuclei with mass numbers close to 56. These are the most stable nuclei of all and no more energy is to be gained from further nuclear reactions. As soon as there is insufficient internal energy being produced, the core cannot support the outer layers against gravitational collapse and the star begins its final contraction. The contraction would stop if the weight of the star could be supported by the ability of the matter in its core to withstand the compressive force. However, only very massive stars reach this stage. They cannot resist the compression and there is a catastrophic collapse. A large amount of gravitational potential energy is converted into kinetic energy and the star becomes a supernova when a large fraction of its mass is thrown out into space. For several days, the star's power output increases enormously by a factor of 10^9 to 10^{10} and many nuclear reactions take place, producing copious numbers of particles including, in particular, neutrons. It is this supply of neutrons which leads to the formation of the heaviest elements.

The principal neutron reactions in a star during this stage are elastic and inelastic scattering and capture (n, γ). Scattering leads back to the original nucleus and only (n, γ) is of consequence in forming new elements. Each neutron capture increases the mass number by one unit and the process can continue to build heavier and heavier nuclei because the neutron does not have to overcome a Coulomb barrier. As noted in the previous section, neutrons are produced during the red giant stage of a star's life as well as in a supernova. However, the neutron fluxes in these two scenarios differ enormously and lead to two different paths by which heavy nuclei are synthesized. These are known as the slow (s) and rapid (r) processes.

In the s process, the mean time between neutron captures is long compared with most β-decay lifetimes, i.e, $t_{\text{capt}} \gg t_\beta$. This means that, in general, neutron capture only takes place on stable nuclei. When a radioactive nucleus is produced, unless its half-life happens to be very long, it will β decay until a stable daughter is reached before a further neutron capture is likely to occur. In a red giant, the neutron flux Φ is estimated to be about 10^{20} m^{-2} s^{-1}. A nucleus with a typical (n, γ) reaction cross section σ of about 100 mb will react at a rate $\sigma\Phi \approx 10^{20} \times 0.1 \times 10^{-28} \approx 10^{-9}$ s^{-1}. Clearly, in this case, nuclei are slowly built up via the s process over a long period of time.

In a supernova, the reverse is the case. The neutron flux is about 10^9–10^{10} times greater than in a red giant, which means that t_β is usually greater than t_{capt} and many neutrons may be captured one after the other, producing very neutron-rich nuclei. This is the r process. Eventually, the resulting nuclei become so unstable that t_β falls below t_{capt} and then β decay transforms them back towards stability until t_β exceeds t_{capt} again.

During the s process, an equilibrium is established when the rate of production of a particular nuclide (mass number A) by neutron capture is equal to its rate of loss by a further neutron capture. The rate of buildup of A is proportional to the product of the number n $(A-1)$ of nuclei with mass number $A-1$ and the capture cross section $\sigma(A-1)$. The rate of loss is proportional to $n(A)\,\sigma\,(A)$. Thus, the rate of change of the number of nuclei with mass number A is given by

$$dn(A)/dt \propto n(A-1)\sigma(A-1) - n(A)\sigma(A).$$

In equilibrium, $dn(A)/dt = 0$ and $n\ (A-1)\sigma(A-1) = n\ (A)\ \sigma\ (A)$. A similar equation relates the next set of nuclei giving $n(A)\sigma(A) = n(A+1)\sigma(A+1)$, and we have the general result that $n(A)\sigma(A)$ should be independent of A within the entire s-process chain of neutron captures. Neutron-capture cross sections have been measured on all stable nuclei and Figure 11.11 shows values of the product of (n, γ)

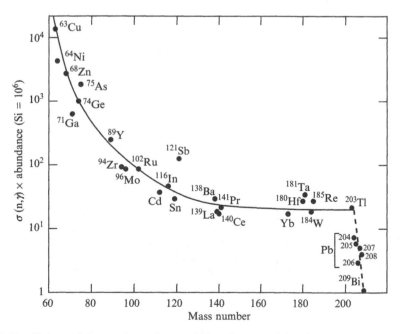

Figure 11.11   Values of the product of natural abundance and (n, γ) capture cross section for various nuclei plotted as a function of mass number. The vertical scale is one for which the value for silicon is arbitrarily set at 10^6. From Clayton *et al.* (1961).

cross section and observed abundance for various stable nuclei plotted as a function of mass number. After decreasing from a peak at the region of greatest stability ($A \approx 60$), the product approaches an approximately constant value beyond $A \approx 100$, consistent with the general expectation for s-process nucleosynthesis.

The s process terminates at ^{209}Bi because heavier nuclei in the continuing sequence are either α-active or β-unstable leading to α-active nuclei. Further build-up could not have continued via the s process because mass would have been lost by α decay as fast as it was gained by neutron capture. Nuclei such as ^{232}Th, ^{235}U and ^{238}U, which we find on earth today, must have been formed by the r process with the time between neutron captures being shorter than any of the α-decay lifetimes of the intermediate nuclei formed on the way.

There are many other nuclei which could not have been formed by the s process and their existence is further evidence for the r process. Examples in the mass region $A = 90$–100 are shown in Figure 11.12. The main s-process route, linking stable and long-lived isotopes of yttrium, zirconium, niobium and molybdenum, is indicated by the arrows. It follows the line of neutron captures on stable (or near stable) nuclides and the β decays when a radioactive product is reached for which t_β is less than t_{capt}. However, there are two stable isotopes: ^{96}Zr and ^{100}Mo, which lie away from the s-process path on the neutron-rich side. They can only have been formed by the r process when reaction rates would have been fast enough to allow neutron capture on the unstable nuclei ^{95}Zr and ^{99}Mo.

Further examination of Figure 11.12 shows that the nucleus ^{92}Mo also lies off the s-process path, but on the proton-rich side. It is an example of many other light isotopes throughout the periodic table which could not have been formed by either r- or s-neutron capture. They are all thought to have been produced via proton-capture (p, γ) reactions in a third process known as the p process. For example, ^{92}Mo could have been formed by proton capture on ^{90}Zr followed by a second proton capture on

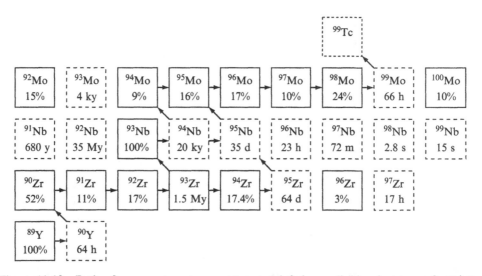

Figure 11.12   Path of s-process neutron captures and β decays linking isotopes of yttrium, zirconium, niobium and molybdenum. Solid rectangles represent stable nuclei (with percentage abundances). Unstable nuclei are shown in dashed rectangles (with decay half-lives).

^{91}Nb. Proton capture is unlikely to have taken place during normal stellar burning, however, because the Coulomb barrier can be high. For example, between a proton and a lead nucleus, it is about 10 MeV. This means that the p process in medium and heavy elements would probably only have occurred at the very high temperatures in a supernova.

In the last three sections, we have been able to give a brief account only of the main mechanisms by which the matter we find on earth was created. A full analysis is complex, requiring detailed knowledge of many nuclear reactions and the different conditions under which they may have taken place. However, when this is done, a good understanding emerges of how virtually all existing atomic nuclei were formed and with reasonable estimates of their observed abundances.

PROBLEMS 11

11.1 If the reaction d(p, γ)^{3}He occurs at a very low initial energy, show that in the centre-of-mass (c–m) system, the energy E_γ of the γ ray is greater than that of the helium ion (mass $m = 3.016$ u) by the factor $(2mc^2/E_\gamma)$. Hence, calculate the energy of the recoiling ^{3}He ion.

11.2 Calculate the energies of both outgoing products in the reaction d(d, n)^{3}He, if the initial kinetic energy is sufficiently small that the momentum of the c–m is negligible.

11.3 What is the distance between two protons at which the Coulomb potential is equal to the Coulomb barrier of 200 keV?

11.4 Estimate the minimum mass of D–T thermonuclear bomb material equivalent to 50×10^6 tonnes of TNT, given that 1 g of TNT liberates about 4 kJ of energy.

11.5 Show that the energy corresponding to the most probable speed in the Maxwell–Boltzmann distribution is kT.
   Calculate the most probable speeds for a deuteron and a triton at $kT = 20$ keV.

11.6 A plasma consists of equal numbers of deuterons and tritons at a temperature for which $kT = 10$ keV. Calculate the plasma density n if, for a confinement time $\tau = 3$s, the Lawson criterion is just satisfied. Determine the reaction rate per particle in the plasma.

11.7 Evaluate the Lawson criterion for a D–D plasma at $kT = 100$ keV. Take the average Q value for the D–D reactions to be 3.6 MeV.

11.8 A deuterium plasma is maintained at a temperature for which the average speed of a deuteron corresponds to $kT = 100$ keV. If the average fusion cross section σ is 0.1 b, estimate $\langle v\sigma \rangle$, where v is the average relative speed between two deuterons, and compare it with the value shown in Figure 11.3. Hence, calculate the power per unit volume. The plasma volume is 1 litre and contains 10^{18} deuterons and 10^{18} electrons.
   How long does the plasma have to burn to regain the input energy?

11.9 If the average magnetic field strength in a tokamak is 3 T, calculate the radii of curvature for 20-keV deuterons and tritons moving in closed orbits.

11.10 What would be the strength of an electric field $\mathbf{E}$, applied at 90° to a magnetic field $\mathbf{B} = 3$ T, which would allow 1-keV tritons to move undeflected in a direction orthogonal to both $\mathbf{E}$ and $\mathbf{B}$? What would be the energy of an undeflected deuteron?

11.11 How many 1-mg pellets of D–T material per second would be required for a 500 MW (thermal) fusion power station if 30% of the material in each pellet were converted in inertial-confinement fusion?

11.12 If the range R' of 3.5-MeV α particles, produced in D–T fusion, is 0.8 mg cm^{-2} in the D–T material, estimate their range R (in units of distance) as a fraction of the radius of a 1-mg D–T pellet compressed to a density of 6×10^{29} atoms m^{-3}.

11.13 A pulse of energy is delivered with an efficiency of 10% to a 1-mg pellet containing equal numbers of deuterium and tritium atoms. Estimate the input energy required to raise the energies of deuterons and tritons to a temperature corresponding to $kT = 20$ keV.

    Neglect the energy of the electrons, assuming they do not have time to reach thermal equilibrium.

11.14 About 3 s after the onset of the Big Bang, the neutron–proton ratio became frozen when the temperature was still as high as 10^{10} K ($kT \approx 0.8$ MeV). About 250 s later, fusion reactions took place converting neutrons and protons into helium-4 nuclei. Show that the resulting ratio of the masses of hydrogen and helium in the Universe was close to 3. The neutron half-life $= 10.24$ min and the neutron–proton mass difference (in energy units) $= 1.29$ MeV.

11.15 Given that the sun (mass $= 2 \times 10^{30}$ kg) was originally composed of 71% hydrogen by weight and assuming it has generated energy at its present rate (3.86×10^{26} W) for about 5×10^9 years by converting hydrogen into helium, estimate the time it will take to burn 10% of its remaining hydrogen. Take the energy release per helium nucleus created to be 26 MeV.

11.16 Explain why the helium-burning phase is likely to be much shorter ($\lesssim 10\%$) than the hydrogen-burning phase of a star's life. Assume that the rate of energy production and the total available mass for burning are the same in both phases.

11.17 Show that the mean reaction time for a nucleus is $1/\sigma\Phi$ in an environment where the interacting particle flux is Φ and σ is the reaction cross section. Hence, determine the neutron flux, during a supernova, assuming an average neutron-capture cross section of 100 mb and an average reaction time per nucleus of 1 s.

11.18 There are eight stable tellurium isotopes between $A = 120$ and 130 (see Appendix F). Indicate the main process (s, r or p) by which each was created inside a star.

Appendix A: Useful Information

A.1 PHYSICAL CONSTANTS[1] AND DERIVED QUANTITIES

Speed of light	c	$= 2.997\,924\,58 \times 10^8\,\mathrm{m\,s^{-1}}$
		$\approx 3.00 \times 10^{23}\,\mathrm{fm\,s^{-1}}$
Avogadro's number	N_A	$= 6.022\,141\,99(47) \times 10^{26}$ molecules per kg–mole
Planck's constant	h	$= 6.626\,068\,76(52) \times 10^{-34}\,\mathrm{J\,s}$
	$\hbar$	$= 1.054\,571\,596(82) \times 10^{-34}\,\mathrm{J\,s}$
		$= 0.65821 \times 10^{-21}\,\mathrm{MeV\,s}$
	$\hbar^2$	$= 41.802\,\mathrm{u\,MeV\,fm^2}$
	$\hbar c$	$= 197.327\,\mathrm{MeV\,fm}$
Elementary charge	e	$= 1.602\,176\,462(63) \times 10^{-19}\,\mathrm{C}$
	$e^2/4\pi\varepsilon_0$	$= 1.4400\,\mathrm{MeV\,fm}$
Fine structure constant	α	$= e^2/4\pi\varepsilon_0\hbar c = 1/137.036$
Boltzmann constant	k	$= 1.380\,6503(24) \times 10^{-23}\,\mathrm{J\,K^{-1}}$
		$= 0.8617 \times 10^{-4}\,\mathrm{eV\,K^{-1}}$

A.2 MASSES AND ENERGIES

Atomic mass unit	m_{u} or u	$= 1.660\,538\,73(13) \times 10^{-27}\,\mathrm{kg}$
	$m_{\mathrm{u}}c^2$	$= 931.494\,\mathrm{MeV}$

---

[1] Source: 1998 CODATA Recommended Values. Uncertainties are given in parentheses.

Electron m_e $= 9.109\,381\,88(72) \times 10^{-31}$ kg
                                 m_e/m_u $= 5.486 \times 10^{-4} = 1/1823$
                                 $m_e c^2$ $= 0.510\,998\,902(21)$ MeV

Proton m_p $= 1.672\,621\,58(13) \times 10^{-27}$ kg
                                 m_p/m_u $= 1.007\,276\,47$
                                 $m_p c^2$ $= 938.272$ MeV

Hydrogen atom m_H $= 1.673\,533 \times 10^{-27}$ kg
                                 m_H/m_u $= 1.007\,825$
                                 $m_H c^2$ $= 938.783$ MeV

Neutron m_n $= 1.674\,927\,16(13) \times 10^{-27}$ kg
                                 m_n/m_u $= 1.008\,664\,915\,78(55)$
                                 $m_n c^2$ $= 939.565$ MeV

Alpha particle m_α $= 6.644656 \times 10^{-27}$ kg
                                 m_α/m_u $= 4.001\,506\,175$
                                 $m_\alpha c^2$ $= 3727.379$ MeV

A.3   CONVERSION FACTORS

Fermi   1fm $= 10^{-15}$ m

Million electron volts   1MeV $= 1.602\,176 \times 10^{-13}$ J
                          $1\text{MeV}/c^2$ $= 1.783 \times 10^{-30}$ kg

Cross section (barn)   1b $= 10^{-28}$ m^2

Year   1y $= 3.1536 \times 10^7$ s

A.4   USEFUL FORMULAE

Energy width of a state of lifetime τ:
$$\Gamma = 6.582\,12 \times 10^{-22}/\tau(\text{s}) \text{ MeV}$$

Non-relativistic speed of mass m with energy E:
$$v = 1.389 \times 10^7 \sqrt{E(\text{MeV})/m(\text{u})} \text{ m s}^{-1}$$

Non-relativistic wave number of mass m with energy E:
$$k \equiv 2\pi/\lambda = 0.21874\sqrt{m(\mathrm{u}) \times E(\mathrm{MeV})}\,\mathrm{fm}^{-1}$$

Wave number for a photon of energy E:
$$k \equiv 2\pi/\lambda = E/\hbar c = E(\mathrm{MeV})/197.327\,\mathrm{fm}^{-1}$$

Appendix B:
Particle in a Square Well

In this appendix, the Schrödinger equation is solved to illustrate the existence of quantum states and discrete energies for a particle confined to a region of space. The simplest examples are those where the confining region is a box with rigid walls represented by a square well in one or more dimensions. This is shown in Figure B.1 for the one-dimensional (1-D) case, which we deal with first.

The potential energy is zero within the region and rises to infinity at the walls $(x = \pm a/2)$. The probability of finding the particle outside the box is zero and, since the wave function $\psi(x)$ must be continuous, this imposes the boundary condition at $x = \pm a/2$:

$$\psi(a/2) = \psi(-a/2) = 0. \tag{B.1}$$

Within the allowed region $(-a/2 < x < a/2)$, the potential energy is zero and the Schrödinger equation may be written as

$$-\frac{\hbar^2}{2m}\frac{d^2\psi}{dx^2} = E\psi \tag{B.2}$$

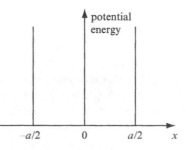

Figure B.1 The one-dimensional square-well potential. The potential rises to infinity for $|x| \geq a/2$.

for which the general solution is

$$\psi(x) = A \cos k_x x + B \sin k_x x \tag{B.3}$$

where A and B are constants and $k_x = \sqrt{2mE/\hbar^2}$. Applying the boundary conditions at $x = \pm a/2$ gives

$$\psi(a/2) = A \cos k_x a/2 + B \sin k_x a/2 = 0$$

and

$$\psi(-a/2) = A \cos k_x a/2 - B \sin k_x a/2 = 0$$

from which we find

$$A \cos k_x a/2 = 0; \quad \text{and} \quad B \sin k_x a/2 = 0.$$

These are satisfied (assuming that A and B are not both zero) for either

$$B = 0 \quad \text{and} \quad k_x a = n_x \pi \quad \text{with} \quad n_x = 1, 3, 5, \cdots \text{ or}$$

$$A = 0 \quad \text{and} \quad k_x a = n_x \pi \quad \text{with} \quad n_x = 2, 4, 6, \cdots.$$

Thus, the final solutions are two classes of wave function:

$$\psi(x) = A \cos k_x x, \quad k_x = n_x \pi / a, \quad n_x = 1, 3, \cdots \tag{B.4}$$

and

$$\psi(x) = B \sin k_x x, \quad k_x = n_x \pi / a, \quad n_x = 2, 4, \cdots. \tag{B.5}$$

The most important consequence of this result is that the energy of the particle can only take on certain discrete values:

$$E = \frac{\hbar^2 k_x^2}{2m} = \frac{h^2}{8ma^2} n_x^2, \quad n_x = 1, 2, 3 \cdots. \tag{B.6}$$

These are called the energy eigenvalues.

It is straightforward to generalize this analysis to three dimensions for the case of a particle in a cubic box of side a. The boundary conditions now constrain the wave function $\Psi(x, y, z)$ to be zero when either x, y or z are $\pm a/2$. Inside the box, the three-dimensional (3-D) Schrödinger equation becomes

$$-\frac{\hbar^2}{2m} \left(\frac{\partial^2 \Psi}{\partial x^2} + \frac{\partial^2 \Psi}{\partial y^2} + \frac{\partial^2 \Psi}{\partial z^2} \right) = E\Psi \tag{B.7}$$

which is solved by choosing a product solution of the form $\Psi(x, y, z) = X(x)Y(y)Z(z)$. Substituting this into Equation (B.7) and dividing by Ψ gives

$$-\frac{\hbar^2}{2m}\left(\frac{1}{X}\frac{\partial^2 X}{\partial x^2} + \frac{1}{Y}\frac{\partial^2 Y}{\partial y^2} + \frac{1}{Z}\frac{\partial^2 Z}{\partial z^2}\right) = E. \qquad (B.8)$$

The three terms on the left-hand side of this equation are, respectively, functions of x, y and z only. Since x, y, and z vary independently of each other, these three terms vary independently of each other, and Equation (B.8) can only be satisfied if each of these terms equals a constant:

$$-\frac{\hbar^2}{2m}\left(\frac{1}{X}\frac{d^2 X}{dx^2}\right) = \text{const.} = E_x \text{ and, similarly, for } Y \text{ and } Z$$

and the three constants E_x, E_y and E_z must add up to E:

$$E_x + E_y + E_z = E.$$

Each 1-D equation is identical in form to Equation (B.2) and we obtain results for each similar to those of Equations (B.4) and (B.5). Thus, the allowed 3-D states can be represented by three positive integers n_x, n_y and n_z, and the total energy follows directly as a generalization of Equation (B.6):

$$E = E_x + E_y + E_z = \frac{h^2}{8ma^2}\left(n_x^2 + n_y^2 + n_z^2\right) \qquad (B.9)$$

with n_x, n_y, $n_z = 1, 2, \cdots$.

Appendix C:
Density of States and the Fermi Energy

C.1 DENSITY OF STATES

Here, we use the results of Appendix B to derive an important quantity for dealing with many problems in physics, such as the distribution of electrons in metals, collisions of particles, and the decay of excited atoms or unstable nuclei. It is called the density of states $\rho(E)$ and it is defined as the number of quantum states per unit energy in a certain volume of space.

We have shown in Appendix B that any allowed quantum state in a box can be identified by a set of positive integers n_x, n_y and n_z. We can obtain a visual representation of these states by plotting the integers as co-ordinates of points in a three-dimensional octant, as shown in Figure C.1. Each point corresponds to an allowed state and, since all positive integer values are allowed, the density of points in this n-space is unity.

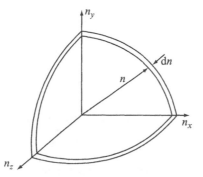

Figure C.1 Representation of allowed states of a particle confined to a cubic volume of space. Each state is represented by a point with co-ordinates n_x, n_y, $n_z = 1, 2, \cdots$. All the states within the spherical shell of the octant between n and $n + \mathrm{d}n$ have approximately the same energy.

Consider a shell of the octant with radius $n = \sqrt{(n_x^2 + n_y^2 + n_z^2)}$ and thickness dn. Assuming that $n \gg 1$, the number of states with radii between n and $n + dn$ is the volume of the shell:

$$\frac{\pi n^2 dn}{2} \tag{C.1}$$

The value of n corresponds to a certain energy, since $E = h^2 n^2 / 8ma^2$ [Equation (B.9)]. Therefore, (C.1) also gives the number of states with energies between E and $E + dE$, which is the density of states at energy E multiplied by dE. It can be written in terms of E by substituting for n and $ndn = (4ma^2/h^2)dE$ in (C.1) to give

$$\rho(E)dE = V \frac{2^{5/2} \pi m^{3/2}}{h^3} \sqrt{E} \; dE \tag{C.2}$$

where $V = a^3$ is the volume of the box within which the particles are contained.

The energy E is the kinetic energy of the particle and so, we can also express the density of states in terms of the magnitude of the momentum $p \; (= \sqrt{2mE})$. Strictly speaking, there is no net momentum, since the states are represented by standing waves, each of which is a superposition of waves travelling in opposite directions. Nevertheless, it is useful to interpret the quantum states in terms of the square of the momentum associated with the to and fro motion in these states, The expression for the density of states in this form is obtained directly from Equation (C.2) by making the substitutions $\sqrt{E} = p/\sqrt{2m}$ and $dE = pdp/m$, whence

$$\rho(p)dp = V \frac{4\pi p^2 dp}{h^3}. \tag{C.3}$$

The entire analysis could have been carried out for travelling waves instead of standing waves. The system is still assumed to be confined to a box of size a but, eventually, the size of the box is allowed to approach infinity, in which case the distinction between a large number of discrete, confined waves and a continuum of free waves can be made arbitrarily small. The boundary condition for a travelling wave is different from that of a standing wave, since a wave of definite momentum cannot vanish anywhere. Instead, periodic boundary conditions are imposed,[1] namely: $\psi(0, y, z) = \psi(a, y, z)$, $\psi(x, 0, z) = \psi(x, a, z)$ and $\psi(x, y, 0) = \psi(x, y, a)$ where 0 and a are the boundaries of the defining region in each of the three dimensions.

The function representing a travelling wave has the form:

$$\psi(\mathbf{r}) \; \propto \; \exp(i\mathbf{k} \cdot \mathbf{r}) = \exp\left[i(k_x x + k_y y + k_z z)\right]$$

and the boundary condition is satisfied if $k_x = 2\pi n_x/a$, $k_y = 2\pi n_y/a$, and $k_z = 2\pi n_z/a$, with $n_x, n_y, n_z = 0, \pm 1, \pm 2, \cdots$. The plus and minus signs indicate that we can have both positive and negative momentum components in each dimension. Therefore, the

---

[1] See for example, Mandl (1992), section. 2.6.1.

points (n_x, n_y, n_z), representing the allowed states, fill the whole n-space of Figure C.1, and the number of states between n and $n + dn$ is the volume of the entire spherical shell:

$$4\pi n^2 dn. \tag{C.4}$$

The magnitude of the particle's momentum $p = \hbar k = hn/a$ and, by substituting for n $(= ap/h)$ in (C.4), we obtain an expression for $\rho(p)dp$, the number of states with momenta between p and $p + dp$, which is identical to that given in Equation (C.3).

C.2   FERMI ENERGY

If A nucleons are packed into a nuclear volume V, we can use Equation (C.2) to determine the *Fermi* energy ε_F, which is the kinetic energy of the highest occupied state, assuming that the particles are arranged to have the minimum total energy. For simplicity, we neglect the Coulomb energy and assume that the neutrons and protons are bound by the same potential well. Since each level can accommodate four nucleons, the relationship between A and ε_F is obtained from Equation (C.2) as

$$A = 4 \int_0^{\varepsilon_F} \rho(E)dE = V\frac{16\pi(2m\varepsilon_F)^{3/2}}{3h^3} \tag{C.5}$$

from which we find

$$\varepsilon_F = \frac{h^2}{2m}\left(\frac{3A}{16\pi V}\right)^{2/3}. \tag{C.6}$$

If A/V is constant, as it is to a good approximation inside a nucleus, we obtain two useful results:

1. The Fermi energy is independent of A.

2. The density of states at the Fermi energy $\rho(\varepsilon_F)$, from Equation (C.2), is proportional to V and, hence, to A.

Appendix D: Spherical Harmonics

For a particle moving in a central (i.e. spherically symmetric) potential $V(\mathbf{r}) = V(r)$, the solutions of the time-independent Schrödinger equation, i.e. the energy eigenstates, are of the form:

$$\psi(r, \theta, \phi) = R(r)\, Y_\ell^m(\theta, \phi) \tag{D.1}$$

where r, θ and ϕ are the usual spherical polar co-ordinates (see Figure D.1). The functions Y_ℓ^m are the *spherical harmonics*. We shall state those of their properties which are required in this book and refer the reader to standard texts on quantum mechanics for a fuller treatment.

The spherical harmonics are given by

$$Y_\ell^m(\theta, \phi) = (-1)^{(m+|m|)/2} \left[\frac{(2\ell+1)(\ell-|m|)!}{4\pi(\ell+|m|)!} \right]^{1/2} P_\ell^{|m|}(\cos\theta)\, e^{im\phi} \tag{D.2}$$

where $\ell = 0, 1, 2, \cdots$ and $m = -\ell, -\ell+1, \cdots \ell$.

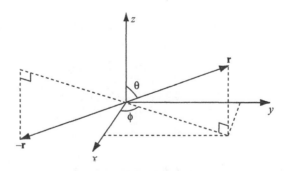

Figure D.1 Co-ordinate framework showing the relationship between the Cartesian and spherical (polar) co-ordinates for a vector $\mathbf{r}$ and its inverse $-\mathbf{r}$.

Table D.1 Examples of associated Legendre functions and spherical harmonics.

ℓ	m	$P_\ell^{|m|}(\cos\theta)$	$Y_\ell^m(\theta,\phi)$
0	0	1	$\sqrt{(1/4\pi)}$
1	0	$\cos\theta$	$\sqrt{(3/4\pi)}\cos\theta$
	± 1	$\sin\theta$	$\mp\sqrt{(3/8\pi)}\sin\theta\,e^{\pm i\phi}$
2	0	$(3\cos^2\theta - 1)/2$	$\sqrt{(5/16\pi)}\,(3\cos^2\theta - 1)$
	± 1	$3\sin\theta\cos\theta$	$\mp\sqrt{(15/8\pi)}\sin\theta\cos\theta\,e^{\pm i\phi}$
	± 2	$\sin^2\theta$	$\sqrt{(15/32\pi)}\sin^2\theta\,e^{\pm 2i\phi}$

The function $P_\ell^{|m|}(\cos\theta)$ is the associated Legendre function. It is a product of $\sin^{|m|}\theta$ and a polynomial of degree $(\ell - |m|)$ in $\cos\theta$. The first few associated Legendre functions and spherical harmonics are listed in Table D.1.

The spherical harmonics are angular momentum eigenfunctions. Explicitly, if $\mathbf{r} \times \mathbf{p}$ is the quantum-mechanical operator for orbital angular momentum, then $Y_\ell^m(\theta,\phi)$ is an eigenfunction of orbital angular momentum squared $(\mathbf{L}^2)$ with eigenvalue $\ell(\ell+1)\hbar^2$ and of ℓ_z with eigenvalue $m\hbar$; ℓ and m are the quantum numbers of orbital angular momentum and of its z component, respectively, and they assume the values:

$$\ell = 0, 1, 2, \cdots \quad \text{and} \quad m = -\ell, \ -\ell+1, \cdots \ell.$$

That ℓ and m are good quantum numbers is a consequence of the fact that in quantum mechanics, as in classical mechanics, angular momentum is conserved for a particle moving in a central potential.

As discussed in Section 2.4.1, *parity* is an important property of a wave function. The operation of *inversion* is defined as the transformation $\mathbf{r} \rightarrow -\mathbf{r}$. Under inversion, the spherical polar co-ordinates change according to $(r, \theta, \phi) \rightarrow (r, \pi - \theta, \phi + \pi)$ (see Figure D.1).

The wave functions [Equation (D.1)] have a definite parity, which we shall now obtain. The associated Legendre function $P_\ell^{|m|}(\cos\theta)$ can be shown to be an even function of $\cos\theta$ if $(\ell - |m|)$ is even and an odd function of $\cos\theta$ if $(\ell - |m|)$ is odd. It follows that

$$P_\ell^{|m|}[\cos(\pi - \theta)] = P_\ell^{|m|}(-\cos\theta) = (-1)^{\ell - |m|}P_\ell^{|m|}(\cos\theta).$$

We also have

$$e^{im(\phi+\pi)} = e^{im\pi}e^{im\phi} = (-1)^m e^{im\phi}.$$

Combining these last two results, it follows from the definition [Equation (D.2)] that

$$Y_\ell^m(\pi - \theta, \phi + \pi) = (-1)^\ell Y_\ell^m(\theta, \phi). \tag{D.3}$$

Correspondingly, under inversion, the wave function [Equation (D.1)] transforms into

$$\psi(r, \pi - \theta, \phi + \pi) = (-1)^\ell \psi(r, \theta, \phi) \tag{D.4}$$

i.e. the energy eigenstates have even or odd parity depending on whether the orbital angular momentum quantum number ℓ is even or odd.

The spherical harmonics have another very important property. They form a complete set of functions, which means that any reasonable function $f(\theta, \phi)$ can be expanded in the form:

$$f(\theta, \phi) = \sum_{\ell=0}^{\infty} \sum_{m=-\ell}^{\ell} C_{\ell m} Y_\ell^m(\theta, \phi). \tag{D.5}$$

The expansion coefficients $C_{\ell m}$ are constants, which are determined by the function $f(\theta, \phi)$ and specify the extent to which the corresponding spherical harmonic is present in the function $f(\theta, \phi)$. (This is analogous to the Fourier expansion of a periodic function as a series of trigonometric functions.)

The spherical harmonics occur in many problems involving angular dependence in three dimensions, both in classical and quantum physics. For example, the angular distribution of particles scattered in a collision process [i.e. the differential cross section $\sigma(\theta, \phi)$ Equation (1.25)] can be represented in the form of Equation (D.5). In Section 2.5.1, the deformation of the surface of a vibrating nucleus is represented in this way [see Equation (2.9)].

Appendix E:
Coulomb Scattering

We outline here a classical derivation of the differential cross section for the elastic scattering of a charged particle by the electrostatic potential given by a point charge. It is also commonly referred to as Rutherford scattering.

In Figure E.1, a particle is shown being deflected through an angle θ by a scattering centre of charge $Z_2 e$. The particle has charge $Z_1 e$, mass m, initial speed v_0, impact parameter b and energy $E = \frac{1}{2} m v_0^2$. We assume, for simplicity, that the scatterer remains at rest; that is, we shall take the scatterer to be infinitely heavy.

The angle of the position vector $\mathbf{r}$ to the y-axis is α and we can use conservation of angular momentum to write

$$mv_0 b = mr^2 \frac{\mathrm{d}\alpha}{\mathrm{d}t}. \tag{E.1}$$

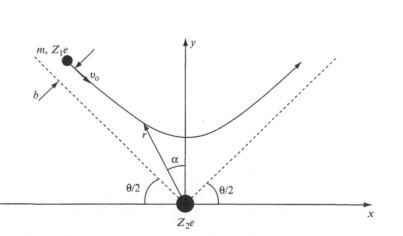

Figure E.1 Trajectory of a particle, mass m, charge $Z_1 e$ being deflected through an angle θ by a charge $Z_2 e$ fixed at the origin. Far from the origin, the particle's speed is v_0 and the impact parameter is b.

The trajectory is symmetric about the y-axis, and, therefore, the scattering brings about a net change in momentum Δp in the y direction. The y component of the momentum is $-mv_o \sin \theta/2$ initially, and $+mv_o \sin \theta/2$ finally. Therefore, we have

$$\Delta p = 2mv_o \sin \theta/2. \tag{E.2}$$

We can also calculate Δp as the integral over the trajectory of the y component of the impulse on the particle due to the Coulomb force F acting on it, as follows:

$$\Delta p = \int_{-\infty}^{\infty} F \cos \alpha \, dt = \int_{-\phi}^{\phi} \frac{Z_1 Z_2 e^2}{4\pi\varepsilon_0 r^2} \cos \alpha \left(\frac{dt}{d\alpha}\right) d\alpha$$

$$= \frac{Z_1 Z_2 e^2}{4\pi\varepsilon_0 v_o b} \int_{-\phi}^{\phi} \cos \alpha \, d\alpha = \frac{2 Z_1 Z_2 e^2 \sin \phi}{4\pi\varepsilon_0 v_o b} = \frac{2 Z_1 Z_2 e^2}{4\pi\varepsilon_0 v_o b} \cos \frac{\theta}{2} \tag{E.3}$$

where $\phi = (\pi - \theta)/2$ and we have substituted for $dt/d\alpha$ from Equation (E.1).

Combining Equations (E.2) and (E.3) gives the key relationship we need between b and θ:

$$b = \frac{Z_1 Z_2 e^2}{8\pi\varepsilon_0 E} \cot \frac{\theta}{2} = A \cot \frac{\theta}{2} \tag{E.4}$$

where we have set $A = Z_1 Z_2 e^2/8\pi\varepsilon_0 E$ for convenience.

Suppose we have an incident flux Φ of particles crossing a plane perpendicular to the beam. The intensity of particles with impact parameters between b and $b + db$ is $\Phi \times 2\pi b \, db$. Using Equation (E.4), this can be written as

$$dR = \pi \Phi A^2 \frac{\cos \theta/2}{\sin^3 \theta/2} d\theta. \tag{E.5}$$

This is equal to the rate at which particles are scattered into the solid angle $d\Omega = 2\pi \sin \theta d\theta$ between θ and $\theta + d\theta$. Thus, from the definition of the differential cross section [Equation (1.25)], we have, finally

$$\frac{d\sigma}{d\Omega} = \frac{1}{\Phi} \frac{dR}{d\Omega} = A^2 \frac{\cos \theta/2 \, d\theta}{\sin^3 \theta/2} \times \frac{1}{2 \sin \theta d\theta}$$

$$= \left(\frac{Z_1 Z_2 e^2}{16\pi\varepsilon_0 E}\right)^2 \mathrm{cosec}^4 \frac{\theta}{2} \tag{E.6}$$

which is the Rutherford formula. If the scatterer is of finite mass, Equation (E.6) still applies with E and θ being interpreted, respectively, as the total kinetic energy and scattering angle in the centre-of-mass system. In this case, a modified form of Equation (E.6) gives the differential cross section in the laboratory system.

Appendix F: Mass Excesses and Decay Properties of Nuclei

The following table lists atomic mass excesses and either abundances or half-lives for a selection of nuclides.[1] For each element, stable isotopes and a number of neighbouring radioisotopes are included. For each stable isotope, the percentage abundance is given relative to other stable isotopes. For unstable isotopes, the main ground-state decay mode and half-life are listed. Abbreviations for the decay modes are: β^- – negative beta decay, β^+ – positive beta decay, α – alpha decay, ε – electron capture, f – spontaneous fission, p – proton decay, and for the half-lives: m – minute, h – hour, y – year, as – 10^{-18} s, zs – 10^{-21} s.

The mass excess is listed in μu where u is an atomic mass unit. Mass excess (in u) is defined as the difference $m_A - A$ between the atomic mass m_A (in u) and the atomic mass number A. Thus, the atomic mass is readily obtained from the mass excess. For example, the mass excess of ^{16}O is listed as -5085 μu, so the mass of ^{16}O (16 + mass excess) is $16 - 0.005085 = 15.994915$ u. For many stable and near-stable nuclides, uncertainties are less than 10^{-6} u but, more typically, are 10^{-5} u (or 10^{-4} u) for nuclides further from stability. Errors in abundances and half-lives are, typically, at or below the level of the least significant digit tabulated.

Nuclide $(_zXA)$	Mass excess (μu)	Abundance or half-life	Nuclide $(_zXA)$	Mass excess (μu)	Abundance or half-life	Nuclide $(_zXA)$	Mass excess (μu)	Abundance or half-life
$_0n1$	8665	β^- 614.6 s	$_3Li6$	15122	7.5%	10	13534	β^- 1.51 My
			7	16004	92.5%	11	21658	β^- 13.81 s
$_1H1$	7825	99.985%	8	22487	β^- 838 ms	12	26920	β^- 21.3 ms
2	14102	0.015%	9	26789	β^- 178.3 ms			
3	16049	β^- 12.33 y				$_5B8$	24607	β^+ 770 ms
			$_4Be7$	16929	ε 53.29 d	9	13329	p 800 zs
$_2He3$	16029	0.000137%	8	5305	α 67 as	10	12937	19.9%
4	2603	99.99986%	9	12182	100%	11	9305	80.1%

[1] Source: Audi *et al.* (1997). Website address: http://csnwww.in2p3.fr/amac

Nuclide $(_zXA)$	Mass excess (μu)	Abundance or half-life
12	14352	β^-20.20 ms
13	17780	β^-17.36 ms
14	25404	β^-12.3 ms
15	31097	β^-9.87 ms
$_6$C9	31040	β^+126.5 ms
10	16853	β^+19.290 s
11	11434	β^+20.39 m
12	0	98.89%
13	3355	1.11%
14	3242	β^-5.73 ky
15	10599	β^-2.449 s
16	14701	β^-747 ms
17	22584	β^-193 ms
18	26753	β^-92 ms
19	35244	β^-46.2 ms
20	40322	β^-16 ms
$_7$N12	18613	β^+11.000 ms
13	5739	β^+9.965 m
14	3074	99.634%
15	109	0.366%
16	6101	β^-7.13 s
17	8450	β^-4.173 s
18	14082	β^-624 ms
19	17026	β^-271 ms
$_8$O14	8595	β^+70.606 s
15	3065	β^+122.24 s
16	-5085	99.762%
17	-868	0.038%
18	-840	0.200%
19	3579	β^-26.464 s
20	4076	β^-13.51 s
21	8655	β^-3.42 s
22	9962	β^-2.25 s
$_9$F17	2095	β^+64.49 s
18	938	β^+109.77 m
19	-1597	100%
20	-19	β^-11.163 s
21	-51	β^-4.158 s
22	2999	β^-4.23 s
23	3575	β^-2.23 s
24	8095	β^-400 ms
25	12099	β^-87 ms

Nuclide $(_zXA)$	Mass excess (μu)	Abundance or half-life
$_{10}$Ne17	17703	β^+109.2 ms
18	5697	β^+1.672 s
19	1880	β^+17.296 s
20	-7560	90.48%
21	-6153	0.27%
22	-8614	9.25%
23	-5533	β^-37.24 s
24	-6385	β^-3.38 m
25	-2212	β^-602 ms
26	462	β^-197 ms
27	7611	β^-32 ms
28	12110	β^-11 ms
$_{11}$Na20	7348	β^+447.9 ms
21	-2345	β^+22.49 s
22	-5563	β^+2.6019 y
23	-10230	100%
24	-9037	β^-14.9590 h
25	-10046	β^-59.1 s
26	-7410	β^-1.077 s
27	-5990	β^-301 ms
28	-1106	β^-30.5 ms
29	2813	β^-44.9 ms
30	9222	β^-48 ms
$_{12}$Mg20	18863	β^+90 ms
21	11715	β^+122 ms
22	-426	β^+3.857 s
23	-5875	β^+11.317 s
24	-14958	78.99%
25	-14163	10.00%
26	-17407	11.01%
27	-15659	β^-9.458 m
28	-16123	β^-20.91 h
29	-11445	β^-1.30 s
30	-9533	β^-335 ms
$_{13}$Al23	7265	β^+470 ms
24	-59	β^+2.053 s
25	-9571	β^+7.183 s
26	-13108	β^+740 ky
27	-18462	100%
28	-18090	β^-2.2414 m
29	-19555	β^-6.56 m
30	-17039	β^-3.60 s
31	-16054	β^-644 ms

Nuclide $(_zXA)$	Mass excess (μu)	Abundance or half-life
32	-11873	β^-31.7 ms
$_{14}$Si25	4106	β^+220 ms
26	-7670	β^+2.234 s
27	-13295	β^+4.16 s
28	-23073	92.23%
29	-23505	4.67%
30	-26230	3.10%
31	-24637	β^-157.3 m
32	-25852	β^-132 y
33	-21999	β^-6.18 s
34	-21425	β^-2.77 s
35	-15416	β^-780 ms
36	-13312	β^-450 ms
$_{15}$P28	-7688	β^+270.3 ms
29	-18199	β^+4.142 s
30	-21686	β^+2.498 m
31	-26238	100%
32	-26093	β^-14.262 d
33	-28275	β^-25.34 d
34	-26364	β^-12.43 s
35	-26686	β^-47.3 s
36	-21740	β^-5.6 s
37	-20387	β^-2.31 s
$_{16}$S29	-3392	β^+187 ms
30	-15097	β^+1.178 s
31	-20446	β^+2.572 s
32	-27929	95.02%
33	-28542	0.75%
34	-32133	4.21%
35	-30968	β^-87.51 d
36	-32919	0.02%
37	-28874	β^-5.05 m
38	-28836	β^-170.3 m
39	-24863	β^-11.5 s
40	-24530	β^-8.8 s
41	-19968	β^-2.6 s
$_{17}$Cl31	-7579	β^+150 ms
32	-14311	β^+298 ms
33	-22548	β^+2.511 s
34	-26238	β^+1.5264 s
35	-31147	75.77%
36	-31693	β^-301 ky

Nuclide ($_zXA$)	Mass excess (μu)	Abundance or half-life	Nuclide ($_zXA$)	Mass excess (μu)	Abundance or half-life	Nuclide ($_zXA$)	Mass excess (μu)	Abundance or half-life
37	−34097	24.23%	45	−43814	β^-162.67 d	53	−55658	β^-1.61 m
38	−31989	β^-37.24 m	46	−46307	0.004%	54	−53556	β^-49.8 s
39	−31992	β^-55.6 m	47	−45454	β^-4.536 d	55	−52765	β^-6.54 s
40	−29587	β^-1.35 m	48	−47467	0.187%	56	−49641	β^-230 ms
41	−29351	β^-38.4 s	49	−44327	β^-8.718 m			
			50	−42481	β^-13.9 s	$_{24}$Cr46	−31638	β^+260 ms
$_{18}$Ar33	−10070	β^+173.0 ms	51	−38530	β^-10.0 s	47	−37093	β^+500 ms
34	−19730	β^+845 ms	52	−34901	β^-4.6 s	48	−45964	β^+21.56 h
35	−24743	β^+1.775 s				49	−48659	β^+42.3 m
36	−32454	0.3365%	$_{21}$Sc 40	−22036	β^+182.3 ms	50	−53950	4.345%
37	−33224	ε35.04 d	41	−30749	β^+596.3 ms	51	−55228	ε27.702 d
38	−37268	0.0632%	42	−34483	β^+681.3 ms	52	−59488	83.789%
39	−35687	β^-269 y	43	−38849	β^+3.891 h	53	−59346	9.501%
40	−37617	99.6003%	44	−40597	β^+3.927 h	54	−61115	2.365%
41	−35499	β^-109.34 m	45	−44090	100%	55	−59156	β^-3.497 m
42	−36951	β^-32.9 y	46	−44830	β^-83.79 d	56	−59355	β^-5.94 m
43	−34332	β^-5.37 m	47	−47592	β^-3.3492 d	57	−56243	β^-21.1 s
44	−34635	β^-11.87 m	48	−47765	β^-43.67 h	58	−55749	β^-7.0 s
45	−31906	β^-21.48 s	49	−49976	β^-57.2 m			
46	−31906	β^-8.4 s	50	−47814	β^-102.5 s	$_{25}$Mn49	−40377	β^+382 ms
			51	−46398	β^-12.4 s	50	−45756	β^+283.9 ms
$_{19}$K35	−11988	β^+190 ms	52	−43350	β^-8.2 s	51	−51785	β^+46.2 m
36	−18707	β^+342 ms				52	−54430	β^+5.591 d
37	−26623	β^+1.226 s	$_{22}$Ti42	−26969	β^+199 ms	53	−58705	ε3.74 My
38	−30920	β^+7.636 m	43	−31476	β^+509 ms	54	−59637	ε312.3 d
39	−36293	93.2581%	44	−40310	ε64.8 y	55	−61950	100%
40	−36001	β^-1.277 Gy	45	−41876	β^+184.8 m	56	−61091	β^-2.5785 h
41	−38174	6.7302%	46	−47370	8.25%	57	−61713	β^-85.4 s
42	−37597	β^-12.360 h	47	−48236	7.44%	58	−60011	β^-3.0 s
43	−39284	β^-22.3 h	48	−52053	73.72%	59	−59553	β^-4.6 s
44	−38444	β^-22.13 m	49	−52129	5.41%	60	−56801	β^-51 s
45	−39300	β^-17.3 m	50	−55208	5.18%			
46	−38024	β^-105 s	51	−53384	β^-5.76 m	$_{26}$Fe50	−37005	β^+150 ms
47	−38322	β^-17.50 s	52	−53102	β^-1.7 m	51	−43175	β^+305 ms
48	−34487	β^-6.8 s	53	−50263	β^-32.7 s	52	−51883	β^+8.275 h
49	−32550	β^-1.26 s	54	−49125	β^-1.5 s	53	−54688	β^+8.51 m
						54	−60385	5.845%
$_{20}$Ca37	−14129	β^+181.1 ms	$_{23}$V45	−34218	β^+547 ms	55	−61702	ε2.73 y
38	−23681	β^+440 ms	46	−39800	β^+422.37 ms	56	−65058	91.754%
39	−29282	β^+859.6 ms	47	−45093	β^+32.6 m	57	−64601	2.119%
40	−37409	96.941%	48	−47746	β^+15.9735 d	58	−66719	0.282%
41	−37722	ε103 ky	49	−51483	ε330 d	59	−65119	β^-44.503 d
42	−41382	0.647%	50	−52837	0.250%	60	−65923	β^-1.5 My
43	−41233	0.135%	51	−56036	99.750%	61	−63250	β^-5.98 m
44	−44519	2.086%	52	−55220	β^-3.743 m	62	−63230	β^-68 s

Continues overleaf

Nuclide ($_zXA$)	Mass excess (μu)	Abundance or half-life	Nuclide ($_zXA$)	Mass excess (μu)	Abundance or half-life	Nuclide ($_zXA$)	Mass excess (μu)	Abundance or half-life
63	−59882	β^-6.1 s	70	−67440	β^-47 s	74	−78822	35.94%
64	−59131	β^-2.0 s				75	−77140	β^-82.78 m
			$_{30}$Zn60	−58168	β^+2.38 m	76	−78597	7.44%
$_{27}$Co53	−45775	β^+240 ms	61	−60486	β^+89.1 s	77	−76451	β^-11.30 h
54	−51536	β^+193.23 ms	62	−65665	β^+9.186 h	78	−77147	β^-88 m
55	−57997	β^+17.53 h	63	−66784	β^+38.47 m	79	−74601	β^-18.98 s
56	−60156	β^+77.27 d	64	−70853	48.6%	80	−74556	β^-29.5 s
57	−63704	ε271.79 d	65	−70755	β^+244.26 d	81	−71176	β^-8 s
58	−64242	β^+70.82 d	66	−73963	27.9%			
59	−66800	100%	67	−72869	4.1%	$_{33}$As68	−63210	β^+151.6 s
60	−66178	β^-5.2714 y	68	−75152	18.8%	69	−67719	β^+15.2 m
61	−67521	β^-1.650 h	69	−73446	β^-56.4 m	70	−69072	β^+52.6 m
62	−65946	β^-1.50 m	70	−74675	0.6%	71	−72885	β^+65.28 h
63	−66385	β^-26.9 s	71	−72273	β^-2.45 m	72	−73247	β^+26.0 h
64	−64186	β^-300 ms	72	−73138	β^-46.5 h	73	−76174	ε80.30 d
65	−63515	β^-1.20 s	73	−70221	β^-23.5 s	74	−76071	β^+17.77 d
			74	−70543	β^-95.6 s	75	−78404	100%
$_{28}$Ni54	−42094	β^+143 ms	75	−67064	β^-10.2 s	76	−77606	β^-1.0778 d
55	−48664	β^+207 ms	76	−66603	β^-5.7 s	77	−79352	β^-38.83 h
56	−57864	β^+5.9 d				78	−78171	β^-90.7 m
57	−60200	β^+35.60 h	$_{31}$Ga63	−60859	β^+32.4 s	79	−79052	β^-9.01 m
58	−64652	68.077%	64	−63162	β^+2.627 m	80	−77422	β^-15.2 s
59	−65648	β^+80 ky	65	−67261	β^+15.2 m	81	−77867	β^-33.3 s
60	−69209	26.223%	66	−68407	β^+9.49 h	82	−75492	β^-19.1 s
61	−68940	1.140%	67	−71795	ε3.2612 d	83	−75019	β^-13.4 s
62	−71651	3.634%	68	−72016	β^+67.629 m			
63	−70327	β^-100.1 y	69	−74419	60.108%	$_{34}$Se71	−67730	β^+4.74 m
64	−72030	0.926%	70	−73973	β^-21.14 m	72	−72887	ε8.40 d
65	−69912	β^-2.5172 h	71	−75295	39.892%	73	−73233	β^+7.15 h
66	−70885	β^-54.6 h	72	−73631	β^-14.10 h	74	−77523	0.89%
67	−68430	β^-21 s	73	−74830	β^-4.86 h	75	−77476	ε119.79 d
68	−68155	β^-17 s	74	−73055	β^-8.12 m	76	−80786	9.36%
69	−64821	β^-11.4 s	75	−73499	β^-126 s	77	−80085	7.63%
			76	−71069	β^-32.6 s	78	−82690	23.78%
$_{29}$Cu57	−50784	β^+196.3 ms	77	−70714	β^-13.2 s	79	−81500	β^-650 ky
58	−55459	β^+3.204 s	78	−68342	β^-5.09 s	80	−83478	49.61%
59	−60496	β^+81.5 s				81	−82007	β^-18.45 m
60	−62632	β^+23.7 m	$_{32}$Ge64	−58422	β^+63.7 s	82	−83300	8.73%
61	−66538	β^+3.333 h	65	−60559	β^+30.9 s	83	−80881	β^-22.3 m
62	−67413	β^+9.74 m	66	−66152	β^+2.26 h	84	−81536	β^-3.1 m
63	−70399	69.17%	67	−67262	β^+18.9 m	85	−77756	β^-31.7 s
64	−70232	β^+12.700 h	68	−71903	ε270.8 d	86	−75729	β^-15.3 s
65	−72206	30.83%	69	−72028	β^+39.05 h			
66	−71127	β^-5.088 m	70	−75750	21.23%	$_{35}$Br74	−70109	β^+25.4 m
67	−72250	β^-61.83 h	71	−75046	ε11.43 d	75	−74224	β^+96.7 m
68	−70360	β^-31.1 s	72	−77924	27.66%	76	−75458	β^+16.2 h
69	−70575	β^-2.85 m	73	−76541	7.73%	77	−78620	β^+57.036 h

Nuclide ($_zXA$)	Mass excess (μu)	Abundance or half-life	Nuclide ($_zXA$)	Mass excess (μu)	Abundance or half-life	Nuclide ($_zXA$)	Mass excess (μu)	Abundance or half-life
78	−78854	β^+6.46 m	88	−88681	β^-17.78 m	92	−94960	17.15%
79	−81662	50.69%	89	−87720	β^-15.15 m	93	−93524	β^-1.53 My
80	−81470	β^-17.68 m	90	−85191	β^-158 s	94	−93684	17.38%
81	−83709	49.31%	91	−83466	β^-58.4 s	95	−91957	β^-64.02 d
82	−83195	β^-35.30 h	92	−80274	β^-4.492 s	96	−91725	2.80%
83	−84820	β^-2.40 h	93	−77967	β^-5.84 s	97	−89049	β^-16.90 h
84	−83496	β^-31.80 m				98	−87253	β^-30.7 s
85	−84392	β^-2.90 m	$_{38}$Sr79	−70292	β^+2.25 m	99	−83488	β^-2.1 s
86	−81203	β^-55.1 s	80	−75476	β^+106.3 m	100	−82233	β^-7.1 s
87	−79289	β^-55.60 s	81	−76787	β^+22.3 m			
88	−75932	β^-16.36 s	82	−81599	ε25.55 d	$_{41}$Nb89	−86506	β^+1.9 h
			83	−82445	β^+32.41 h	90	−88736	β^+14.60 h
$_{36}$Kr73	−61074	β^+27.0 s	84	−86575	0.56%	91	−93010	ε680 y
74	−66742	β^+11.50 m	85	−87068	ε64.84 d	92	−92807	β^+34.7 My
75	−68967	β^+4.3 m	86	−90738	9.86%	93	−93622	100%
76	−74052	β^+14.8 h	87	−91121	7.00%	94	−92717	β^-20.3 ky
77	−75332	β^+74.4 m	88	−94386	82.58%	95	−93165	β^-34.975 d
78	−79614	0.35%	89	−92547	β^-50.53 d	96	−91900	β^-23.35 h
79	−79917	β^+35.04 h	90	−92262	β^-28.84 y	97	−91903	β^-72.1 m
80	−83622	2.25%	91	−89790	β^-9.63 h	98	−89669	β^-2.86 s
81	−83408	ε229 ky	92	−88970	β^-2.71 h	99	−88382	β^-15.0 s
82	−86515	11.6%	93	−85978	β^-7.423 m			
83	−85864	11.5%	94	−84640	β^-75.3 s	$_{42}$Mo88	−78048	β^+8.0 m
84	−88493	57.0%	95	−80641	β^-23.90 s	89	−80519	β^+2.04 m
85	−87473	β^-10.756 y				90	−86064	β^+5.56 h
86	−89390	17.3%	$_{39}$Y83	−77649	β^+7.08 m	91	−88250	β^+15.49 m
87	−86646	β^-76.3 m	84	−79614	β^+4.6 s	92	−93189	14.84%
88	−85553	β^-2.84 h	85	−83573	β^+2.68 h	93	−93188	ε4.0 ky
89	−82362	β^-3.15 m	86	−85113	β^+14.74 h	94	−94912	9.25%
90	−80476	β^-32.32 s	87	−89122	β^+79.8 h	95	−94159	15.92%
91	−76554	β^-8.57 s	88	−90497	β^+106.65 d	96	−95321	16.68%
92	−73847	β^-1.840 s	89	−94152	100%	97	−93979	9.55%
93	−68739	β^-1.286 s	90	−92849	β^-64.10 h	98	−94592	24.13%
			91	−92697	β^-58.51 d	99	−92288	β^-65.94 h
$_{37}$Rb77	−69594	β^+3.80 m	92	−91053	β^-3.54 h	100	−92522	9.63%
78	−71859	β^+17.66 m	93	−90418	β^-10.18 h	101	−89654	β^-14.61 m
79	−76004	β^+22.9 m	94	−88406	β^-18.7 m	102	−89703	β^-11.3 m
80	−77481	β^+33.4 s	95	−87176	β^-10.3 m	103	−86796	β^-67.5 s
81	−81005	β^+4.576 h				104	−86238	β^-60 s
82	−81792	β^+1.273 m	$_{40}$Zr86	−83532	β^+16.5 h			
83	−84888	ε86.2 d	87	−85184	β^+1.68 h	$_{43}$Tc92	−84740	β^+4.23 m
84	−85615	β^+32.77 d	88	−89774	ε83.4 d	93	−89752	β^+2.75 h
85	−88211	72.165%	89	−91111	β^+78.41 h	94	−90344	β^+293 m
86	−88833	β^-18.631 d	90	−95296	51.45%	95	−92343	β^+20.0 h
87	−90816	27.835%	91	−94355	11.22%	96	−92129	β^+4.28 d

Continues overleaf

Nuclide ($_zXA$)	Mass excess (μu)	Abundance or half-life	Nuclide ($_zXA$)	Mass excess (μu)	Abundance or half-life	Nuclide ($_zXA$)	Mass excess (μu)	Abundance or half-life
97	−93636	ε2.6 My	106	−96517	27.33%	110	−92832	β^+4.9 h
98	−92784	β^-4.2 My	107	−94871	β^-6.5 My	111	−94890	ε2.8047 d
99	−93745	β^-211.1 ky	108	−96106	26.46%	112	−94467	β^+14.97 m
100	−92342	β^-15.8 s	109	−94047	β^-13.7012 h	113	−95938	4.29%
101	−92686	β^-14.22 m	110	−94848	11.72%	114	−95083	β^-71.9 s
102	−90787	β^-5.28 s	111	−92357	β^-23.4 m	115	−96122	95.71%
			112	−92687	β^-21.03 h	116	−94740	β^-14.10 s
$_{44}$Ru93	−82953	β^+59.7 s	113	−89845	β^-93 s	117	−95484	β^-43.2 m
94	−88640	β^+51.8 m	114	−89634	β^-2.42 m	118	−93645	β^-5.0 s
95	−89587	β^+1.643 h				119	−94154	β^-2.4 m
96	−92402	5.52%	$_{47}$Ag101	−87193	β^+11.1 m			
97	−92445	β^+2.9 d	102	−87998	β^+12.9 m	$_{50}$Sn106	−83125	β^+1.92 m
98	−94712	1.88%	103	−91028	β^+65.7 m	107	−84338	β^+2.90 m
99	−94061	12.7%	104	−91371	β^+69.2 m	108	−88031	β^+10.30 m
100	−95780	12.6%	105	−93471	β^+41.29 d	109	−88713	β^+18.0 m
101	−94418	17.0%	106	−93334	β^+23.96 m	110	−92148	ε4.11 h
102	−95651	31.6%	107	−94907	51.839%	111	−92265	β^+35.3 m
103	−93676	β^-39.26 d	108	−94047	β^-2.37 m	112	−95179	0.97%
104	−94570	18.7%	109	−95245	48.161%	113	−94826	β^+115.09 d
105	−92250	β^-4.44 h	110	−93890	β^-24.6 s	114	−97218	0.65%
106	−92673	β^-373.59 d	111	−94705	β^-7.45 d	115	−96654	0.34%
107	−90092	β^-3.75 m	112	−92996	β^-3.130 h	116	−98256	14.54%
108	−89813	β^-4.55 m	113	−93434	β^-5.37 h	117	−97046	7.68%
			114	−91192	β^-4.6 s	118	−98394	24.22%
$_{45}$Rh95	−84101	β^+5.02 m	115	−91241	β^-20.0 m	119	−96691	8.58%
96	−85482	β^+9.90 m				120	−97803	32.59%
97	−88664	β^+30.7 m	$_{48}$Cd102	−85218	β^+5.5 m	121	−95763	β^-27.06 h
98	−89283	β^+8.7 m	103	−86581	β^+7.3 m	122	−96560	4.63%
99	−91867	β^+16.1 d	104	−90152	β^+57.7 m	123	−94278	β^-129.2 d
100	−91884	β^+20.8 h	105	−90532	β^+55.5 m	124	−94725	5.79%
101	−93836	ε3.3 y	106	−93542	1.25%	125	−92215	β^-9.64 d
102	−93157	β^+206.0 d	107	−93385	β^+6.50 h	126	−92346	β^-207 ky
103	−94496	100%	108	−95817	0.89%	127	−89650	β^-2.10 h
104	−93345	β^-42.3 s	109	−95014	ε462.6 d	128	−89465	β^-59.1 m
105	−94308	β^-35.36 h	110	−96994	12.49%			
106	−92716	β^-29.80 s	111	−95818	12.80%	$_{51}$Sb116	−93203	β^+15.8 m
107	−93249	β^-21.7 m	112	−97243	24.13%	117	−95160	β^+2.80 h
			113	−95599	12.22%	118	−94468	β^+3.6 m
$_{46}$Pd97	−83522	β^+3.10 m	114	−96642	28.73%	119	−96053	ε38.19 h
98	−87279	β^+17.7 m	115	−94569	β^-53.46 h	120	−94926	β^+15.89 m
99	−88232	β^+21.4 m	116	−95245	7.49%	121	−96182	57.21%
100	−91495	ε3.63 d	117	−92782	β^-2.49 h	122	−94825	β^-2.7238 d
101	−91711	β^+8.47 h	118	−93086	β^-50.3 m	123	−95784	42.79%
102	−94392	1.02%	119	−90081	β^-2.69 m	124	−94062	β^-60.20 d
103	−93913	ε16.991 d				125	−94752	β^-2.7582 y
104	−95965	11.14%	$_{49}$In108	−90285	β^+58.0 m	126	−92754	β^-12.46 d
105	−94916	22.33%	109	−92845	β^+4.2 h	127	−93086	β^-3.85 d

Nuclide ($_zXA$)	Mass excess (μu)	Abundance or half-life	Nuclide ($_zXA$)	Mass excess (μu)	Abundance or half-life	Nuclide ($_zXA$)	Mass excess (μu)	Abundance or half-life
128	−90833	β^-9.01 h	124	−94104	0.10%	141	−85594	β^-18.27 m
129	−90850	β^-4.40 h	125	−93602	β^+16.9 h	142	−83552	β^-10.6 m
			126	−95731	0.09%			
$_{52}$Te115	−88417	β^+5.8 m	127	−94821	ε36.4 d	$_{57}$La133	−91606	β^+3.912 h
116	−91584	β^+2.49 h	128	−96470	1.91%	134	−91510	β^+6.45 m
117	−91366	β^+62 m	129	−95221	26.4%	135	−93029	β^+19.5 h
118	−94175	ε6.00 d	130	−96492	4.1%	136	−92346	β^+9.87 m
119	−93592	β^+16.03 h	131	−94918	21.2%	137	−93538	ε60 ky
120	−95980	0.096%	132	−95845	26.9%	138	−92893	0.0902%
121	−95070	β^+19.40 d	133	−94094	β^-5.243 d	139	−93652	99.9098%
122	−96953	2.603%	134	−94605	10.4%	140	−90528	β^-1.6781 d
123	−95727	0.908%	135	−92793	β^-9.14 h	141	−89043	β^-3.92 h
124	−97181	4.816%	136	−92780	8.9%	142	−85925	β^-91.1 m
125	−95575	7.139%	137	−88438	β^-3.818 m	143	−83942	β^-14.2 m
126	−96694	18.952%	138	−86012	β^-14.08 m			
127	−94783	β^-9.35 h				$_{58}$Ce132	−88514	β^+3.51 h
128	−95539	31.687%	$_{55}$Cs127	−92582	β^+6.25 h	133	−88449	β^+97 m
129	−93405	β^-69.6 m	128	−92252	β^+3.640 m	134	−90972	ε3.16 d
130	−93777	33.799%	129	−93936	β^+32.06 h	135	−90854	β^+17.7 h
131	−91478	β^-25.0 m	130	−93294	β^+29.21 m	136	−92862	0.19%
132	−91477	β^-3.204 d	131	−94540	ε9.689 d	137	−92217	β^+9.0 h
133	−89061	β^-12.5 m	132	−93570	β^+6.479 d	138	−94015	0.25%
134	−88460	β^-41.8 m	133	−94553	100%	139	−93353	ε137.640 d
			134	−93287	β^-2.0648 y	140	−94566	88.48%
$_{53}$I120	−89952	β^+81.0 m	135	−94029	β^-2.3 My	141	−91729	β^-32.501 d
121	−92634	β^+2.12 h	136	−92694	β^-13.16 d	142	−90761	11.08%
122	−92407	β^+3.63 m	137	−92916	β^-30.07 y	143	−87618	β^-33.039 h
123	−94402	β^+13.27 h	138	−88989	β^-33.41 m	144	−86357	β^-284.893 d
124	−93789	β^+4.1760 d	139	−86643	β^-9.27 m	145	−82770	β^-3.01 m
125	−95376	ε59.408 d				146	−81310	β^-13.52 m
126	−94381	β^+13.11 d	$_{56}$Ba126	−88756	β^+100 m			
127	−95531	100%	127	−88879	β^+12.7 m	$_{59}$Pr137	−89319	β^+1.28 h
128	−94195	β^-24.99 m	128	−91691	ε2.43 d	138	−89251	β^+1.45 m
129	−95013	β^-15.7 My	129	−91326	β^+2.23 h	139	−91068	β^+4.41 h
130	−93326	β^-12.36 h	130	−93689	0.106%	140	−90929	β^+3.39 m
131	−93876	β^-8.02070 d	131	−93069	β^+11.50 d	141	−92353	100%
132	−92006	β^-2.295 h	132	−94944	0.101%	142	−89960	β^-19.12 h
133	−92194	β^-20.8 h	133	−93997	ε10.51 y	143	−89188	β^-13.57 d
134	−90123	β^-52.5 m	134	−95497	2.417%	144	−86699	β^-17.28 m
135	−89950	β^-6.57 h	135	−94317	6.592%	145	−85493	β^-5.984 h
			136	−95430	7.854%			
$_{54}$Xe120	−87848	β^+40 m	137	−94179	11.23%	$_{60}$Nd138	−88074	β^+5.04 h
121	−88614	β^+40.1 m	138	−94759	71.70%	139	−88074	β^+29.7 m
122	−91455	ε20.1 h	139	−91165	β^-83.06 m	140	−90690	ε3.37 d
123	−91529	β^+2.08 h	140	−89400	β^-12.752 d	141	−90396	β^+2.49 h

Continues overleaf

Nuclide $(_zXA)$	Mass excess (μu)	Abundance or half-life	Nuclide $(_zXA)$	Mass excess (μu)	Abundance or half-life	Nuclide $(_zXA)$	Mass excess (μu)	Abundance or half-life
142	−92281	27.13%	$_{63}$Eu145	−83739	β^+5.93 d	154	−75578	α3.0 My
143	−90190	12.18%	146	−82800	β^+4.59 d	155	−74251	β^+9.9 h
144	−89917	23.80%	147	−83259	β^+24.1 d	156	−75721	0.06%
145	−87431	8.30%	148	−81846	β^+54.5 d	157	−74538	β^+8.14 h
146	−86888	17.19%	149	−82074	ε93.1 d	158	−75596	0.10%
147	−83904	β^-10.98 d	150	−80302	β^+36.9 y	159	−74265	ε144.4 d
148	−83112	5.76%	151	−80154	47.8%	160	−74807	2.34%
149	−79856	β^-1.728 h	152	−78260	β^+13.537 y	161	−73071	18.9%
150	−79114	5.64%	153	−78774	52.2%	162	−73205	25.5%
151	−76175	β^-12.44 m	154	−77025	β^-8.593 y	163	−71273	24.9%
152	−75320	β^-11.4 m	155	−77111	β^-4.7611 y	164	−70829	28.2%
153	−72305	β^-28.9 s	156	−75249	β^-15.19 d	165	−68300	β^-2.334 h
154	−70521	β^-25.9 s	157	−74580	β^-15.18 h	166	−67196	β^-81.6 h
			158	−72153	β^-45.9 m	167	−64348	β^-6.20 m
$_{61}$Pm141	−86393	β^+20.90 m				168	−62770	β^-8.7 m
142	−87054	β^+40.5 s	$_{64}$Gd148	−81890	α74.6 y			
143	−89072	β^+265 d	149	−80664	β^+9.28 d	$_{67}$Ho161	−72149	ε2.48 h
144	−87414	β^+363 d	150	−81345	α1.79 My	162	−70908	β^+15.0 m
145	−87257	ε17.7 y	151	−79656	ε124 d	163	−71269	ε4.570 ky
146	−85308	ε5.53 y	152	−80212	0.20%	164	−69770	ε29 m
147	−84866	β^-2.6234 y	153	−78254	ε241.6 d	165	−69681	100%
148	−82532	β^-5.370 d	154	−79138	2.18%	166	−67719	β^-26.83 h
149	−81671	β^-53.08 h	155	−77381	14.80%	167	−66874	β^-3.1 h
150	−79020	β^-2.68 h	156	−77880	20.47%	168	−64504	β^-2.99 m
151	−78797	β^-28.40 h	157	−76043	15.65%	169	−63132	β^-4.7 m
152	−76512	β^-4.12 m	158	−75899	24.84%	170	−60387	β^-2.76 m
153	−75887	β^-5.250 m	159	−73615	β^-18.479 h			
			160	−72949	21.86%	$_{68}$Er158	−70092	ε2.29 h
$_{62}$Sm140	−81009	β^+14.82 m	161	−70334	β^-3.646 m	159	−69319	β^+36 m
141	−81531	β^+10.2 m	162	−69019	β^-8.4 m	160	−70918	ε28.58 h
142	−84807	β^+72.49 m				161	−69998	β^+3.21 h
143	−85377	β^+8.83 m	$_{65}$Tb153	−76569	β^+2.34 d	162	−71225	0.14%
144	−88005	3.1%	154	−75309	β^+21.5 h	163	−69970	β^+75.0 m
145	−86594	ε340 d	155	−76500	ε5.32 d	164	−70803	1.61%
146	−86964	α103 My	156	−75257	β^+5.35 d	165	−69277	ε10.36 h
147	−85107	15.0%	157	−75979	ε71 y	166	−69710	33.6%
148	−85182	11.3%	158	−74590	β^+180 y	167	−67954	22.95%
149	−82820	13.8%	159	−74657	100%	168	−67632	26.8%
150	−82728	7.4%	160	−72836	β^-72.3 d	169	−65412	β^-9.40 d
151	−80072	β^-90 y	161	−72434	β^-6.88 d	170	−64539	14.9%
152	−80272	26.7%	162	−70510	β^-7.60 m	171	−61975	β^-7.516 h
153	−77906	β^-46.27 h	163	−69356	β^-19.5 m	172	−60648	β^-49.3 h
154	−77795	22.7%	164	−66656	β^-3.0 m	173	−57596	β^-1.434 m
155	−75364	β^-22.3 m				174	−55663	β^-3.3 m
156	−74474	β^-9.4 h	$_{66}$Dy152	−75287	ε2.38 h			
157	−71648	β^-8.03 m	153	−74239	β^+6.4 h	$_{69}$Tm165	−67568	β^+30.06 h

Nuclide ($_zXA$)	Mass excess (μu)	Abundance or half-life	Nuclide ($_zXA$)	Mass excess (μu)	Abundance or half-life	Nuclide ($_zXA$)	Mass excess (μu)	Abundance or half-life
166	−66447	β^+7.70 h	182	−49447	β^-9 My	191	−39072	β^-15.4 d
167	−67151	ε9.25 d	183	−46474	β^-1.067 h	192	−38521	41.0%
168	−65830	β^+93.1 d	184	−44552	β^-4.12 h	193	−35852	β^-30.5 h
169	−65789	100%				194	−34820	β^-6.0 y
170	−64202	β^-128.6 d	$_{73}$Ta176	−55255	β^+8.09 h			
171	−63574	β^-1.92 y	177	−55528	β^+56.56 h	$_{77}$Ir185	−43414	β^+14.4 h
172	−61604	β^-63.6 h	178	−54246	β^+9.31 m	186	−42049	β^+16.64 h
173	−60400	β^-8.24 h	179	−54066	ε1.82 y	187	−42639	β^+10.5 h
			180	−52534	ε8.152 h	188	−41148	β^+41.5 h
$_{70}$Yb166	−66121	ε56.7 h	181	−52004	99.988%	189	−41283	ε13.2 d
167	−65054	β^+17.5 m	182	−49848	β^-114.43 d	190	−39410	β^+11.78 d
168	−66106	0.13%	183	−48627	β^-5.1 d	191	−39409	37.3%
169	−64813	ε32.026 d	184	−45991	β^-8.7 h	192	−37398	β^-73.831 d
170	−65241	3.05%				193	−37076	62.7%
171	−63678	14.3%	$_{74}$W178	−54150	ε21.6 d	194	−34924	β^-19.28 h
172	−63622	21.9%	179	−52928	β^+37.05 m	195	−34023	β^-2.5 h
173	−61793	16.12%	180	−53294	0.120%			
174	−61142	31.8%	181	−51802	ε121.2 d	$_{78}$Pt186	−40569	β^+2.08 h
175	−58727	β^-4.185 d	182	−51794	26.498%	187	−39442	β^+2.35 h
176	−57432	12.7%	183	−49776	14.314%	188	−40605	ε10.2 d
177	−54743	β^-1.911 h	184	−49067	30.642%	189	−39168	β^+10.87 h
178	−53356	β^-74 m	185	−46579	β^-75.1 d	190	−40070	0.01%
			186	−45638	28.426%	191	−38316	ε2.802 d
$_{71}$Lu170	−61528	β^+2.012 d	187	−42842	β^-23.72 h	192	−38965	0.79%
171	−62091	β^+8.24 d	188	−41513	β^-69.4 d	193	−37015	ε50 y
172	−60918	β^+6.70 d				194	−37336	32.9%
173	−61073	ε1.37 y	$_{75}$Re181	−49936	β^+19.9 h	195	−35226	33.8%
174	−59667	β^+3.31 y	182	−48793	β^+64.0 h	196	−35065	25.3%
175	−59232	97.41%	183	−49179	ε70.0 d	197	−32677	β^-19.8915 h
176	−57318	2.59%	184	−47475	β^+38.0 d	198	−32124	7.2%
177	−56245	β^-6.734 d	185	−47044	37.40%	199	−29424	β^-30.80 m
178	−54049	β^-28.4 m	186	−45013	β^-3.7183 d	200	−28576	β^-12.5 h
179	−52676	β^-4.59 h	187	−44249	62.60%	201	−25507	β^-2.5 m
			188	−41888	β^-16.98 h	202	−24262	β^-44 h
$_{72}$Hf171	−59507	β^+12.1 h	189	−40772	β^-24.3 h			
172	−60537	ε1.87 y				$_{79}$Au192	−35190	β^+4.94 h
173	−59346	β^+23.6 h	$_{76}$Os182	−47814	ε22.10 h	193	−35868	β^+17.65 h
174	−59960	0.162%	183	−46892	β^+13.0 h	194	−34662	β^+38.02 h
175	−58497	ε70 d	184	−47509	0.020%	195	−34983	ε186.10 d
176	−58598	5.206%	185	−45957	ε93.6 d	196	−33448	β^+6.183 d
177	−56780	18.606%	186	−46162	1.58%	197	−33448	100%
178	−56302	27.297%	187	−44252	1.6%	198	−31775	β^-2.69517 d
179	−54185	13.629%	188	−44164	13.3%	199	−31252	β^-3.139 d
180	−53451	35.100%	189	−41855	16.1%	200	−29286	β^-48.4 m
181	−50901	β^-42.39 d	190	−41555	26.4%	201	−28359	β^-26 m

Continues overleaf

Nuclide ($_zXA$)	Mass excess (μu)	Abundance or half-life	Nuclide ($_zXA$)	Mass excess (μu)	Abundance or half-life	Nuclide ($_zXA$)	Mass excess (μu)	Abundance or half-life
$_{80}$Hg192	−34429	ε4.85 h	209	−19617	100%	224	23231	β^-3.30 m
193	−33356	β^+3.80 h	210	−15895	β^-5.013 d	225	25607	β^-4.0 m
194	−34619	ε440 y	211	−12742	α2.14 m	226	29340	β^-49 s
195	−33366	β^+9.9 h	212	−8728	β^-60.55 m	227	31831	β^-2.47 m
196	−34185	0.15%	213	−5625	β^-45.59 m			
197	−32804	ε64.14 h	214	−1301	β^-19.9 m	$_{88}$Ra222	15361	α38.0 s
198	−33248	9.97%				223	18497	α11.435 d
199	−31738	16.87%	$_{84}$Po204	−19693	β^+3.53 h	224	20202	α3.66 d
200	−31691	23.10%	205	−18834	β^+1.66 h	225	23604	β^-14.9 d
201	−29715	13.18%	206	−19535	β^+8.8 d	226	25403	α1.600 ky
202	−29374	29.86%	207	−18422	β^+5.80 h	227	29171	β^-42.2 m
203	−27142	β^-46.612 d	208	−18769	α2.898 y	228	31064	β^-5.75 y
204	−26524	6.87%	209	−17585	α102 y			
205	−23944	β^-5.2 m	210	−17143	α138.376 d	$_{89}$Ac223	19126	α2.10 m
206	−22501	β^-8.15 m	211	−13363	α516 ms	224	21708	β^+2.9 h
			212	−11148	α299 ns	225	23221	α10.0 d
$_{81}$Tl198	−29533	β^+5.3 h				226	26090	β^-29.37 h
199	−30188	β^+7.42 h	$_{85}$At208	−13417	β^+1.63 h	227	27747	β^-21.773 y
200	−29054	β^+26.1 h	209	−13841	β^+5.41 h	228	31015	β^-6.15 h
201	−29196	ε72.912 h	210	−12869	β^+8.1 h	229	32926	β^-62.7 m
202	−27909	β^+12.23 d	211	−12520	ε7.214 h	230	36028	β^-122 s
203	−27671	29.524%	212	−9266	α314 ms	231	38551	β^-7.5 m
204	−26151	β^-3.78 y	213	−7079	α125 ns			
205	−25588	70.476%	214	−3644	α558 ns	$_{90}$Th226	24891	α30.57 m
206	−23905	β^-4.199 m	215	−1359	α100 μs	227	27699	α18.72 d
207	−22592	β^-4.77 m	216	2409	α300 μs	228	28731	α1.9131 y
208	−17995	β^-3.053 m	217	4710	α32.3 ms	229	31755	α7.34 ky
			218	8682	α1.5 s	230	33127	α75.38 ky
$_{82}$Pb200	−28185	ε21.5 h	219	11294	α56 s	231	36297	β^-25.52 h
201	−27150	β^+9.33 h				232	38050	100%
202	−27856	ε52.5 ky	$_{86}$Rn212	−9311	α23.9 m	233	41577	β^-22.3 m
203	−26625	ε51.873 h	213	−6132	α25.0 ms	234	43596	β^-24.10 d
204	−26971	1.4%	214	−4654	α270 ns			
205	−25533	ε15.3 My	215	−1271	α2.30μs	$_{91}$Pa229	32088	ε1.50 d
206	−25551	24.1%	216	258	α45 μs	230	34533	β^+17.4 d
207	−24119	22.1%	217	3914	α540 μs	231	35879	α32.76 ky
208	−23364	52.4%	218	5587	α35 ms	232	38582	β^-1.31 d
209	−18926	β^-3.253 h	219	9475	α3.96 s	233	40240	β^-26.967 d
210	−15827	β^-22.3 y	220	11384	α55.6 s	234	43302	β^-6.70 h
211	−11269	β^-36.1 m	221	15459	β^-25 m	235	45432	β^-24.5 m
212	−8112	β^-10.64 h	222	17570	α3.8235 d			
						$_{92}$U230	33927	α20.8 d
$_{83}$Bi204	−22194	β^+11.22 h	$_{87}$Fr219	9241	α20 ms	231	36289	ε4.2 d
205	−22625	β^+15.31 d	220	12312	α27.4 s	232	37146	α68.9 y
206	−21517	β^+6.243 d	221	14246	α4.9 m	233	39628	α159.2 ky
207	−21545	β^+31.55 y	222	17544	β^-14.2 m	234	40946	0.0055%
208	−20273	β^+368 ky	223	19731	β^-21.8 m	235	43923	0.720%

Nuclide ($_zXA$)	Mass excess (μu)	Abundance or half-life	Nuclide ($_zXA$)	Mass excess (μu)	Abundance or half-life	Nuclide ($_zXA$)	Mass excess (μu)	Abundance or half-life
236	45562	α23.42 My	249	75947	β^-64.15 m	$_{102}$No252	88966	α2.30 s
237	48724	β^-6.75 d	250	78350	f9 ky	253	90650	α? 1.7 m
238	50783	99.2745%				254	90949	α? 55 s
239	54288	β^-23.45 m	$_{97}$Bk245	66355	ε4.94 d	255	93232	α3.1 m
			246	68664	β^+1.80 d			
$_{93}$Np234	42888	β^+4.4 d	247	70299	α1.38 ky	$_{103}$Lr257	99603	α646 ms
235	44056	ε396.1 d	248	73076	α9 y	258	101879	α3.9 s
236	46560	ε154 ky	249	74980	β^-320 d	259	102996	α6.3 s
237	48167	α2.144 My				260	105572	α3.0 m
238	50940	β^-2.117 d	$_{98}$Cf250	76400	α13.08 y	261	106946	f? α? 39 m
239	52931	β^-2.3565 d	251	79580	α900 y	262	109695	β^+? 3.6 h
			252	81619	α2.645 y			
$_{94}$Pu236	46048	α2.858 y	253	85127	β^-17.81 d	$_{104}$Rf257	103071	α4.7 s
237	48404	ε45.2 d	254	87317	f60.5 d	258	103565	f12 ms
238	49553	α87.7 y	255	91037	β^-85 m	259	105626	α2.7 s
239	52157	α24.11 ky	256	93441	f12.3 m	260	106431	f20.1 ms
240	53807	α6.564 ky				261	108750	α65 s
241	56845	β^-14.35 y	$_{99}$Es251	79983	ε33 h	262	109920	f2.06 s
242	58737	α373.3 ky	252	82974	α471.7 d			
243	61997	β^-4.956 h	253	84818	α20.47 d	$_{105}$Db261	112110	α1.8 s
244	64198	α80.8 My	254	88016	α275.7 d	262	114150	f34 s
			255	90267	β^-39.8 d	263	115073	f29 s
$_{95}$Am240	55288	β^+50.8 h						
241	56823	α432.2 y	$_{100}$Fm251	81567	β^+5.30 h	$_{106}$Sg 265	121064	α16 s
242	59543	β^-16.02 h	252	82460	α25.39 h	266	121933	α20 s
243	61373	α7.37 ky	253	85176	ε3.00 d			
244	64279	β^-10.1 h	254	86848	α3.240 h	$_{107}$Bh266	127011	α? 1 s
			255	89955	α20.07 h	267	127741	α? 15 s
$_{96}$Cm242	58829	α162.8 d	256	91767	f157.6 m			
243	61382	α29.1 y	257	95099	α100.5 d	$_{108}$Hs268	132153	α? 2 s
244	62746	α18.10 y				269	134118	α13 s
245	65486	α8.5 ky	$_{101}$Md255	91075	β^+27 m			
246	67218	α4.73 ky	256	94053	β^+78.1 m	$_{109}$Mt270	140720	α? 2 s
247	70347	α15.6 My	257	95535	ε5.52 h			
248	72342	α340 ky	258	98426	α51.50 d			

Appendix G:
Answers and Hints to Problems

CHAPTER 1

1.1 Density $= A/(\frac{4}{3}\pi \times (1.2)^3 \times A) = 0.138\,\mathrm{u\,fm^{-3}} \to 2.29 \times 10^{17}\,\mathrm{kg\,m^{-3}}$.

1.2

Energy	$\lambda_\gamma(\mathrm{m})$	$\lambda_{el}(\mathrm{m})$	$\lambda_n(\mathrm{m})$	$\lambda_A(\mathrm{m})$
1 MeV	1.24×10^{-12}	8.7×10^{-13}	2.86×10^{-14}	2.87×10^{-15}
10 cV	1.24×10^{-7}	3.9×10^{-10}	9.0×10^{-12}	9.1×10^{-13}
1/40 eV	4.96×10^{-5}	7.8×10^{-9}	1.81×10^{-10}	1.82×10^{-11}

Ratio: $p_{13}/p_\gamma = p_{13}c/E_\gamma = \sqrt{2m_{13}c^2 \times (KE)}/E_\gamma = \sqrt{2m_{13}c^2/E_\gamma} \approx 50$.

1.3 The uncertainty principle gives electron momentum $p_{el} \sim \Delta p \approx \hbar/\Delta x$.
Relativistic energy: $E_{el} \approx p_{el}c \sim \hbar c/\Delta x = 197\ (\mathrm{MeV\,fm})/10\ (\mathrm{fm}) = 20\,\mathrm{MeV}$.

1.4 (a) Since the number of nucleons is conserved in the reaction, the Q value can be evaluated using either atomic masses (given) or mass excesses (me) (see Appendix F). Thus, $Q = [m_{^4\mathrm{He}} + m_{^9\mathrm{Be}} - m_{^{13}\mathrm{C}}]c^2 = [me(^4\mathrm{He}) + me(^9\mathrm{Be}) - me(^{13}\mathrm{C})]m_u c^2$
$= [0.002603 + 0.012182 - 0.003355]m_u c^2 = 10.6\,\mathrm{MeV}$, where $m_u c^2 = 931.494\,\mathrm{MeV}$ (from Appendix A.2).
(b) From momentum conservation (neglecting the γ ray): $p_\alpha = p_{^{13}\mathrm{C}}$, which gives $KE_{^{13}\mathrm{C}} = 1.54\,\mathrm{MeV}$. We also know that $Q = KE_{^{13}\mathrm{C}} + E_\gamma - KE_\alpha$. Whence, $E_\gamma = Q + KE_\alpha - KE_{^{13}\mathrm{C}} = 14.1\,\mathrm{MeV}$.

1.5 (a) Use Equation (1.4): $E_{111} = 3h^2/8m_n a^2 = 10\,\mathrm{MeV}$, which gives a $= 7.8\,\mathrm{fm}$.
(b) $E_{112} = 2E_{111} = 20\,\mathrm{MeV}$, $E_{122} = 30\,\mathrm{MeV}$ and $E_{113} = 36.7\,\mathrm{MeV}$.

1.6 The decay Q value [Equation (1.6)] can be written in terms of mass excesses (me), since the number of nucleons in the parent is the same as the number of nucleons contained in the parent plus daughter. Therefore, using data for $^{234}\mathrm{U}$, $^4\mathrm{He}$ and $^{230}\mathrm{Th}$ from Appendix F, we obtain $Q_\alpha = (0.040946 - 0.002603 - 0.033127)m_u c^2$

$= 4.86\,\text{MeV}$, where $m_u c^2 = 931.494\,\text{MeV}$ (from Appendix A.2).
$$Q_d = [me(^{234}\text{U}) - me(^2\text{H}) - me(^{232}\text{Pa})]m_u c^2 = -10.9\,\text{MeV}.$$
Similarly, $Q_t = -10.2\,\text{MeV}$ and $Q_{^3\text{He}} = -10.6\,\text{MeV}$.

1.7 Equation (1.7) gives $E_{^{208}\text{Pb}} = 8.784 \times 4/208 = 0.169\,\text{MeV}$. $Q = E_{^{208}\text{Pb}} + E_\alpha = 8.953$ MeV $= (m_{^{212}\text{Po}} - m_{^{208}\text{Pb}} - m_{^4\text{He}})c^2$, whence, $m_{^{212}\text{Po}} = 211.98885\,\text{u}$.

1.8 $m_{^{36}\text{Cl}} = m_{^{36}\text{Sl}} + 1.142/m_u c^2 = 35.968307\,\text{u}$; $m_{^{36}\text{Ar}} = m_{^{36}\text{Cl}} - 0.709/m_u c^2 = 35.967546$ u. $^{36}\text{Cl}\ \beta^-$ decays to ^{36}Ar and decays by β^+ emission or electron capture to ^{36}S.

1.9 Equation (1.17) gives $R = N_B/N_A(0) = e^{-\lambda_A t} - e^{-2\lambda_A t}$.
$dR/dt = -\lambda_A e^{-\lambda_A t} + 2\lambda_A e^{-2\lambda_A t} = 0$ for a maximum. Whence, $e^{-\lambda_A t} = 1/2$, which gives t (for maximum N_B) $= \ln 2/\lambda_A$ and $R = 0.25$.

1.10 $\dfrac{dN}{dt} = P - \lambda N \longrightarrow e^{\lambda t}\left(\dfrac{dN}{dt} + \lambda N\right) = \dfrac{d}{dt}(Ne^{\lambda t}) = Pe^{\lambda t}$.

Integrate to obtain $Ne^{\lambda t} = (P/\lambda)e^{\lambda t} + C$ and then use $N = 0$ at $t = 0$ to obtain $C = -P/\lambda$.
NiCl_2: mol. wt. $= 129.6$ and $^{35}\text{Cl}/\text{Cl} = f = 0.758$.
Number of ^{35}Cl atoms $N_{^{35}\text{Cl}} = 2fmN_A/\text{mol. wt.}\ (m = 1\,\text{g})$.
Activity $A = \lambda N = P(1 - e^{-\lambda t}) \approx P\lambda t$, since $\lambda t \ll 1$.
$t = A/P\lambda = At_{1/2}/\ln 2(\sigma\Phi N_{^{35}\text{Cl}})$, where we have used Equation (1.14) to substitute for λ and Equation (1.22) for the rate of production. The required activity $A = 100\ \mu\text{Ci} = 3.7\times 10^6\,\text{Bq}$. Using this, and substituting other values, gives $t = 19.3\,\text{days}$.

1.11 If there are n nuclei initially, the number transformed in time t is $n\sigma\Phi t$, which, after substituting values, gives the fraction: $\sigma\Phi t = 3.7 \times 10^{-4}$.

1.12 Since $\sigma(\theta)$ is isotropic, the total number of protons emitted into the full (4π) solid angle is $15 \times 4\pi/(2 \times 10^{-3}) = I$. Use $I = I_o\sigma mN_A/A$ [from Equation (1.24)] to obtain $\sigma = IA/(I_o mN_A)$, where m is the mass of the target in units of mass per unit area. Substituting values gives $\sigma = 241\,\text{mb}$.

CHAPTER 2

2.1 Work done to move a charge dq from infinity to the surface of a spherical charge q, radius r, is $\Delta V = qdq/4\pi\varepsilon_0 r$. The Coulomb energy is the total work done:
$V = \displaystyle\int_0^Q \dfrac{qdq}{4\pi\varepsilon_0 r}$. If the radius of the total charge Q is R, $q/Q = (r/R)^3$ and substituting for r in the integral gives

$$V = \frac{1}{4\pi\varepsilon_0}\int_0^Q \frac{qdq}{R(q/Q)^{1/3}} = \frac{3}{5}\left(\frac{Q^2}{4\pi\varepsilon_0 R}\right).$$

Difference in binding energies: $B(^{41}\text{Sc}) - B(^{41}\text{Ca}) = \frac{3}{5}(e^2/4\pi\varepsilon_0 R)[Z^2 - (Z-1)^2]$ $= 7.28\,\text{MeV}$. Solving (with $Z = 21$) gives $R = 4.87\,\text{fm}$.

2.2 Use Equations (1.1) and (2.1) to write the nuclear mass as

$$m(A, Z) = Zm_H + (A - Z)m_n - [a_v A - a_s A^{2/3} - \frac{a_c Z^2}{A^{1/3}} - \frac{a_a(A - 2Z)^2}{A} \pm \Delta]/c^2.$$

Evaluate relative mass energies: $[m(A, Z) - Am_u]c^2$, using $m_u c^2 = 931.494\,\text{MeV}$ and $\Delta = 2\,\text{MeV}$, to obtain values of $-33.23,\ -44.56,\ -43.46,\ -45.91$ and $-35.93\,\text{MeV}$, for $A = 46$ and $Z = 19$ to 23 inclusive.

The minima (second and fourth values) correspond to the β-stable nuclei $^{46}_{20}\text{Ca}$ and $^{46}_{22}\text{Ti}$.

2.3 Differentiate the equation for $m(A, Z)$ in Problem 2.2, with A constant:
$dm/dZ = m_H - m_n + [2a_c Z/A^{1/3} - 4a_a(1 - 2Z/A)]/c^2 = 0$ for a minimum.

Whence $Z_{min} = \dfrac{[m_n - m_H]c^2 + 4a_a}{2a_c A^{-1/3} + 8a_a A^{-1}} = 43.8$ for $A = 101$.

2.4 Use Equation (2.1) to obtain: $B(^{238}_{92}\text{U}) = 1802\,\text{MeV}$ and $B(^{119}_{46}\text{Pd}) = 991\,\text{MeV}$, whence the energy release in fission (SEMF): $E_f = 2B(^{119}_{46}\text{Pd}) - B(^{238}_{92}\text{U}) = 180\,\text{MeV}$.
Estimating from Figure 2.3 gives $\Delta B \approx (8.5 - 7.6) \times 238 = 214\,\text{MeV}$.

Figure 2.3 gives B/A for the most stable $A = 119$ nucleus (not $^{119}_{46}\text{Pd}$) and, therefore, predicts a higher energy release than the SEMF does.

2.5 Energy changes due to the spin-orbit force are: $\Delta E_{j=\ell+\frac{1}{2}} \propto \ell$ and $\Delta E_{j=\ell-\frac{1}{2}} \propto -(\ell + 1)$ [see after Equation (2.6)]. A state of spin j can hold up to $(2j + 1)$ neutrons (and $(2j + 1)$ protons). Therefore, the weighting factor for $j = \ell + 1/2$ is $2\ell + 2$ and for $j = \ell - 1/2$, it is 2ℓ. Using these factors, the weighted mean energy change is zero, independent of the spin-orbit force.

2.6 A state with quantum number $j\ (= \ell \pm 1/2)$ can contain a maximum number $N_j = 2(2j + 1)$ nucleons. Therefore, if $N_j = 16$, $j = 7/2$ and $\ell = 3$ or 4, but since the parity is odd, $\ell = 3$. I^π values: $3/2^-(^7\text{Li})$, $3/2^-(^{11}\text{B})$, $1/2^+(^{31}\text{P})$, $3/2^+(^{39}\text{K})$, $7/2^-(^{59}\text{Co})$ and $5/2^+(^{127}\text{I})$.

2.7 In the ground state of ^{16}O, both protons and neutrons have the closed-shell configurations $[(1s_{1/2})^2(1p_{3/2})^4(1p_{1/2})^2]_{0^+}$ with resultant spin 0 and positive parity, as indicated by the suffix 0^+. The additional neutron in ^{17}O goes into the $1d_{5/2}$ shell in the ground state, which, therefore, is a $5/2^+$ state.

Possible configurations for the negative-parity excited states of ^{17}O are with a neutron promoted from either the $1p_{1/2}$ or $1p_{3/2}$ shell-model orbitals to a higher state. The two extra-core neutrons and the neutron hole can then couple together to form different spin–parity configurations. Note that, as described in Section 2.4.2 (and see footnote), the spin–parity assignments (I^π) of a vacancy in a shell-model state (quantum numbers j and ℓ) are $I = j, \pi = (-1)^\ell$. Furthermore, the two extra-core neutrons are subject to the restriction (as noted in Section 2.4.2) that, if they are in the same shell-model state, they must be coupled with a net even spin and parity, i.e. two $d_{5/2}$ neutrons (or protons) can be coupled to give $I^\pi = 0^+, 2^+$ or 4^+ only.

The simplest way to obtain a $1/2^-$ state is to promote one of the $1p_{1/2}$ neutrons to the $1d_{5/2}$ shell, leaving one hole in the $1p_{1/2}$ shell and two neutrons in the $1d_{5/2}$ shell. Omitting closed shells, the neutron configuration is $[(1p_{1/2})^{-1}(1d_{5/2})^2_{0^+}]_{1/2^-}$. The

$(1p_{1/2})^{-1}$ represents a vacancy in the $1p_{1/2}$ shell-model state (with $I^{\pi} = 1/2^-$). $(1d_{5/2})^2_{0+}$ represents the two extra-core particles in the $1d_{5/2}$ state coupled to $I^{\pi} = 0^+$. Coupling the $1p_{1/2}$ hole with 0^+ gives a net spin–parity of $1/2^-$ (as indicated by $]_{1/2-}$).

Similarly, other possible configurations for the ^{17}O $1/2^-$ state are:

$$[(1p_{1/2})^{-1}(2s_{1/2})^2_{0+}]_{1/2-}, \ [(1p_{1/2})^{-1}(1d_{3/2}2s_{1/2})_{1+}]_{1/2-}, \text{ etc.}$$

In the latter example, the extra-core neutrons are in different shell-model states ($1d_{3/2}$ and $2s_{1/2}$) and are coupled with $I^{\pi} = 1^+$. The configuration spin ($1/2^-$) is one result of coupling $1/2^-$ (for the hole) with 1^+.

With this interpretation in mind, we can write possible configurations for the ^{17}O $3/2^-$ state:

$$[(1p_{3/2})^{-1}(1d_{5/2})^2_{0+}]_{3/2-}, \ [(1p_{1/2})^{-1}(2s_{1/2}1d_{5/2})_{2+}]_{3/2-}, \ [(1p_{1/2})^{-1}(1d_{5/2})^2_{2+}]_{3/2-}, \text{ etc.}$$

and for the ^{17}O $5/2^-$ state:

$$[(1p_{1/2})^{-1}(1d_{5/2})^2_{2+}]_{5/2-}, \ [(1p_{3/2})^{-1}(1d_{5/2})^2_{2+}]_{5/2-}, \ [(1p_{1/2})^{-1}(1d_{5/2}2s_{1/2})_{2+}]_{5/2-}, \text{ etc.}$$

2.8 (a) Equation (2.10) predicts $E(I) \propto I(I+1)$ for the rotational model. For the states in Figure 2.15, we find $E(I)/I(I+1) = 15.5, 15.4, 15.3$ and 15.1, for $I = 2, 4, 6$ and 8, respectively [where $E(I)$ is in keV], i.e. good agreement with the model.
(b) The moment of inertia of a (rigid) uniform sphere $\mathcal{I}_{\text{rigid}} = \frac{2}{5}MR^2$, where M and R are the mass and radius, respectively. Using Equation (2.10) for the nucleus, we write $E(2^+) = 3\hbar^2/\mathcal{I}$. Whence,

$$\frac{\mathcal{I}}{\mathcal{I}_{\text{rigid}}} = \frac{3(\hbar c)^2}{0.4Mc^2R^2E(2^+)} = \frac{3 \times 197.3^2}{0.4 \times 180 \times 931.5 \times (1.25A^{1/3})^2 \times 0.093} = 0.38.$$

2.9 Equation (2.10) gives: $E(2^+) = \mathcal{I}\omega^2/2 = 3\hbar^2/\mathcal{I}$. Eliminate $\mathcal{I}$ to get $\omega = 2E(2^+)/\sqrt{6}\hbar = 1.15 \times 10^{20}\,\text{s}^{-1}$. Rotation period $T = 2\pi/\omega = 5.4 \times 10^{-20}\,\text{s}$.
Speed of a $50\,\text{MeV}$ nucleon $v = 9.8 \times 10^{22}\,\text{fm s}^{-1}$ (non rel.). Nuclear radius $R = 1.25A^{1/3} = 7\,\text{fm}$. Therefore, transit time $2R/v \approx 1.4 \times 10^{-22}\,\text{s}$ (or about $T/400$).

2.10 Use the fact that the energy difference E_{γ} between sequential γ transitions is $47\,\text{keV}$.

$$\text{Equation (2.11) gives } \mathcal{I}_{\text{exp}} = \frac{4\hbar^2}{\Delta E_{\gamma}} = \frac{4 \times 41.802(\text{u MeV fm}^2)}{0.047\,(\text{MeV})} = 3558\,\text{u fm}^2.$$

$\mathcal{I}_{\text{rigid}}(\text{SD}) \approx 1.3 \times \mathcal{I}_{\text{rigid}}(\text{sphere}) = 1.3 \times \frac{2}{5}MR^2$. $\quad \mathcal{I}_{\text{rigid}}(\text{SD}) = 3516\,\text{u fm}^2$, i.e. $\mathcal{I}_{\text{exp}} \approx \mathcal{I}_{\text{rigid}}(\text{SD})$.

CHAPTER 3

3.1 Momentum of the recoiling nucleus: $p_r = p_{\gamma} = E_{\gamma}/c$. Therefore, recoil energy $E_r = p_r^2/2m = E_{\gamma}^2/2mc^2$. For the ^{16}O decay, $m = 16\,\text{u}$, giving $E_r = 1.26\,\text{keV}$. Excitation energy $E^* = E_{\gamma} + E_r = 6128.63 + 1.26 = 6129.89\,\text{keV}$.

3.2 Conservation of angular momentum and parity in radiative transitions leads to the selection rules given in Section 3.2.1. For the transitions in the question, the possible multipoles are: (i) **E1**, $M2$, $E3$, $M4$, $E5$; (ii) **E2**, $M3$, $E4$, $M5$, $E6$, $M7$; (iii) **E1**; (iv) **E2**, $M3$, $E4$, $M5$.

3.3 $1/2^+$, $3/2^+$, $5/2^+$, $7/2^+$. If M1 does *not* occur, $J^\pi = 7/2^+$.

3.4

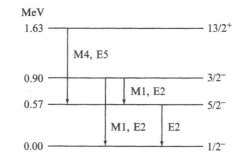

Using Equations (3.10) with half-life $t_{1/2} = \ln 2/T$, one obtains:
$5/2^- \to 1/2^-$: $T(E2) = 5.4 \times 10^9\,\mathrm{s}^{-1}$, which gives $t_{1/2} \sim 10^{-10}\,\mathrm{s}$.
$3/2^- \to 1/2^-$: $T(M1) = 2.3 \times 10^{13}\,\mathrm{s}^{-1} \gg T(E2) = 5.3 \times 10^{10}\,\mathrm{s}^{-1}$.   Verify   that $3/2^- \to 5/2^-$ is weaker than $3/2^- \to 1/2^-$. Therefore, $t_{1/2} \sim 3 \times 10^{-14}\,\mathrm{s}$.
$13/2^+ \to 5/2^-$: $T(M4) = 0.24\,\mathrm{s}^{-1} \gg T(E5) = 2.4 \times 10^{-4}\,\mathrm{s}^{-1}$. Thus, we predict that $t_{1/2} \sim 3\,\mathrm{s}$ (isomer). (Actual value 0.8 s.)
Estimates, based on Equations (3.10) are very approximate and the difference between the estimate and actual value, noted here, is not unexpected.

3.5 (a) β decay: $^{22}\mathrm{Na} \to {}^{22}\mathrm{Ne} + \beta^+ + \nu$. EC: $\mathrm{e}^- + {}^{22}\mathrm{Na} \to {}^{22}\mathrm{Ne} + \nu$.
     In terms of nuclear (m_N) and atomic (m_A) masses, the Q values are:
(b) $Q_{\beta^+} = m_N(\mathrm{Na}) - m_N(\mathrm{Ne}) - m_e = m_A(\mathrm{Na}) - m_A(\mathrm{Ne}) - 2m_e = 1.820\,\mathrm{MeV}$.
(c) $Q_{\mathrm{EC}} = m_e + m_N(\mathrm{Na}) - m_N(\mathrm{Ne}) = m_A(\mathrm{Na}) - m_A(\mathrm{Ne}) = 2.842\,\mathrm{MeV}$.

3.6 Decay energy $Q_n = (m_n - m_p - m_e)c^2 = (m_n - m_H)c^2 = 0.782\,\mathrm{MeV}$. The mirror transition $n \to p$ is a superallowed transition for which $\log_{10} ft_{1/2} \approx 3.5$ (from Table 3.3). From Figure 3.8, $\log_{10} f \approx 0.2$ for $Q_n = 0.782\,\mathrm{MeV}$, giving $\log_{10} t_{1/2} \approx 3.3$ and half-life $t_{1/2}$ (estimated) $\approx 30\,\mathrm{min}$. (Observed half-life $= 10.24\,\mathrm{min}$.)

3.7 Maximum proton momentum (p_p) is when all the decay energy ($Q_n = 0.782\,\mathrm{MeV}$) is carried by the electron, i.e. $p_p = p_e$. Relativistically: $(p_e c)^2 = E_e^2 - m_e^2 c^4$ $= (m_e c^2 + Q_n)^2 - m_e^2 c^4 = Q_n^2 + 2m_e c^2 Q_n$.

Proton energy: $E_p = \dfrac{p_p^2}{2m_p} = \dfrac{Q_n^2 + 2m_e c^2 Q_n}{2m_p c^2} = \dfrac{1.41}{2 \times 938} = 7.5 \times 10^{-4}\,\mathrm{MeV}$.

3.8 Note: (1) As explained in the paragraph after Equation (3.21), the half-life for a decay branch $= t_{1/2}({}^{137}\mathrm{Cs}\ 7/2^+$ state) divided by the decay-branch fraction. (2) End-point energy $T_o = Q_\beta - E^*$ ($E^* =$ final-state energy).
     Estimate (approximate) $\log_{10} f$ values from Figure 3.8.

Transition	$t_{1/2}$ (branch)	$\log t_{1/2}$	T_0	$\log f$	$\log ft_{1/2}$
$7/2^+ \to 3/2^+$	538 y	10.2	1.176	1.8	12.0
$7/2^+ \to 11/2^-$	31.9 y	9.0	0.514	0.5	9.5

The type of transition follows from the selection rules listed in Table 3.2:

Transition	L	$\Delta\pi$	Type of transition
$7/2^+ \to 3/2^+$	2	no	2nd-forbidden Fermi and G–T
$7/2^+ \to 11/2^-$	1	yes	1st-forbidden G–T

3.9 Use Equation (3.25): $B = [1.44 \,(\text{MeV fm}) \times 2 \times 90]/[1.4 \times 234^{1/3}(\text{fm})] = 30.0 \,\text{MeV}$. Outer distance $b = [1.44(\text{MeV fm}) \times 2 \times 90]/4.268(\text{MeV}) = 61 \,\text{fm}$.

3.10 From Equation (3.28), half-life ratio $(^{209}\text{Bi}/^{211}\text{Bi})$: $\dfrac{t_{1/2}^{209}}{t_{1/2}^{211}} = \dfrac{\exp(-2G^{211})}{\exp(-2G^{209})}$ (assuming that $f^{211} = f^{209}$ and $P^{211} = P^{209}$).

Use Equation (3.29) to determine G^{211} and G^{209}; obtain $B^{211} = 25.910 \,\text{MeV}$ and $B^{209} = 25.976 \,\text{MeV}$ from Equation (3.25), taking $r = R = 1.2(A_1^{1/3} + 4^{1/3}) \,\text{fm}$. Whence, Equation (3.29) gives $G^{211} = 23.428$ and $G^{209} = 51.391$.

From the α-decay rate λ, Equation (3.28), which is inversely proportional to the half-life, we obtain the estimate $t_{1/2}^{209} = t_{1/2}^{211}\exp[-2(G^{211} - G^{209})] = 2.4 \times 10^{26} \,\text{s}$ $\approx 8 \times 10^{18}$ y, i.e. ^{209}Bi is predicted to be essentially stable.

3.11 Use Equation (3.29) to calculate G^{92} $(Z = 92)$ and G^{94} $(Z = 94)$ for $Q = 5 \,\text{MeV}$ and $A_1 = 234$. Obtain $B^{92} = 27.872 \,\text{MeV}$ and $B^{94} = 28.492 \,\text{MeV}$ from Equation (3.25), taking $r = R = 1.2(A_1^{1/3} + 4^{1/3}) \,\text{fm}$. Whence, $G^{92} = 38.027$ and $G^{94} = 39.308$. Therefore, $\log_{10} t_{1/2}^{94}/\log_{10} t_{1/2}^{92} = \log_{10}[\exp(2G^{94} - 2G^{92})] = 1.1$.

CHAPTER 4

4.1 Using Equation (4.2) with $\sigma = 1 \,\text{b}$ $(100 \,\text{fm}^2)$ and $B/E = 0.5$, gives $R = 8 \,\text{fm}$.

4.2 Important partial waves include $0 \le \ell \le R/\lambda = kR$. $R(A = 125) = 6 \,\text{fm}$; k (9-MeV neutrons) $= 0.66 \,\text{fm}^{-1}$ (using the formula in Appendix A), whence, $kR \approx 4$. Therefore about five partial waves are important.

4.3 Summing Equation (4.4) gives $\sigma_r(\text{max}) = \sum_0^{\ell_{max}} (2\ell + 1)\pi\lambda^2 = \pi(\ell_{max} + 1)^2 \lambda^2$.

The classical angular momentum $L(\approx \ell_{max}\hbar) = pR = \hbar kR = \hbar R/\lambda$; ℓ_{max} corresponds to the grazing angular momentum. In the classical limit, $\ell_{max} \gg 1$; therefore, $\sigma_r(\text{max}) \to \pi\ell_{max}^2\lambda^2 \to \pi R^2$.

4.4 $q^2 = |\mathbf{k}_{in} - \mathbf{k}_{out}|^2 = 2k^2 - 2k^2\cos\theta = 2k^2(1 - \cos\theta) = 4k^2\sin^2\frac{\theta}{2}$ (since $|\mathbf{k}_{in}| = |\mathbf{k}_{out}| = k$), i.e. $q = 2k\sin\theta/2$.

450-MeV electrons are fully relativistic and so $k \approx E/\hbar c = 2.28 \,\text{fm}^{-1}$.

In Figure 4.6, minima occur for $q \approx 1.0, 1.8, 2.7$ and 3.35, which correspond to θ values of about 25°, 46°, 73° and 95°, respectively.

4.5

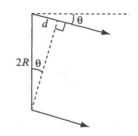

Successive minima (or maxima) occur when the path difference d for electrons (wavelength λ) emerging from opposite sides of a nucleus (diameter $2R$) changes by λ. This leads to the condition $\lambda = 2R\Delta\theta$ or $\Delta\theta = \pi\lambda/R = \pi/kR$.

Taking $\Delta\theta \approx 23°$ and $k = 2.28\,\text{fm}^{-1}$, from the results of Problem 4.4, gives $R \approx 3.4\,\text{fm}$.

4.6 Use Equations (4.25): $k_r^\alpha/k_r^p = \sqrt{(2m_\alpha \times 140)/(2m_p \times 90)} = 2.49$. $k_i^p = k_r^p/18$, $k_i^\alpha = 30k_r^\alpha/280 = 2.49k_r^p \times 30/280 = 0.267k_r^p$.

Mean free path $d \propto 1/k_i$ [see comment below Equation (4.25)]; therefore: $d_p/d_\alpha = k_i^\alpha/k_i^p = 0.267 \times 18 = 4.8$.

4.7 For a given momentum p, kinetic energy $E \propto 1/\text{mass}$. Therefore, $E_d/E_p = m_p/m_d = 0.5$; but $E_p = E_d + Q$, therefore, $E_d = Q = 8\,\text{MeV}$.

From Equation (4.27), $\sqrt{\ell(\ell+1)} \approx 2pcR \sin\frac{\theta}{2}/\hbar c$; $pc = \sqrt{2m_pc^2E_p} = 173\,\text{MeV}$. Using this and $\theta = 40°$, gives $\ell(\ell+1) \approx 6$, i.e. $\ell = 2$. Therefore, the excited state is 2^+, since the ground state is 0^+ (for an even–even nucleus).

4.8 Half-life $= \ln 2 \times$ mean lifetime $= \hbar \ln 2/\Gamma = 3.4 \times 10^{-15}\,\text{s}$.
v_n $(0.17\,\text{eV}) = 5.7 \times 10^3\,\text{m}\ \text{s}^{-1} = 5.7 \times 10^{18}\,\text{fm}\ \text{s}^{-1}$. Radius $R = 1.2 \times 113^{1/3}\,\text{fm} = 5.8\,\text{fm}$. Therefore, collision time $\approx 2R/v_n = 2 \times 10^{-18}\,\text{s}$.

4.9 Using Equations (4.29) and (4.30), with $E = E_r$, gives $\sigma_c(\text{peak}) = \dfrac{4\pi}{k^2}\dfrac{\Gamma_n}{\Gamma} = 19\,200\,\text{b}$;
$4\pi/k^2 = 22\,640\,\text{b}$ (for $E_n = 115\,\text{eV}$); whence, $\Gamma_n/\Gamma = 0.848$, $\Gamma_n = 79.71\,\text{meV}$ and $\Gamma_\gamma = \Gamma - \Gamma_n = 14.29\,\text{meV}$; $\sigma(n,\gamma) = \sigma_c\Gamma_\gamma/\Gamma = 2919\,\text{b}$. Note that the spin factor, Equation (4.31), is unity in this example, since $i_a = 1/2$, $i_A = 0$ and $\ell = 0$.

4.10 In the laboratory system: $E_{\text{lab}} = \frac{1}{2}mv_0^2$ (where $v_0 = $ projectile speed).

In the c-m system, the total momentum is zero, i.e. $mv = Mw$, where v and w are the respective c-m speeds of m and M. We also have $v + w = v_0$, whence, $v = Mv_0/(M+m)$ and $w = mv_0/(M+m)$.   Thus, $E_{\text{cm}} = \frac{1}{2}mv^2 + \frac{1}{2}mw^2 = E_{\text{lab}}M/(M+m)$.

E_{cm} (for $n+{}^6\text{Li}$) $= 244.5 \times 6/7 = 209.6\,\text{keV}$. The Q value for the $n+{}^6\text{Li} \rightarrow {}^7\text{Li}$ reaction (using data from Appendix F) $= 7250\,\text{keV}$. Excitation energy $= Q + E_{\text{cm}} = 7460\,\text{keV}$.

4.11 A graph of σ versus $1/E$ gives a straight line whose intercepts on the x and y axes are $0.0494\,\text{MeV}^{-1}$ and $2800\,\text{mb}$ ($280\,\text{fm}^2$), respectively. Equating the line to Equation (4.2) gives $R = 9.44\,\text{fm}$ and $B = 20.2\,\text{MeV}$.

4.12 Two results from classical mechanics you need are: (1) $E_{\text{cm}}/E_{\text{lab}} = M/(M+m)$ (see Problem 4.10); (2) the angular momentum (L) of two colliding particles about an axis through the centre of mass is conserved in the collision.

Thus, $L = mv_0bM/(M+m)$,   where b is   the   impact   parameter   and v_0 ($= \sqrt{2E_{\text{lab}}/m}$) is the projectile speed in the laboratory system. Taking the cross section $\sigma = \pi b^2$ (geometric limit), we obtain directly:

$$L^2 = m^2 \left(\frac{2E_{\text{lab}}}{m}\right) \frac{\sigma}{\pi} \left(\frac{M}{M+m}\right)^2 = \frac{2E_{\text{cm}} \sigma m M}{\pi(M+m)}$$

4.13 Using $E_{\text{cm}} = 33 \text{ MeV}$, obtain $\sigma \approx 1$ b $(= 100 \text{ fm}^2)$ from Figure 4.16; then, using $L^2 = \ell(\ell+1)\hbar^2$ and the result of Problem 4.12, gives $\ell(\ell+1) = 505$ or $\ell = 22$.

4.14 Write Equation (4.20) as $\eta = \left(\dfrac{K-k}{K+k}\right) e^{-2ikR} = a e^{-2ikR} \approx a - i\,2akR$, since

$kR \ll 1$; then, Equation (4.13) gives $\sigma_{\text{sc}} = \dfrac{\pi}{k^2}|\eta - 1|^2 = \dfrac{\pi}{k^2}\left[(a-1)^2 + (2akR)^2\right]$

$$= \frac{\pi}{k^2}\left[\frac{4k^2}{(K+k)^2} + 4a^2k^2R^2\right] \approx 4\pi(K^{-2} + R^2), \text{ since } K \gg k.$$

This is essentially independent of k, since $K \approx$ constant when $k \ll K$ [see discussion after Equation (4.21)].

CHAPTER 5

5.1 Maximum energy transfer occurs in a head-on collision. Speed of approach = speed of separation. Therefore, since the α-particle speed v is unchanged $(m_\alpha \gg m_e)$, electron speed $= 2v$.

Maximum electron energy $= \frac{1}{2}m_e(2v)^2 = 4E_\alpha m_e/m_\alpha = 2.7 \text{ keV}$.

5.2

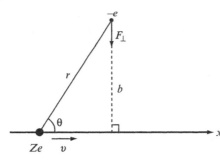

Given the assumptions, the net impulse is perpendicular to the motion of the charged particle. Therefore, momentum transfer to electron

$p = \int_{-\infty}^{\infty} F_\perp \, dt = (ze^2/4\pi\varepsilon_0) \int \sin\theta \, dt/r^2$
$= (ze^2/4\pi\varepsilon_0) \int \sin^3\theta \, dt/b^2$.   Use $x = vt$
$= b\cot\theta$, whence, $dt = -bd\theta/(v\sin^2\theta)$.
Substituting gives:
$p = (ze^2/4\pi\varepsilon_0) \int \sin\theta d\theta \, vb = ze^2/2\pi\varepsilon_0 vb$.
E(electron):
$p^2/2m = (e^2/4\pi\varepsilon_0)^2 \, (2z^2/mv^2b^2)$.

5.3 Momentum transferred to a nucleus scales as Z, i.e. $p_{\text{nuc}}^2 = Z^2 p_{\text{el}}^2$.

Therefore, $\dfrac{E_{\text{nuc}}}{E_{\text{el}}} = \dfrac{p_{\text{nuc}}^2/2m_{\text{nuc}}}{p_{\text{el}}^2/2m_{\text{el}}} = \dfrac{Z^2 m_{\text{el}}}{m_{\text{nuc}}} = \dfrac{36m_{\text{el}}}{12m_{\text{u}}} \approx \dfrac{1}{600}$.

5.4 From Equations (5.5) and (5.6), $dE/\rho dx \propto z^2$ and $R' = \rho R \propto m/z^2$ (for a given v). Hence:

	p	d	t	^{3}He	α
$-dE/\rho dx$	59	59	59	236	236
$R' = \rho R$	50	100	150	37.5	50

5.5 (a) If $dE/dx = -k/E^n$, we obtain $R = -\int_E^0 E^n dE/k = E^{1+n}/(1+n)k$. Therefore, $R \times (-dE/dx) = E/(1+n)$. Evaluating at the initial energy, we obtain $(50 \times 59) = 5000/(1+n)$ keV, whence $n = 0.69$.

(b) The distance R^* a particle of energy E travels to half its energy is given by $R^* = R(E) - R(E/2)$. Whence, the fractional distance $R^*/R(E) = 1 - (1/2)^{1+n} = 0.69$.

5.6 From Equation (5.2), note that $dE/dx \propto (z^2/mv^2) \ln(2mv^2/I) = K$, where $m = $ electron mass. Calculate K, setting $I = 11Z$ (see Section 5.2.1) = 0.154 keV and $mv^2 = 2E_{ion} \times (m/m_{ion})$, then scale results for p, d and ^{3}He relative to the α particle.

Ion	mv^2 (keV)	$\ln(2mv^2/I)$	z^2/mv^2 (keV)$^{-1}$	K (keV)$^{-1}$	$-dE/\rho dx$ (keV mg^{-1}cm^2)
α	2.74	3.57	1.46	5.21	233
p	10.97	4.96	0.091	0.452	20
d	5.48	4.27	0.182	0.780	35
^{3}He	3.66	3.86	1.09	4.224	189

5.7 Using Equation (5.13), with $\theta = 60°$ and $E'_\gamma = 0.5E_\gamma$, gives $E_\gamma = 2mc^2 = 1.022$ MeV.

5.8 Use Equation (5.13): $E'(180°) = E/(1 + 2E/mc^2)$, which is maximum when $E \to \infty$. $E'_{max}(180°) = mc^2/2 = 255.5$ keV.

5.9 Use Equation (5.13) with $E_\gamma = hc/\lambda = 1.24$ cV $\ll mc^2$. Electron energy $E_{el} = E_\gamma - E'_\gamma = E_\gamma - E_\gamma/(1 + E_\gamma/mc^2) \approx E_\gamma^2/mc^2$.

Electron speed $v = \sqrt{2E_{el}/m} = \sqrt{2}c(E_\gamma/mc^2) = 1030$ m s^{-1}.

5.10 Use Equations. (5.16) and (5.17): $\mu = \rho\mu_m = 0.141$cm^{-1}; $e^{-\mu x} = 10^{-6} \to x = 98$ cm.

5.11 Use Equations. (5.16) and (5.17): For iron, $\mu = \rho\mu_m = 314.8$ m^{-1}. Fraction of photons absorbed in the slab $= 1 - e^{-\mu x} = 0.793$.

Energy absorbed (per m^2) $= 10^{18} \times 100 \times 1.6 \times 10^{-16} \times 0.793 = 1.27 \times 10^4$ J. Mass of iron (per m^2) $M = 39.35$ kg. Therefore, the increase in temperature $= E/(M \times$ heat capacity) $= 3°$ C.

5.12 Use Equation (5.16): $\dfrac{\ln(0.62)}{\ln(0.65)} = \dfrac{\mu_{1.17}x}{\mu_{1.33}x} = \dfrac{\mu_{1.17}}{\mu_{1.33}} = 1.11$ (where x is the slab thickness).

With two slabs, the attenuation is $e^{-2\mu x}$: i.e. 0.38 (for $E_\gamma = 1.17$ MeV) and 0.42 (for $E_\gamma = 1.33$ MeV).

5.13 Use Equation (5.18): $x = \ln(10^3)/N\sigma = A \ln(10^3)/\rho N_A \sigma = 0.50$ mm.

5.14 (a) Use Equations (5.18) and (5.19): $\lambda = 1/\Sigma_t = 1/N\sigma_t \to \sigma_t = 1/N\lambda = 10.5$ b.

(b) $\lambda_a = \lambda\Sigma_t/\Sigma_a = \lambda(\Sigma_s + \Sigma_a)/\Sigma_a = 7\lambda = 14$ cm.

5.15 Use Equations (5.34), (5.35) and (5.37):

		ξ	n_{th}	$n_{1/2}$
n_{th} = No. of collisions to thermalize	Be	0.207	88	3.4
$n_{1/2}$ = No. of collisions to halve the	Fe	0.035	516	20
incident energy	Pb	0.0096	1890	72

CHAPTER 6

6.1 The curie (Ci) and becquerel (Bq) are units of activity with $1\,Bq = 1$ decay per second and $1\,Ci = 3.7 \times 10^{10}$ Bq (see end of Section 1.5.4). Hence, $50\,\mu Ci = 1.85 \times 10^6$ Bq.

A 5-MeV α particle produces $(5 \times 10^6\,eV)/(34\,eV)$ ion pairs, each leading to a collected charge of $e = 1.6 \times 10^{-19}$ C.

$$\text{Hence, current} = 1.85 \times 10^6 (s^{-1}) \times \frac{5 \times 10^6 (eV)}{34(eV)} \times 1.6 \times 10^{-19}(C) = 43.5 \text{ nA}.$$

6.2 $$C = \frac{\Delta Q}{\Delta V} = \frac{4 \times 10^6 (eV)}{34(eV)} \times \frac{1.6 \times 10^{-19}(C)}{2 \times 10^{-3}(V)} = 9.4 \text{ pF}.$$

6.3 If $\mathcal{E}$ is the electric field strength, ionization begins when $e\mathcal{E}\lambda = 10$ eV, i.e. when $\mathcal{E} = 2\,kV/mm$.

A multiplication $M = 1024 = 2^{10}$ is obtained after 10 ionizing collisions, which requires an average of 10 mean free paths $= 0.05\,mm$.

Therefore, ionization begins at radius $r = 0.05 + 0.05 = 0.1$ mm, whence, Equation (6.1) gives $V = \mathcal{E} r \ln b/a = 1060$ V.

6.4 (a) Since total dead time in one second $= m\tau$ s, it follows that the total 'live time' in one second $= (1 - m\tau)$ s, and the true count rate $n = m/(1 - m\tau)$, giving $n = 1250\ s^{-1}$.
(b) Rewriting the last equation: $m = n/(1 + n\tau) = 3571$ for $\tau = 200\ \mu s$ and $n = 12\,500\ s^{-1}$.

6.5 (a) For γ-ray energies E greater than twice the electron rest-mass energy $(2mc^2)$, pair production can occur leading to three peaks at energies E, $E - mc^2$ and $E - 2mc^2$ (see Section 6.5.1). A larger crystal absorbs more of the radiation emitted following positron annihilation, thus enhancing the full-energy peak (see discussion in Section 6.5.3).

(b) The channel numbers in the spectrum are proportional to the γ-ray energies E, $E - mc^2$ and $E - 2mc^2$. Therefore, $E/1650 = (E - 511)/1252 = (E - 1022)/854$. From the equality of the first two expressions one finds $E = 2118\,keV$. Using the peak at channel number 854, we can check our interpretation and, indeed, equating the first and last expressions gives the same value for E.

6.6 $(E/4.5\ eV) \times 10\% \times 90\% \times 15\% = 1$ (photoelectron) $\longrightarrow E = 333$ eV.

PMT multiplication $M = 4^{10}$. Output voltage $V = \dfrac{Q}{C} = \dfrac{E_\gamma}{E} \times \dfrac{Me}{C} = 2.02$ V.

6.7 Detector efficiency $\epsilon =$ (probability of the neutron interacting in the detector)$\times$ (fraction of interactions with hydrogen) $= [1 - \exp -(\Sigma_C + \Sigma_H)t] \times \Sigma_H/(\Sigma_H + \Sigma_C)$. $\Sigma_C = \Sigma_H = N_H \sigma_H = \rho N_A \sigma_H/13 = 0.102\,\mathrm{cm}^{-1}$, whence, $\epsilon = 0.17$.

6.8 Conservation of momentum gives $m_1 v_1 = m_2 v_2$ (where 1 and 2 refer to f_1 and f_2, respectively). Thus, ratio of energies $E_2/E_1 = v_2/v_1$. Ratio of flight times: $t_1/t_2 = v_2/v_1 = m_1/m_2 = 1.2$. Therefore f_2 has the most energy.

6.9 $t_y = 0.3(\mathrm{m})/c = 1$ ns. Neutron speed $v_n = 0.3(\mathrm{m})/15(\mathrm{ns}) = 2 \times 10^7$ m s^{-1}. Whence, E_n (non-relativistic) $= 2.1$ MeV.

6.10

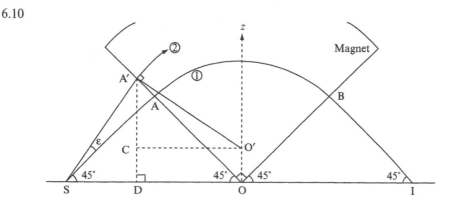

Consider first the central ray (ray 1). By symmetry, and since $A\hat{O}B = 90°$ and $SA = AO = IB = OB = r$, SOI is a straight line with the centre of curvature O lying on the z-axis. Also, $A'O' = r$, since O' is the centre of curvature for ray 2. If O' also lies on the z-axis, ray 2 passes through I (as required), i.e. we need to show that $SD + CO' = SO = 2r\cos45°$. $SA' = r/\cos\varepsilon$, $SD = SA'\cos(45° + \varepsilon)$, $CO' = A'O'\cos(45° - \varepsilon)$. Therefore, $SD + CO' = \dfrac{r}{\cos\varepsilon}\cos(45° + \varepsilon) + r\cos(45° - \varepsilon)$. To first order in ε, $\cos\varepsilon \approx 1$, and $SD + CO' \approx r[\cos(45° + \varepsilon) + \cos(45° - \varepsilon)]$ $= 2r\cos45°\cos\varepsilon \approx 2r\cos45° = SO$ (QED).

6.11 (a) Path length (1 m) $= \pi r/2 \to r = 0.637$ m. $B = mv/qr = \sqrt{2mE}/qr = 0.72$ T.
(b) Since $B \propto \sqrt{E}$, $dB/B = dE/2E$. With $dB/B = 0.01$, $dE = 0.2$ MeV and $E' = E - dE = 9.8$ MeV.

6.12 In a Sloan–Lawrence linear accelerator (discussed in Section 6.8.2), the ion must travel the length of the drift tube in half a r.f. cycle, i.e. in a time $1/2f$. Hence, $L = v/2f$, where v is the speed of the oxygen ion.
From a formula in Appendix A.4, the speed v of an 80-MeV ^{16}O ion $= 3.1 \times 10^7$ m s^{-1}. Hence, length L of drift tube $= v/2f = (3.1 \times 10^7$ m s$^{-1})\div (2 \times 30 \times 10^6$ s$^{-1}) = 0.52$ m.

6.13 Use Equation (6.7): $R = \sqrt{2mE}/Bq = 12.7$ cm.

6.14 Use Equation (6.5): $f = qB/2\pi m = 30.7$ MHz.

CHAPTER 7

7.1 See Section 7.3: Absorbed dose $D =$ absorbed energy per unit mass; equivalent dose $H = D\times$ (radiation weighting factor w_R); effective dose $E = H\times$ (tissue weighting factor w_T). The weighting factors w_R and w_T are dimensionless. D is measured in units of Gy (gray); H and E in units of Sv (sievert). All three units correspond to an energy absorption of 1 J kg^{-1}. For $H = 2$ mSv, energy absorbed by whole body (70 kg) $= 0.14$ J (w_R =1). Doses to the liver ($w_T = 0.05$) are:
(1) 17.6 μGy, 17.6 μSv, 0.88 μSv; (2) 17.6 μGy, 176 μSv, 8.8 μSv.

7.2 Number of photons $= \dfrac{\text{dose} \times \text{mass}}{E_{\text{photon}}} = \dfrac{(0.5 \times 10^{-3} \times 5)\text{J}}{(5 \times 10^4\text{V}) \times (e \text{ in coulomb})} = 3.12 \times 10^{11}$.

7.3 Substituting $\mu = 1$ into Equation (7.7), gives survival probability $S_v = 0.736$.
(1) For a single '10 times' dose, $\mu = 10$ in Equation (7.7) gives $S_v = 0.00050$.
(2) For 10 single doses (i.e. $\mu = 1$ for each), with enough time between irradiations, survival probability $S_v = (0.736)^{10} = 0.046$.

7.4 Total (effective) probability of loss from the system λ_e(effective) $= \lambda$(decay) $+\lambda_b$(biological).

Using Equation (1.14) gives $\dfrac{1}{t^e_{1/2}} = \dfrac{1}{t_{1/2}} + \dfrac{1}{t^b_{1/2}}$ or $t^e_{1/2} = \dfrac{t_{1/2}t^b_{1/2}}{(t_{1/2} + t^b_{1/2})}$.

7.5 Dose rate $\dfrac{dD}{dt} = \left[\dfrac{dD}{dt}\right]_0 e^{-\lambda_e t} \longrightarrow D = \int_0^\infty \dfrac{dD}{dt}\,dt = \dfrac{1}{\lambda_e}\left[\dfrac{dD}{dt}\right]_0 = \dfrac{t^e_{1/2}}{\ln 2}\left[\dfrac{dD}{dt}\right]_0.$

7.6 To find D from the result of Problem 7.5, we require $[dD/dt]_0 =$ initial rate of energy absorption per unit mass = (initial activity $\mathcal{A}$) $\times$ (average energy per decay)/(mass M). [Activity $\mathcal{A}$ is defined following Equation (1.14).]
    With $\mathcal{A} = 1$ Ci $= 3.7 \times 10^{10}$ Bq, $E_{av} = 6.2$ keV, $M = 70$ kg, $t^e_{1/2} \approx t^b_{1/2} = 3$ d,

$D = \dfrac{t^b_{1/2}}{\ln 2} \times \dfrac{\mathcal{A}E_{av}}{M} = 0.196$ Gy (1 Gy $= 1$ J kg^{-1}; see Section 7.3).

7.7 Dose = absorbed energy/mass $= t\,\mathcal{A}E_{av}/M_{\text{body}}$, where $t = 70$ y, $E_{av} = 0.5$ MeV and ^{40}K activity $\mathcal{A} = N_K \ln 2/t_{1/2}$. Number of ^{40}K atoms per kg: N_K/M_{body} $= 0.27\% \times 0.012\%/A m_u = 4.88 \times 10^{18}$ kg^{-1}. Substituting values gives an absorbed dose of 0.0152 Gy.

7.8 Assume 50% of α particles ($w_R = 20$) are absorbed in the hand. The range of these α particles $R' = 3.7$ mg cm^{-2}. [Range R' is defined after Equation (5.4).] Therefore, since the area of the source is 1 cm^2, 50% of the emitted energy will be absorbed in a mass $M = 3.7 \times 10^{-6}$ kg of tissue. Total energy absorbed $E_{\text{abs}} = 50\% \times$ activity (10μCi) $\times E_\alpha$(5 MeV)$\times 30$s $= 4.45 \times 10^{-6}$ J. Equivalent dose [Equation (7.2)] $= E_{\text{abs}}w_R/M = 24$ Sv.

7.9 Calculate the energy absorbed in 1 h by a piece of tissue, density $\rho = 1000\,\mathrm{kg\,m^3}$ and $x = 20\,\mathrm{cm}$ thick, presenting an area A to a 1-MBq source of 1-MeV γ rays located 1 m from the tissue. Incident energy flux at the tissue $\Phi_E = 1\,(\mathrm{MBq}) \times 1\,(\mathrm{MeV})/4\pi \times (1\,\mathrm{m})^2 = 1.27 \times 10^{-8}\,\mathrm{J\,m^{-2}\,s^{-1}}$. Fraction of energy incident on tissue, which is absorbed $f = (1 - \text{attenuation in tissue}) = (1 - e^{-\mu x}) = 0.753$. Total energy absorbed (in time t) $= E_{\mathrm{abs}} = \Phi_E A f t$. Absorbing mass $M = \rho A x$. Therefore, dose (in $t = 1\,\mathrm{h}$) $= E_{\mathrm{abs}}/M = 1.7 \times 10^{-7} \approx 1/6\,\mu\,\mathrm{Gy\,h^{-1}} = 1/6\,\mu\mathrm{Sv\,h^{-1}}$, since the weighting factor for γ rays is unity. This result is given by Equation (7.8) (QED).

7.10 Using Equation (7.8) with dD/dt (background) $= 0.297\,\mu\mathrm{Sv\,h^{-1}}$ and emitted γ-ray energy $E_\gamma = 1.17 + 1.33 = 2.5\,\mathrm{MeV}$ gives $A = 1.6\,\mathrm{MBq}$.

7.11 Sea-level dose from natural sources (see Table 7.3) $= 2210/52 = 42.5\,\mu\mathrm{Sv}$ per week. Excess dose at $10\,\mathrm{km} = 4.97\,\mu\mathrm{Sv\,h^{-1}}$. Thus, 8.6 flying hours per week doubles the sea-level dose.

7.12 A working year $= 2080\,\mathrm{h}$; 10% of the recommended dose rate limit (see Table 7.6) $= (10\% \times 20\,\mathrm{mSv/y}) = 0.96\,\mu\mathrm{Sv/h}$. Equation (7.8) gives $3062\,\mu\mathrm{Sv\,h^{-1}}$ for the unshielded source at 2 m. Therefore, attenuation needed $= 3.14 \times 10^{-4} = e^{-\mu x}$. For lead $\mu = \rho\mu_{\mathrm{m}} = 1.1\,\mathrm{cm^{-1}}$, whence $x = 7.33\,\mathrm{cm}$.

7.13 Dose $= 13\,\mu\mathrm{Sv} = \dfrac{t^b_{1/2}}{\ln 2} \times \dfrac{A \times E}{M}$. Substituting: $A = 10^3\,\mathrm{Bq}$, $E = (0.75 \times 0.662) + 0.2 = 0.697\,\mathrm{MeV}$ and $M = 70\,\mathrm{kg}$, gives $t^b_{1/2} = 5.656 \times 10^6\,\mathrm{s} \approx 65\,\mathrm{d}$.

7.14 Equivalent dose to breast $= 50\,\mathrm{Sv}$. Using information from Table 7.5 gives a cancer risk $= 50 \times 2 \times 10^{-3} = 0.1$.

7.15 Using information from Table 7.5, the risk per capita $\mathrm{y}^{-1} = 2.6 \times 10^{-3} \times 5 \times 10^{-2} = 1.3 \times 10^{-4}\,\mathrm{y^{-1}}$, which implies 7.8×10^5 deaths y^{-1} worldwide. This is approximately 2.6% of all cancer deaths (taking an average lifespan of 60 y).

7.16 Use Table 7.4 for effective doses. Bone marrow: $37(\mathrm{kBq}) \times 28 = 1036\,\mu\mathrm{Sv}$; thyroid: $37(\mathrm{kBq}) \times 22 = 814\,\mu\mathrm{Sv}$; whole body: $74(\mathrm{kBq}) \times 13 = 962\,\mu\mathrm{Sv}$. Total $= 2812\,\mu\mathrm{Sv}$. Taking the whole-body risk factor from Table 7.5 gives the number of deaths $= 2812 \times 10^{-6} \times 5 \times 10^{-2} \times 200 \times 10^6 = 2.8 \times 10^4$ (0.014% of the population).

7.17 Absorbed dose rate $\dot{D}_{\mathrm{abs}} = (\text{activity/kg}) \times E = 4.21 \times 10^{-10}\,\mathrm{Gy\,s^{-1}} = 0.0132\,\mathrm{Gy\,y^{-1}}$. The equivalent dose rate $\dot{D}_{\mathrm{eq}} = 0.265\,\mathrm{Sv\,y^{-1}}$ (using $w_R = 20$ from Table 7.1). From Table 7.2, w_T(bone + marrow) $= 0.13$. Therefore, the effective dose rate $\dot{D}_{\mathrm{eff}} = \dot{D}_{\mathrm{eq}} \times w_T$(bone + marrow) $= 34\,\mathrm{mSv\,y^{-1}}$, which exceeds the limit of $20\,\mathrm{mSv\,y^{-1}}$ (see Table 7.6).

7.18 Effective doses: (1) neutron: $2 \times 10^{-3} \times 5 = 10\,\mathrm{mSv}$ (using Table 7.1); (2) $^{90}\mathrm{Sr}$: $100 \times 28 \times 10^{-6} = 2.8\,\mathrm{mSv}$ (from Table 7.4). Thus (using data in Table 7.6), his $^{123}\mathrm{I}$ dose must not exceed $20 - 12.8 = 7.2\,\mathrm{mSv}$. Therefore, using information in Table 7.4, he could ingest up to $0.33\,\mathrm{MBq}$ of $^{131}\mathrm{I}$.

CHAPTER 8

8.1 (a) Using Equation (5.16), attenuation $= 0.5 = \exp(-\mu x_{1/2})$, whence, $\mu = (\ln 2)/x_{1/2}$.
(b) If $x = 1.05x_{1/2}$, $I = 1000\exp(-\mu x) = 1000\exp(-1.05\ln 2) = 483\,\text{s}^{-1}$.
   Fractional error in a count N (due to Poisson statistics): $\sigma_N/N = 1/\sqrt{N}$. This should equal the result of a 5% change in x, i.e. $1/\sqrt{N} = |dI/I| = \mu dx = \ln 2|dx/x_{1/2}| = 0.05\ln 2$. Solving this gives $N = 833$ counts, which is obtained in about 1.7 s at a counting rate of $500\,\text{s}^{-1}$.

8.2 Thermal power $P = AQ_\alpha$, where activity $A = N\ln 2/t_{1/2}$ [see after Equation (1.14)]. Therefore, $\ln 2 \times NQ_\alpha/t_{1/2} = 66.7\,\text{W}$, whence, the number of ^{210}Po atoms $N = 1.33 \times 10^{21}$. Mass of $^{210}\text{Po} = N \times 210m_\text{u} = 0.46\,\text{g}$.

8.3 Atomic mass of copper: $A = 63 \times 0.69 + 65 \times 0.31 = 63.62$. Number of ^{63}Cu atoms in the foil $N_{63} = 0.69 \times (0.1/Am_\text{u}) = 6.53 \times 10^{20}$.
   The activity A of a sample, irradiated for a time t_1 in a reactor, after a time interval t_2 following its removal from the reactor is, from Equations (8.1) and (8.2), given by $A = \Phi\sigma N_{63}(1 - e^{-\lambda t_1})e^{-\lambda t_2}$. Substituting $t_1 = 90\,\text{h}$, $t_2 = 8\,\text{h}$ and $\lambda = \ln 2/12.7(\text{h})$, gives $A = 5.66 \times 10^9\,\text{Bq} = 153\,\text{mCi}$.

8.4 Use Equations (8.1) and (8.2): A count $C = K(1 - e^{-\lambda t_1})(1 - e^{-\lambda t_2})$ where $\lambda = \ln 2/t_{1/2}$ and K is independent of the irradiation and counting times t_1 and t_2. Given that $C\,(t_1 = t_2 = t_{1/2}) = 1000$, we deduce that $K = 4000$, whence, (1) $C\,(t_1 = t_2 = 0.1t_{1/2}) = 18$, and (2) $C\,(t_1 = t_2 = 10t_{1/2}) = 3992$.

8.5 Use Equations (8.1) and (8.2) with $t_1 = 12\,\text{h}$, $f = 0.95$, $\varepsilon = 2 \times 10^{-2}$, $t_2 = 6\,\text{h}$. The count rate of $10\,\text{s}^{-1} = Af\varepsilon$, where the activity $A = \Phi\sigma N(1 - e^{-\lambda t_1})e^{-\lambda t_2}$. The number of gold atoms in the foil $N = 0.1/197m_\text{u} = 3.057 \times 10^{20}$ atoms.
   Rearranging the equation and solving gives the flux $\Phi = 1.54 \times 10^5\,\text{cm}^{-2}\,\text{s}^{-1}$.

8.6 Use Equation (8.3): $E_\text{b}(A) = E_0[(A - 4)/(A + 4)]^2$, with $E_0 = 5000\,\text{keV}$.
   $E_\text{b}(20) = 2222\,\text{keV}$, $E_\text{b}(50) = 3628\,\text{keV}$, $E_\text{b}(100) = 4260\,\text{keV}$.
   Resolution required: $\Delta E \approx E_\text{b}(A + 1) - E_\text{b}(A)$. Whence, $\Delta E = 90\,\text{keV}$ (for $A = 20$); $\Delta E = 23\,\text{keV}$ (for $A = 50$); $\Delta E = 7\,\text{keV}$ (for $A = 100$).

8.7 Energy spread: $20\,\text{keV} = (\rho x) \times (-dE/\rho dx)$. This gives (ρx) the source thickness (mass per unit area) $= 9.2 \times 10^{-5}\,\text{g cm}^{-2}$ – or the number of ^{210}Po atoms per unit area $N = (\rho x)/210\,m_\text{u}$. Therefore, area activity $= \lambda N = (\ln 2/t_{1/2}) \times (\rho x)/210\,m_\text{u} \rightarrow$ area activity $= 1.537 \times 10^{10}\,\text{Bq cm}^{-2} = 0.415\,\text{Ci cm}^{-2}$.

8.8 Use Equation (8.3) and refer to Section 5.5.2. $E_\text{b} = E_0/4$ (for α particles scattered at $180°$ from ^{12}C).
   In the c-m frame, speed of protons $= V_\text{cm}$. In the lab. frame, speed of knock-on protons at $0° = 2V_\text{cm}$. Therefore, forward proton energy $E_\text{f} = \frac{1}{2}m_\text{p}\,(2V_\text{cm})^2$. From Equation (5.22), $V_\text{cm} = 4v_0/5$ and, since $E_0 = \frac{1}{2}m_\alpha v_0^2$, $E_\text{f} = 16E_0/25$. Whence, $E_\text{f}/E_\text{b} = 64/25 = 2.56$.

8.9 From Equation (1.25), the event rate (into $d\Omega$):

$R(180°) = 0.1\,\text{s}^{-1} = \sigma_R(180°)\,N\Phi d\Omega$. This can be written as $I\left(\frac{(\rho t)N_A}{M_A}\right)\sigma_R(180°)d\Omega$, [see Equation (1.24)]. Equation (8.4) gives $\sigma_R(180°) = 0.348\,\text{b sr}^{-1}$.

Beam intensity $I = 10\,\text{nA} = 3.1 \times 10^{10}\,\text{s}^{-1}$. Rearranging the equation and solving gives $(\rho t) = 6.3 \times 10^{-8}\,\text{g cm}^{-2}$.

8.10

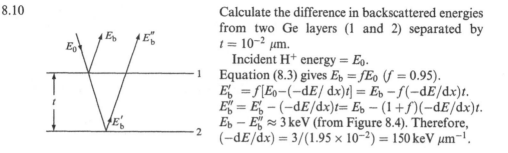

Calculate the difference in backscattered energies from two Ge layers (1 and 2) separated by $t = 10^{-2}\,\mu\text{m}$.

Incident H^+ energy $= E_0$.

Equation (8.3) gives $E_b = fE_0$ $(f = 0.95)$.

$E'_b = f[E_0 - (-dE/\,dx)t] = E_b - f(-dE/dx)t$.

$E''_b = E'_b - (-dE/dx)t = E_b - (1+f)(-dE/dx)t$.

$E_b - E''_b \approx 3\,\text{keV}$ (from Figure 8.4). Therefore,

$(-dE/dx) = 3/(1.95 \times 10^{-2}) = 150\,\text{keV}\,\mu\text{m}^{-1}$.

8.11 Use Equation (8.7): number of counts $N = I\sigma_X(mN_A/A)t\varepsilon$, with $\sigma_X = 800 \times 0.5 = 400\,\text{b}$, $m = 10^{-9}\,\text{g cm}^{-2}$, $t = 600\,\text{s}$. Substituting values gives $N = 9021$.

8.12 Use Equation (8.7): count rate $= N/t = I\sigma_X(mN_A/A)\varepsilon = 0.6\,\text{s}^{-1}$. Rearranging and substituting values gives $\sigma_X = 383\,\text{b}$.

Count in 25 m $= 900 \pm 30$ (Poisson statistics). Therefore, accuracy $= 3.3\%$.

8.13 ^{14}C rate $= 3.33\,\text{s}^{-1}$. A 10 μA current of $^{12}\text{C}^{3+}$ ions corresponds to a particle intensity of $(10^{-5}\,\text{A})/[3 \times (e\ \text{in coulomb})] = 2.08 \times 10^{13}\,\text{s}^{-1}$. Therefore, $(^{14}\text{C rate})/(^{12}\text{C rate}) = 1.6 \times 10^{-13}$. $M \equiv$ mass of ^{12}C in sample totally consumed in a 30-min run. Therefore, $M = (^{12}\text{C rate} \times 1800\,\text{s} \times 12m_u)/\varepsilon \to 37\ \mu\text{g}$.

8.14 To achieve the required accuracy, we need to record 100 counts, which, according to Poisson statistics, has a standard deviation of 10 (10%). If N ^{26}Al atoms are ingested, the number recorded $100 = Nf\varepsilon \times (5/5500)$.

Therefore, activity ingested $= N\ln 2/t_{1/2} = 0.033\,\text{Bq}$ (using $t_{1/2} = 7.4 \times 10^5$ y).

8.15 In 1 h, background count $B = 18$. From Equation (8.15), the critical limit $= 1.645\sqrt{2B} = 9.87$, where we have used $k = 1.645$ for a 95% confidence level. From Equation (8.19), the detection limit $L_d = 2.71 + 3.29\sqrt{2B} = 22.45$.

8.16 Detection limit $L_d =$ minimum activity $\times 20$ h $= 36$. From Equation (8.19) with $k = 1.645$, $L_d = 2.71 + 3.29\sqrt{2B}$, whence $B = 51$. Background rate $= B/20$ h $= 7.1 \times 10^{-4}\,\text{s}^{-1}$.

8.17 Use Equation (8.11). If $T = 4$ weeks (2.42×10^6 s), true detected rate $< 9.5 \times 10^{-7}\,\text{s}^{-1}$ (90% confidence). An upper limit to the cross section is obtained by equating this to $I(mN_A/M_A)\sigma\varepsilon$ [using Equation (1.24) with $m = (\rho t)$]. Substituting values gives $\sigma < 0.82$ pb.

CHAPTER 9

9.1 (a) At $E_\gamma = 60\,\text{keV}$: transmission $= \exp[-\mu_m(\rho x)] = e^{-4} = 0.018$ (1.8%).
At $E_\gamma = 511\,\text{keV}$: transmission $= e^{-1.94} = 0.143$ (14.3%).
(b) Additional reductions $= 0.26$ (for $E_\gamma = 60\,\text{keV}$); $= 0.93$ (for $E_\gamma = 511\,\text{keV}$).

9.2 Let $d =$ total thickness. Intensity ratio: $\dfrac{I_1}{I_2} = 2 = \dfrac{\exp(-\mu^t d)}{\exp[-(\mu^t(d-b) + \mu^b b)]}$

$= \exp(\mu^b - \mu^t)b$, where t and b refer to tissue and bone, respectively. Note that $\mu^b = \rho\mu_m^b = 1.8\mu_m^b$.
Bone thickness $b = \ln 2/(1.8 \times 0.32 - 0.2) = 1.8\,\text{cm}$.

9.3 Number of photons detected (to form an image in a given time) is proportional to activity ($\mathcal{A}$) $\times \gamma$ branch (f_γ) $\times$ detection efficiency (ε).
For ^{131}I: $E_\gamma = 364\,\text{keV}$, $f_\gamma = 0.81$ and for ^{123}I: $E_\gamma = 159\,\text{keV}$, $f_\gamma = 0.83$ (see Section 9.2.2).

Therefore, $\dfrac{\mathcal{A}_{131\text{I}}}{\mathcal{A}_{123\text{I}}} = \dfrac{(\varepsilon f_\gamma)_{123\text{I}}}{(\varepsilon f_\gamma)_{131\text{I}}} = \dfrac{0.83}{0.81} \times \left(\dfrac{1 - \exp[-\mu\,(159\,\text{keV})x]}{1 - \exp[-\mu\,(364\,\text{keV})x]}\right) = 2.30$.

Using data in Table 7.4, we obtain the ratio of doses $= 2.30 \times (22/0.21) = 241$.

9.4 (a) Transmission $T =$ solid angle/$4\pi \approx$ area of channel/$4\pi l^2 = A/4\pi l^2$.
(i) $A = \pi d^2/4 \to T = d^2/16 l^2$; (ii) $A = d^2 \to T = d^2/4\pi l^2$; (iii) $A = \sqrt{3}d^2/2 \to T = \sqrt{3}d^2/8\pi l^2$.
(b)

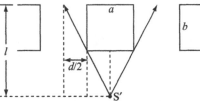

Rays from S' access two adjacent channels; 50% of each is blocked if limiting rays are as shown.
By geometry, $\dfrac{a/2}{l-b} = \dfrac{d/2}{b}$. Therefore, if $l = 2b$, $a = d$.

9.5 Overall efficiency $\eta = Tf\varepsilon$, where $T =$ transmission for (unhindered) γ rays, $f = \gamma$-ray attenuation (brain + skull) and $\varepsilon =$ detector efficiency: $[1 - \exp(-\mu_{\text{NaI}}x)]$.
Transmission: for SPECT: $T = d^2/4\pi l^2$ (see Problem 9.4) $= 0.25/5027 = 5 \times 10^{-5}$; for PET: T (for a disc, $r = 7.5\,\text{cm}$, at $l = 20\,\text{cm}$) $\approx \pi r^2/4\pi l^2 = 0.035$ (a more exact calculation gives 0.032).
Attenuation in brain and skull: f_{spect} (140 keV) $\approx \exp[-(0.16 \times 1.8 + 0.15 \times 9)]$
$= 0.194$; f_{pet} (511 keV) $\approx \exp[-(0.09 \times 1.8 + 0.097 \times 9)] = 0.355$.
Detector efficiency: $\varepsilon_{\text{spect}}$ (140 keV) $\approx [1 - \exp(-2.42)] = 0.91$; ε_{pet} (511 keV)
$\approx [1 - \exp(-0.33)] = 0.28$.
Combining figures gives $\eta_{\text{pet}}/\eta_{\text{spect}} \approx 360$ (using $T_{\text{pet}} = 0.032$).

9.6 Maximum activity is achieved when the bombardment time is so long that equilibrium is reached. This corresponds to the minimum amount of ^{15}O gas required in the cell (i.e. minimum pressure). In equilibrium, activity $A(= 10^6$ Bq$)$ = production rate = $I\sigma m N_A / A$ [using Equation (1.24)].

$I = 6.24 \times 10^{12}$ protons s^{-1} and $\sigma = 50$ mb, which give $m = 0.080$ mg cm^{-2} of ^{15}N. Hence, the density of gas in the cell = 0.080 mg cm^{-3}.

At STP, 1 mole of gas occupies 22.4 litres. Therefore, density of ^{15}N gas at STP = 1.34 mg cm^{-3}. The pressure in the gas cell = $0.080/1.34 = 0.060$ bar.

9.7 Using Equation (1.17) and noting that 0.85 of ^{99}Mo (parent) decays lead to $^{99}Tc^m$ (daughter), we find:

$$N_{Tc}(t) = \frac{0.85\lambda_{Mo}}{\lambda_{Tc} - \lambda_{Mo}} N_{Mo}(0)[\exp(-\lambda_{Mo}t) - \exp(-\lambda_{Tc}t)]$$

Since $t^{Mo}_{1/2} \gg t^{Tc}_{1/2}$, $\lambda_{Tc} \gg \lambda_{Mo}$ and we can substitute $\exp(-\lambda_{Mo}t^{Tc}_{1/2}) \approx 1$ and $\exp(-\lambda_{Tc}t^{Tc}_{1/2}) = 1/2$ into the above equation and obtain the $^{99}Tc^m$ activity (at $t = t^{Tc}_{1/2}$): $A_{Tc}(t^{Tc}_{1/2}) = \lambda_{Tc}N_{Tc}(t^{Tc}_{1/2}) = 0.85A_{Mo}/2$, where $A_{Mo} = \lambda_{Mo}N_{Mo}$ is the ^{99}Mo activity. If $A_{Mo}(0)$ is the ^{99}Mo activity initially, after $t = 1$ week it will have decayed to $A_{Mo} = A_{Mo}(0)e^{-\lambda_{Mo}t} = 0.171A_{Mo}(0)$.

Therefore, $A_{Mo}(0) = 5$(MBq) $\times 2/(0.171 \times 0.85) = 69$ MBq.

9.8 See Section 9.5.1: $f = 2\mu_p B/h = 42.6$ MHz.

9.9 Following the discussion in Section 9.5.1, the population ratio of the two spin states $N_-/N_+ = \exp(-\Delta E/kT) = \exp(-2\mu_p B/kT) \approx 1 - (2\mu_p B/kT)$, since $\mu_p B \ll kT$. For $T = 293$ K and $B = 1$ T, $\mu_p B/kT = 3.49 \times 10^{-6}$; $\Delta N = N_+ - N_- \approx (1 - N_-/N_+)N/2 \approx \mu_p NB/kT$.

1 m^3 of H_2O contains $N = 6.69 \times 10^{28}$ protons. Therefore, $M = \mu_p \Delta N = \mu_p N(\mu_p B/kT) = 3.29 \times 10^{-3}$ J T^{-1} m^{-3}.

9.10

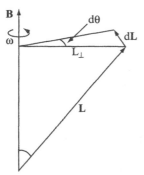

The magnetic moment of a proton is aligned with its spin ($\frac{1}{2}\hbar$). Therefore, the magnetization (**M**) of a sample $\propto$ net angular momentum (**L**) of the protons in the sample, i.e. $\mathbf{M} = \gamma\mathbf{L}$.

$M = \Delta N\mu_p$ (from Problem 9.9) and $L = \Delta N s_z = \Delta N \times \frac{1}{2}\hbar \rightarrow M/L = \gamma = 2\mu_p/\hbar$.

Torque on **M** (due to **B**) = $\mathbf{M} \times \mathbf{B} = d\mathbf{L}/dt$, using the classical result.

$|\mathbf{M} \times \mathbf{B}| = \gamma L_\perp B = dL/dt = L_\perp \, d\theta/dt = L_\perp\omega$.

Frequency $= \omega/2\pi = \gamma B/2\pi = 2\mu_p B/h$.

9.11 (a) A component $B_{ex}/2$ of the applied field maintains a torque on **M** (see Section 9.5.1) causing **M** to precess with angular frequency $\gamma B_{ex}/2 = \mu_p B_{ex}/\hbar$ (using the result of Problem 9.10).

In time $t (= 10^{-5}$ s$)$, $\omega t = \mu_p B_{ext} t / \hbar = \pi/2 \rightarrow B_{ex} = \hbar\pi/2\mu_p t = 1.17 \times 10^{-3}$ T $= 11.7$ gauss.

9.12 Use dose $\propto \exp(-\mu_{en}x)$, where $x =$ depth of penetration. Doses at A, B and tumour are proportional to $\exp(-\mu_{en}x_A)$, $\exp(-\mu_{en}x_B)$ and $\exp(-\mu_{en}x_t)$, respectively, where $x_A = 4.5$ cm, $x_B = 13.5$ cm and $x_t = 9$ cm. Normalizing the dose to the tumour to be unity, we find

$$\frac{\text{Average dose at A and B (with } \mu_{en} = 0.1)}{\text{Average dose at A and B (with } \mu_{en} = 0.2)} = \frac{1.10}{1.43} = 0.77$$

i.e. Dose to normal tissue is less when μ_{en} is less.

9.13 Let $S_1 =$ survival fraction for a single-dose fraction $= (1 + \mu)e^{-\mu}$ [using Equation (7.7)]. Since $S_{10} = (S_1)^{10} = 0.5$, $S_1 = 0.933 \rightarrow \mu = 0.42$ (graphically or by trial and error).
    For $\mu = 4.2$, survival fraction $= 0.078$.

CHAPTER 10

10.1 Coulomb energy (168 MeV) $= Z_1 Z_2 e^2 / 4\pi\varepsilon_0 r = 1.44$(MeV-fm)$Z_1 Z_2 / r$(fm), using information from Appendix A1. Whence, $r = 18.1$ fm.

10.2 Using equations in Section 10.3.2 and data in Table 10.2:

$$\eta = v\frac{6\sigma_f(235)}{6\sigma_a(235) + 94\sigma_c(238)} = 1.94; \ \sigma_a(U) = 0.06\sigma_a(235) + 0.94\sigma_c(238) = 43.4 \text{ b.}$$

10.3 The resonance escape probability p is given by Equation (10.7). For clarity, assemble the relevant data, given in Tables 10.2, 10.3 and 5.1, in the form of a table:

Nuclide	N	σ_s (b)	$N\sigma_s$	ξ	$N\sigma_s\xi$
^{235}U	0.016	10	0.16	0.0084	-
^{238}U	0.984	8.3	8.17	0.0084	0.07
C	600	4.7	2820	0.158	445.56
Totals			2828.3		445.6

Equation (10.8) gives $\langle\xi\rangle = 0.1576$. Substituting these data in Equation (10.7) gives:

$$p = \exp\left[-\frac{2.73}{0.1576}\left(\frac{0.984}{2828.3}\right)^{0.514}\right] = 0.749.$$

10.4 The mean free path for absorption (λ_a) is inversely proportional to the macroscopic absorption cross section Σ_a [see the discussion following Equation (5.20)]. Also, from Equation (10.4): $\Sigma_a(U + C) = \Sigma_a(U) + \Sigma_a(C)$ and $\sigma_a(U) = 0.02\sigma_a(235) + 0.98\sigma_c(238) = 16.27$ b.

Therefore, $\dfrac{\lambda_a(U+C)}{\lambda_a(C)} = \dfrac{\Sigma_a(C)}{\Sigma_a(U)+\Sigma_a(C)} = \dfrac{401\sigma_a(C)}{\sigma_a(U)+400\sigma_a(C)} = 0.1.$

10.5 If $\Phi(\mathbf{r}) = \Phi(z)$, Equation (10.21) becomes: $\partial^2\Phi/\partial z^2 = \Phi(z)/L^2$.
   Solution: $\Phi(z) = ae^{-z/L} \rightarrow \langle z \rangle = \int_0^\infty ze^{-z/L}dz / \int_0^\infty e^{-z/L}dz = L.$

10.6 Begin by assembling the relevant data, given in Tables 10.2 and 10.3, in the form of a table:

Nuclide	N	σ_a	σ_s	$N\sigma_a$	$N\sigma_s$	ξ	$N\sigma_s\xi$
^{nat}U	1	7.60	8.3	7.60	8.3	0.0084	0.07
O	6	0.0	3.8	0	22.8	0.120	2.74
S	1	0.52	1.1	0.52	1.1	0.061	0.07
D_2O	750	0.001	10.6	0.75	7950	0.509	4046.6
Totals				8.87	7982.2		4049.4

We calculate the neutron multiplication factor k_∞ for an infinite reactor from the four-factor formula [Equation (10.11)]: $k_\infty = \eta\varepsilon pf$. We obtain the four factors in turn:

   For ^{nat}U: $\eta = 1.328$ [see Equation (10.6)].

   The fast fission factor $\varepsilon = 1$, as discussed in Section 10.3.2.

   The resonance escape probability is given by Equation (10.7), with $\langle\xi\rangle = 4049.4/7982.2 = 0.507$, from Equation (10.8); 98.28% of natural uranium consists of ^{238}U, therefore, $N_{238}/N(^{nat}U) = 0.9928$ and we have $N_{238}/\Sigma_s = 0.9928/7982.2 = 1/8040$. Equation (10.7) gives:

$$p = \exp\left[-\dfrac{2.73}{0.507}\left(\dfrac{1}{8040}\right)^{0.514}\right] = 0.948.$$

   The thermal utilization factor f is given by Equation (10.9):
$f = N\sigma_a(^{nat}U)/\sum N\sigma_a = 7.6/8.87 = 0.8568.$
   Combining these results in Equation (10.11) gives $k_\infty = 1.0787$.

10.7 From Section 10.4.5, the radius R of a spherical reactor is given by $R = \pi/B$, where B is defined by Equation (10.32) with L_c^2 replaced by $(L_c^2 + L_s^2)$, as explained at the end of Section 10.4.4.

   From Equation (10.34) and Table 10.5, and the data obtained in Problem 10.6, $L_c^2 = (1 - 0.8568)L^2 = 4296\,\text{cm}^2$, $L_s^2 = 131\,\text{cm}^2$, whence $B^2 = (k_\infty - 1)/(L_c^2 + L_s^2) = 1.78 \times 10^{-5}\,\text{cm}^{-2}$.

   Hence, $R = \pi/B = 745\,\text{cm}$, and diameter of reactor $\approx 15\,\text{m}$.

10.8 Fission consumption: $N_f = 2\,(\text{GW})/200\,(\text{MeV/fission}) \rightarrow 6.24 \times 10^{19}$ atoms s^{-1}.
   Total ^{235}U consumption $= N_f\sigma_a(235)/\sigma_f(235) = N_f \times 680/579 = 7.33 \times 10^{19}$ atoms s$^{-1} = 902\,\text{kg y}^{-1}$ (using σ_a and σ_f for ^{235}U from Table 10.2).

10.9 Use Equations (10.37) and (10.38). Power $P = N(235)\,\sigma_f(235)\Phi E$; ^{239}Pu production rate $R = N(238)\sigma_c(238)\Phi$. Substituting for Φ gives $R = N(238)\sigma_c(238)P/[N(235)\sigma_f(235)E] = 4.04 \times 10^{19}$ atoms s$^{-1} = 506\,\text{kg y}^{-1}$.

10.10 Atomic wt. (1.5% enriched uranium) $= 238.0$; ^{235}U atoms per tonne $N(235)$ $= 0.015 \times 10^3\,(\text{kg})/238m_u = 3.80 \times 10^{25}$. Equations (10.37) and (10.38) give

power $P = N(235)\sigma_f(235)\Phi E \rightarrow \Phi = P/(N(235)\sigma_f(235)E) = 5.68 \times 10^{12}\,\mathrm{cm^{-2}\,s^{-1}}$.
For f, use Equation (10.9) with $N_M/N_F = 500$, $\sigma(M) = 0.0045$ b and $\sigma_a(F) = 0.015\sigma_a(235) + 0.985\sigma_c(238) = 12.88$ b. Whence $f = 0.851$.

10.11 Use Equation (10.40). $\Sigma_a(M) = 3.61 \times 10^{-4}\,\mathrm{cm^{-1}}$ (see Section 10.5.2).
Using $f = 0.851$ (see Problem 10.10) gives the prompt neutron lifetime $t_p \approx (1-f)/v\Sigma_a(M) = 1.9$ ms.

10.12 Let the time between collisions at thermal energies $= t$. Collision probability $\propto$ cross section. Therefore, diffusion time $t_d = $ (number of scattering collisions before capture)$\times t \approx (\sigma_s/\sigma_a)t \approx 1000t$ (using data in Table 10.3 for graphite values).
Slowing down: number of collisions $= 115$, but time between collisions varies. $\xi = \ln\langle E_0/E_1\rangle \rightarrow \langle E_0/E_1\rangle = e^\xi = 1.17$ (for graphite) $= \langle v_0/v_1\rangle^2$. So, the fractional decrease in average speed per collision $\langle v_1/v_0\rangle = 1/\sqrt{1.17} = 0.92 = F$.
Collision time for the $(n-1)$th collision $= F \times$ collision time for the nth collision and so on. Therefore, slowing-down time $t_s = t + Ft + F^2t + \cdots + F^{115}t \approx t/(1-F) = 12.5t$. Whence, $t_s/t_d \approx 0.01$.

10.13 (a) Use Equations in Section 10.5.3 to obtain $\dfrac{N_X}{N_I} = \dfrac{(\gamma_I + \gamma_X)\lambda_I}{(\lambda_X + \sigma_c\Phi)\gamma_I} = 0.737$ after substituting $\Phi = 7.5 \times 10^{12}\,\mathrm{cm^{-2}\,s^{-1}}$ and using the data given in Section 10.5.3 and Figure 10.8.
(b) Use Equation (10.46): $q_X = -\dfrac{(\gamma_I + \gamma_X)\eta f}{(1 + \lambda_X/\sigma_c\Phi)v} = -0.148$, i.e. about -1.5%.

10.14 Following the discussion in Section 10.5.3, the rate of production of ^{149}Sm is $\gamma\Sigma_f\Phi$, where Φ is the neutron flux in the reactor. In equilibrium, this is equal to the rate of loss $N_{Sm}\sigma_c\Phi$; whence, $N_{Sm} = \gamma\Sigma_f/\sigma_c$.
The fractional change in the neutron multiplicity factor q_{Sm} is given by Equation (10.44), with $q_X = q_{Sm}$ and $f' = \Sigma_a(F)/(\Sigma_a(C) + N_{Sm}\sigma_c)$. From the first part of Equation (10.45), we can write $\dfrac{q_{Sm}}{f} = -\dfrac{N_{Sm}\sigma_c}{\Sigma_a(F)} = -\dfrac{\gamma\Sigma_f}{\Sigma_a(F)}$.
Since $\Sigma_f/\Sigma_a(F) = \eta/v$ [see discussion after Equation (10.45)], we then obtain $q_{Sm} = -\gamma\eta f/v$. Substituting values gives $q_{Sm} = -0.0055$.

10.15 Mean free path $\lambda = 1/N_{nat\,U}\sigma_a(^{nat}U)$ (see Section 5.5.1). $N_{nat\,U} = \rho N_A/238 = 4.78 \times 10^{22}\,\mathrm{cm^{-3}}$. Using given data: $\sigma_a(^{nat}U)$ (2 MeV) $= 0.61$ b; $\sigma_a(^{nat}U)$ (thermal) $= 7.6$ b, whence, $\lambda(2\,\mathrm{MeV}) = 34$ cm and λ (thermal) $= 2.8$ cm.
Fraction absorbed by ^{238}U:

$$F = \frac{\Sigma(238)}{\Sigma(235) + \Sigma(238)} = 1 - \frac{\Sigma(235)}{\Sigma(235) + \Sigma(238)} = 1 - \frac{0.0072\sigma_a(235)}{\sigma_a(^{nat}U)}.$$

Whence, $F(2\,\mathrm{MeV}) = 0.97$ and $F(\mathrm{thermal}) = 0.36$.

10.16 The doubling time T_d is given by Equation (10.47): $T_d = [(B-1)\sigma_a\Phi]^{-1}$.
From Section 10.7.1, the breeding ratio $B = \eta - 1 - (C + L)$. Since $C + L = 0.1$ (10% neutron loss) and $\eta = 2.4$ for ^{233}U (from Table 10.6), $B = 1.3$. The neutron

flux is found from Equations (10.37) and (10.38): power $P = N(233)\sigma_f(^{233}U)\Phi E$, giving $\Phi = 3.35 \times 10^{15}$ cm^{-2}s^{-1}.

Substituting these values for B and Φ, and $\sigma_a(^{233}U) = 2.2$ b in Equation (10.47), gives the doubling time $T_d = 4.52 \times 10^8$ s ≈ 14 y.

10.17 From the definition of the neutron multiplication factor k (see beginning of Section 10.3), neutrons produced in the nth generation per neutron in the $(n-1)$th generation $= Fv = k = 0.9$, whence, $F = 0.9/2.5 = 0.36$. Therefore, nine neutrons (of the initial 25) lead to fission.

From the discussion in Section 10.7.2, the neutron gain factor $= 1/(1-k)$, whence, energy gain $= \dfrac{9}{(1-k)} \times \dfrac{200(\text{MeV/fission})}{800(\text{MeV})} = 22.5$.

10.18 For the subcritical assembly, $k = 0.38 \times 2.5 = Fv = 0.95$.

Fraction of neutrons absorbed by fissile fuel $= F\sigma_a/\sigma_f = 0.44$. Fraction captured by ^{232}Th: $F_{232} = 0.9(1 - 0.44) = 0.504 \rightarrow$ breeding ratio $B = 0.504/0.44 > 1$.

Neutron gain factor $= 1/(1-k)$ (see Section 10.7.2); therefore, neutron production rate $N = I\,(10\,\text{mA}) \times 30/(1-k) = 3.745 \times 10^{19}$s^{-1}.

^{233}U production rate $= N \times F_{232} \times 233\,m_u = 7.303 \times 10^{-6}$ kg s^{-1} = 230 kg y^{-1}.

CHAPTER 11

11.1 In the c-m system, $\mathbf{p}_\gamma + \mathbf{p}_{He} = 0$. Therefore, $p_\gamma^2 = E_\gamma^2/c^2 = 2m_{He}E_{He}$, whence, $E_\gamma/E_{He} = 2m_{He}c^2/E_\gamma$. If the initial kinetic energy is negligible, $Q = E_\gamma + E_{He}$. Eliminating E_γ from these equations, we obtain $(Q - E_{He})^2 = 2m_{He}c^2 E_{He}$. Solving this equation (with $Q = 5.49$ MeV) gives $E_{He} = 5.354$ keV.

11.2 If the c-m momentum is negligible, momentum conservation gives $p_n = p_{He}$, whence, $E_n/E_{He} = m_{He}/m_n = 3.0$. If the initial kinetic energy is negligible, $Q = E_n + E_{He} = 3.27$ MeV. Solving these two equations gives: $E_n = 2.45$ MeV and $E_{He} = 0.82$ MeV.

11.3 Coulomb potential: 0.2 MeV $= e^2/4\pi\varepsilon_0 r = 1.44$ (MeV fm)$/r$(fm), using information from Appendix A1. Whence $r = 7.2$ fm.

11.4 Each d–t pair (mass $= 5.03 m_u$) liberates 17.6 MeV; therefore, the energy of D–T fuel per kg $= 17.6$ (MeV)$/5.03 m_u = 3.376 \times 10^{14}$ J kg^{-1}.

Energy of 50 megatonnes of TNT $= 2 \times 10^{17}$ J, which is equivalent to 592 kg of D–T material.

11.5 The maximum of the Maxwell–Boltzmann distribution, $p(v) \propto v^2 \exp(-\tfrac{1}{2}mv^2/kT)$ (see Section 11.3.1), corresponds to the most probable speed. Differentiating and setting $dp/dv = 0$ (corresponding to the maximum) gives $\tfrac{1}{2}mv^2 = kT$.

At $kT = 20$ keV, v(deuteron) $= 1.38 \times 10^6$ m s^{-1}; v(triton) $= 1.13 \times 10^6$ m s^{-1}.

11.6 The Lawson criterion being just satisfied corresponds to Equation (11.6) with the inequality sign being replaced by an equal sign. We calculate the right-hand side of this equation: with $Q = 17.59$ MeV (from Section 11.2.1) and $\langle v\sigma \rangle \approx 1.2 \times 10^{-22}$ $m^3 \, s^{-1}$ (from Figure 11.3), we obtain $n\tau = 12kT/[\langle v\sigma \rangle Q] = 5.68 \times 10^{19} \, s \, m^{-3}$ and for $\tau = 3$ s, $n = 1.9 \times 10^{19} \, m^{-3}$.

Equation (11.2) gives the total reaction rate. Therefore, since $n_1 = n_2 = n/2$, the reaction rate per particle $= \frac{n}{2} \langle v\sigma \rangle = 1.14 \times 10^{-3} \, s^{-1}$.

11.7 If there are n deuterons per unit volume $(n \gg 1)$, the number of independent ways any two of them can interact with each other is $n^2/2$. Therefore, following a similar derivation to that of Equation (11.3), the fusion energy output is $E_f = n^2 \langle v\sigma \rangle Q\tau/2 > 3nkT$ to satisfy the Lawson criterion. Therefore, $n\tau > 6kT/\langle v\sigma \rangle Q$. $\langle v\sigma \rangle \approx 5 \times 10^{-23} \, m^3 \, s^{-1}$ (from Figure 11.3), whence, $n\tau > (6 \times 0.1)/(5 \times 10^{-23} \times 3.6) \approx 3 \times 10^{21} \, s \, m^{-3}$.

11.8 Speed of a 100-keV d: $v_d = 3.1 \times 10^6 \, m \, s^{-1}$ (using a formula in Appendix A.4). On average, any two deuterons will be moving at right angles to each other, which gives an average relative speed $v_{rel} = \sqrt{2} v_d$. This gives $\langle v_{rel} \sigma \rangle = 4.38 \times 10^{-23} \, m^3 \, s^{-1}$ (from Figure 11.3, $\langle v_{rel} \sigma \rangle \approx 5 \times 10^{-23} \, m^3 \, s^{-1}$).

If we have a large number n of deuterons per unit volume, the number of independent ways any two of them can interact with each other is $n^2/2$. Therefore, following a similar derivation to that of Equation (11.3), the power output per unit volume $= n^2 \langle v_{rel} \sigma \rangle Q/2$. Substituting $n = 10^{21} \, m^{-3}$ and a mean Q value of 3.6 MeV, gives $1.26 \times 10^7 \, W \, m^{-3}$.

From Equation (11.5), plasma energy $E_p = 3nkT = 4.8 \times 10^7 \, J \, m^{-3}$. Therefore, the plasma has to burn for about 3.8 s.

11.9 Use radius $r = \sqrt{2mE}/qB$ [e.g. from Equation (6.4)]. At $B = 3$ T, $r = 0.96$ cm for a 20-keV deuteron, and 1.2 cm for a 20-keV triton.

11.10 Since the charged particle is undeflected, the magnitudes of the electric and magnetic forces are equal, i.e. $qE = Bqv$ or $E = Bv$. The speed of a 1-keV triton $= 2.529 \times 10^5 \, m \, s^{-1}$; therefore, for $B = 3$ T, $E = 0.76$ MV m^{-1}.

For fixed v, kinetic energy $\propto m$. Therefore, deuteron energy $= 668$ eV.

11.11 Each d–t pair (mass $= 5.03 m_u$) liberates 17.6 MeV; therefore, energy per pellet $= 17.6 \, (\text{MeV}) \times 0.3 \times 10^{-6} (\text{kg})/5.03 \, m_u = 101$ MJ. A power output of 500 MW requires about five pellets s^{-1}.

11.12 D–T density $\rho = 6 \times 10^{29}$ atoms $m^{-3} = 2.5$ g cm^{-3}. Radius of a 1-mg sphere $= 0.0457$ cm. Range $R = $ range $R'(\text{g } cm^{-2})/\rho = 3.2 \times 10^{-4}$ cm $\approx 0.7\%$ of the radius.

11.13 Mass $m = 10^{-3}$ g. Number of deuterons (tritons) $N_d(N_t) = m/5.03 \, m_u$. For an energy efficiency of 10%, input energy $= 10(N_d + N_t) \times 20$ keV $= 7.7$ MJ.

11.14 Use Equation (11.11). At $kT = 0.8$ MeV, n/p ratio (at time $t = 0$): $n_0/p_0 = \exp(-\Delta mc^2/kT) = 0.2$. Neutron half-life $= 10.24$ m. So, at $t = 250$ s, $n_t = n_0 e^{-\lambda t} = 0.75 n_0$, whence $p_t/n_t = [p_0 + (n_0 - n_t)]/n_t = 7$. For every 14 protons, there are two neutrons and these convert into 12 protons and one α particle in the Big Bang, i.e. a mass ratio of 3:1.

11.15 Originally, the sun contained $= 0.71 \times M_\odot/m_H = 8.48 \times 10^{56}$ protons. In the sun, the consumption of four protons generates 26 MeV. Note that this is less than the Q value by the neutrino energy (see the discussion in the example near the end of Section 11.6.1). Therefore, the solar luminosity: 3.86×10^{26} W $\equiv 3.71 \times 10^{38}$ protons s^{-1}. In 5×10^9 y, 0.58×10^{56} protons were consumed, leaving 7.9×10^{56} protons now. Time to consume a further 10% $\approx 6.8 \times 10^9$ y.

11.16 H burning: $4p \rightarrow$ He, $Q \approx 26$ MeV; $Q/\text{mass} = 26$ (MeV)/4 (u).
He burning: $3\alpha \rightarrow {}^{12}$C, $Q \approx 7.3$ MeV; $Q/\text{mass} = 7.3$ (MeV)/12 (u).
   Therefore, mass is consumed in He burning about 10 times as fast as in H burning.

11.17 If there are N nuclei at a given time t that have not undergone a reaction, the rate of change $dN/dt = -N\sigma\Phi$, where $N\sigma\Phi$ is the reaction rate [Equation (1.22)]. Solving the equation gives $N(t) = N_0 \exp(-\sigma\Phi t)$, where N_0 is the number at $t = 0$. Thus, nuclei are depleted in a mean time given by $\tau = \int_0^\infty tN(t)dt / \int_0^\infty N(t)dt = 1/\sigma\Phi$.
   If $\tau = 1$ s (in a supernova), $\Phi = 1/\sigma = 10^{25}$ s^{-1} cm^{-2}.

11.18 ^{122}Te is populated by the s process via: ^{121}Sb$(n, \gamma)^{122}$Sb $\xrightarrow{\beta^-}$ ^{122}Te.
The s process populates the line of stable isotopes to ^{126}Te.
^{128}Te and ^{130}Te are populated by the r process (^{127}Te and ^{129}Te are too unstable).
^{120}Te is populated via the p process by double proton capture on ^{118}Sn in a supernova.

References

Publications specifically referred to in the text.

Audi, A., Bersillon, O., Blachot, J. and Wapstra, A. H. (1997) The Nubase evaluation of nuclear and decay properties, *Nuclear Physics A*, **624**, 1.

Back, B. B. *et al.* (1977) *Nuclear Physics A*, **285**, 317.

Barlow, R. (1988) *Statistics*, Wiley, p. 29.

Barnett, A. R. and Lilley, J. S. (1974) *Physical Review*, **C9**, 2010.

Bennet, D. J. and Thomson, J. R. (1989) *The Elements of Nuclear Power*, Longman, p. 79.

Blatt, J. M. and Weisskopf, V. F. (1952) *Theoretical Nuclear Physics*, Wiley, pp. 623–627.

Bohr, A. and Mottleson, B. R. (1975) *Nuclear Structure*, Vol. II, Benjamin, p. 48.

Bormann, M. *et al.* (1970) *Nuclear Instruments and Methods*, **88**, 245.

Breeze, M. B. H., Grime, G. W. and Watt, F. (1992) The nuclear microprobe, *Annual Review of Nuclear and Particle Science*, **42**, 8.

Cahill, T. A. (1980) Proton microprobes and particle-induced X-ray analytical systems, *Annual Review of Nuclear and Particle Science*, **30**, pp. 219 and 220.

Clayton, D. *et al.* (1961) *Annals of Physics*, **12**, 331.

Dauk, J. *et al.* (1975) *Nuclear Physics A*, **241**, 170.

Eisen, Y. *et al.* (1977) *Nuclear Physics A*, **291**, 459.

Ellegaard, C. and Vedelsby, P. (1968) *Physics Letters*, **26B**, 155.

Evans, R. D. (1955) *The Atomic Nucleus*, McGraw-Hill, pp. 538, 560.

Farwell, G. *et al.* (1954) *Physical Review*, **95**, 1212.

Goldsmith, H. H. *et al.* (1947) *Reviews of Modern Physics*, **19**, 259.

Hendee, W. R. (1999) Physics and applications of medical imaging, *Reviews of Modern Physics*, **71**, s444.

Hobbie, R. K. (1997) *Intermediate Physics for Medicine and Biology*, 3rd edition, AIP Press, Ch. 17.

Hofmann *et al.* (1996) *Z. Phys.*, **A354**, 229.

Hughes, D. J. and Schwartz, R. B. (1958) Neutron cross sections, Brookhaven National Laboratory Report BNL-325 (unpublished).

Katz, L. and Penfold, A. S. (1952) *Reviews of Modern Physics*, **24**, 28.

Keefe, D. (1982) Inertial confinement fusion, *Annual Review of Nuclear and Particle Science*, **32**, 395.

Keeley, N. *et al.* (1995) *Nuclear Physics A*, **582**, 314.

Kelly, E. L. and Segrè, E. (1949) *Physical Review*, **75**, 999.

Kozub, R. L. *et al.* (1975) *Physical Review*, **C11**, 1497.

Larson, J. S. *et al.* (1972) *Physics Letters*, **42B**, 205.

Lederer, C. M. and Shirley, V. (eds) (1978) *Table of Isotopes*, Wiley, Appendix V.

Litttmark, V. and Ziegler, J. F. (1980) *Handbook of Range Distributions for Energetic Ions in all Elements*, Pergamon Press.

Mandl, F. (1992) *Quantum Mechanics*, Wiley.

Marion, J. and Young, F. C. (1968) *Nuclear Reaction Analysis*, North-Holland, Amsterdam, p. 99.

Metzger, F. (1952) *Physical Review*, **88**, 1360.

Okada, S. (1957) *Archives of Biochemistry and Biophysics*, **67**, 102.

Park, J. Y. and Satchler, G. R. (1971) *Particles and Nuclei*, **1**, 233.

Porter, F. T. *et al.* (1957) *Physical Review*, **107**, 135.

Rasey, J. S. and Nelson, N. J. (1981) *Radiation Research*, **85**, 69.

Satchler, G. R. (1967) *Nuclear Physics A*, **92**, 273.

Sick, I. (1975) *Physical Review Letters*, **35**, 910.

Siegbahn, K. (ed.) (1966) α-, β- and γ-*Ray Spectroscopy*, Vol. 1, North-Holland, pp. 51, 832, 838.

Stehn, J. R. *et al.* (1958) Neutron Cross Sections, Brookhaven National Laboratory Report BNL-325 (unpublished).

Till, J. E. and McCulloch, E. A. (1963) *Radiation Research*, **18**, 96.

Twin, P. J. *et al.* (1986) *Physical Review Letters*, **57**, 811.

Ungrin, J. *et al.* (1971) *Mat. Fys. Medd. Dan. Vid. Selsk.*, **38**, 8.

Bibliography

Undergraduate nuclear physics textbooks

Burcham, W. E. (1979) *Elements of Nuclear Physics*, Longman Scientific & Technical, Harlow, England.

Burcham, W. E. and Jobes, M. (1995) *Nuclear and Particle Physics*, Longman Scientific & Technical, Harlow, England.

Cohen, B. L. (1971) *Concepts of Nuclear Physics*, McGraw-Hill, New York.

Heyde, K. (1999) *Basic Ideas and Concepts in Nuclear Physics*, Institute of Physics Publishing, Bristol and Philadelphia.

Hodgson, P. E., Gadioli, E. and Gadioli Erba, E. (1997) *Introductory Nuclear Physics*, Clarendon Press, Oxford.

Jelley, N. A. (1990) *Fundamentals of Nuclear Physics*, Cambridge University Press, Cambridge.

Krane, K. S. (1987) *Introductory Nuclear Physics*, Wiley, New York.

Pearson, J. M. (1986) *Nuclear Physics: Energy and Matter*, Adam Hilger, Bristol and Boston.

Satchler, G. R. (1980) *Introduction to Nuclear Reactions*, Macmillan, London.

Williams, W. S. C. (1992) *Nuclear and Particle Physics*, Oxford University Press, Oxford.

Graduate nuclear physics textbooks

Blatt, J. M. and Weisskopf, V. F. (1952) *Theoretical Nuclear Physics*, Wiley, New York.

Bohr, A. and Mottleson, B. R. (1969) *Nuclear Structure*, Vol. I, Benjamin, New York.

Bohr, A. and Mottleson, B. R. (1975) *Nuclear Structure*, Vol. II, Benjamin, New York.

Evans, R. D. (1955) *The Atomic Nucleus*, McGraw-Hill, New York.

Preston, M. A. and Bhaduri, R. K. (1975) *Structure of the Nucleus*, Addison-Wesley, Reading MA, U.S.A.

Satchler, G. R. (1983) *Direct Nuclear Reactions*, Clarendon Press, Oxford.

Siegbahn, K., ed. (1966) α-, β- and γ-Ray Spectroscopy, Vols 1 and 2, North-Holland, Amsterdam.

Books and articles related to topics in Part II

Audi, A., Bersillon, O., Blachot, J. and Wapstra, A. H. (1997) *The Nubase evaluation of nuclear and decay properties, Nuclear Physics A*, **624**, 1.

Barlow, R. (1989) *Statistics*, Wiley, Chichester, England.

Bennet, D. J. and Thomson, J. R. (1989) *The Elements of Nuclear Power*, 3rd edition, Longman Scientific & Technical, Harlow, England.

Bowman, C. D. (1998) *Accelerator-driven systems for waste transmutation*, Annual *Review of Nuclear and Particle Science* **48**, 505.

Breeze, M. B. H., Grime, G. W. and Watt, F. (1992) *The nuclear microprobe*, Annual Review of *Nuclear and Particle Science* **42**, p. 1.

Bromley, D. (1984) *Neutrons in science and technology, Nuclear Instruments and Methods*, **225**, 240.

Cahill, T. A. (1980) *Proton microprobes and particle-induced X-ray analytical systems, Annual Review of Nuclear and Particle Science* **30**, 211.

Coggle, J. E. (1983) *Biological Effects of Radiation*, 2nd edition, Taylor & Francis, New York.

England, J. B. A. (1974) *Techniques in Nuclear Structure Physics*, Vols I and II, Macmillan, London.

Filippone, B. W. (1986) *Nuclear reactions in stars, Annual Review of Nuclear and Particle Science*, **36**, 717.

Firestone, R. B. and Shirley, V. S. (Ed.) (1996) *Table of Isotopes*, Vols I and II, 8th edition with CD-ROM, Wiley, New York.

Gilmore, G. and Hemingway, J. (1995) *Practical Gamma-Ray Spectroscopy*, Wiley, Chichester, England.

Gove, N. B. and Martin, M. J. (1971) *log f tables, Nuclear Data Tables A*, **10**, 205.

Guinn, V. P. (1974) *Applications of nuclear science in crime investigation, Annual Review of Nuclear Science* **24**, 561.

Hall, E. J. (1994) *Radiobiology for the Radiologist*, 4th edition, Lippincott, Philadelphia.

Hendee, W. R. (1999) *Physics and applications of medical imaging, Reviews of Modern Physics.* **71**, s444.

Hobbie, R. K. (1997) *Intermediate Physics for Medicine and Biology*, 3rd edition, AIP Press, New York.

Hubell, J. H. (1982) *Photon mass attenuation and energy-absorption coefficients, International Journal of Applied Radiation and Isotopes*, **33**, 1269.

Johansson, S. A. E. and Johansson, T. B. (1976) *Analytical applications of particle-induced X-ray emission, Nuclear Instruments and Methods*, **137**, 473.

Keefe, D. (1982) *Inertial confinement fusion, Annual Review of Nuclear and Particle Science*, **32**, 391.

Knoll, G. F. (2000) *Radiation Detection and Measurement*, 3rd edition, Wiley, New York.

Kutschera, W. and Paul, M. (1990) *Accelerator mass spectrometry in nuclear physics and astrophysics, Annual Review of Nuclear and Particle Science*, **40**, 411.

Leo, W. R. (1994) *Techniques in Nuclear and Particle Physics Experiments*, 2nd edition, Springer-Verlag, New York and Berlin.

Litherland, A. E. (1980) *Ultrasensitive mass spectrometry with accelerators, Annual Review of Nuclear and Particle Science*, **30**, 437.

Littmark, U. and Ziegler, J. F. (1980) *Handbook of Range Distributions for Energetic Ions in all Elements*, Pergamon Press, New York.

Mandelkern, M. A. (1995) *Nuclear techniques for medical imaging: positron emission tomography, Annual Review of Nuclear and Particle Science*, **45**, 205.

Mandl, F. (1992) *Quantum Mechanics*, Wiley, Chichester, England.

Marion, J. and Young, F. C. (1968) *Nuclear Reaction Analysis*, North-Holland, Amsterdam.

Phillips, A. C. (1999) *The Physics of Stars*, 2nd edition, Wiley, Chichester, England.

Post, R. F. (1970) *Controlled Fusion Research and High-Temperature Plasmas, Annual Review of Nuclear Science*, **20**, 509.

Rubbia, C. (1995) *Conceptual Design of a Fast Neutron Operated High Power Energy Amplifier*, Rapport CERN/AT/95–44.

Truran, J. W. (1984) *Nucleosynthesis, Annual Review of Nuclear and Particle Science*, **34**, 53.

Tsipenyuk, Yu. M. (1997) *Nuclear Methods in Science and Technology*, Institute of Physics Publishing, Bristol.

Ziegler, J. F. (1980) *Handbook of Stopping Cross Sections for Energetic Ions in all Elements*, Pergamon Press, New York.

Index